ENSURING THE QUALITY, CREDIBILITY, AND RELEVANCE OF
U.S. JUSTICE STATISTICS

Panel to Review the Programs of the Bureau of Justice Statistics

Robert M. Groves and Daniel L. Cork, *Editors*

Committee on National Statistics
Committee on Law and Justice

Division of Behavioral and Social Sciences and Education

NATIONAL RESEARCH COUNCIL
OF THE NATIONAL ACADEMIES

THE NATIONAL ACADEMIES PRESS
Washington, DC
www.nap.edu

THE NATIONAL ACADEMIES PRESS   500 Fifth Street, NW   Washington, DC 20001

NOTICE: The project that is the subject of this report was approved by the Governing Board of the National Research Council, whose members are drawn from the councils of the National Academy of Sciences, the National Academy of Engineering, and the Institute of Medicine. The members of the committee responsible for the report were chosen for their special competences and with regard for appropriate balance.

Support of the work of the Committee on National Statistics is provided by a consortium of federal agencies through a grant from the National Science Foundation (No. SES-0453930). The project that is the subject of this report was supported by an allocation from the U.S. Department of Justice to the National Science Foundation under this grant. Any opinions, findings, conclusions, or recommendations expressed in this publication are those of the author(s) and do not necessarily reflect the views of the organizations or agencies that provided support for the project.

International Standard Book Number 13: 978-0-309-13910-6
International Standard Book Number 10: 0-309-13910-4

Additional copies of this report are available from the National Academies Press, 500 Fifth Street, NW, Washington, DC 20001; (202) 334-3096; Internet, http://www.nap.edu.

Printed in the United States of America

Suggested citation: National Research Council. (2009). *Ensuring the Quality, Credibility, and Relevance of U.S. Justice Statistics.* Panel to Review the Programs of the Bureau of Justice Statistics. Robert M. Groves and Daniel L. Cork, eds. Committee on National Statistics and Committee on Law and Justice, Division of Behavioral and Social Sciences and Education. Washington, DC: The National Academies Press.

THE NATIONAL ACADEMIES
Advisers to the Nation on Science, Engineering, and Medicine

The **National Academy of Sciences** is a private, nonprofit, self-perpetuating society of distinguished scholars engaged in scientific and engineering research, dedicated to the furtherance of science and technology and to their use for the general welfare. Upon the authority of the charter granted to it by the Congress in 1863, the Academy has a mandate that requires it to advise the federal government on scientific and technical matters. Dr. Ralph J. Cicerone is president of the National Academy of Sciences.

The **National Academy of Engineering** was established in 1964, under the charter of the National Academy of Sciences, as a parallel organization of outstanding engineers. It is autonomous in its administration and in the selection of its members, sharing with the National Academy of Sciences the responsibility for advising the federal government. The National Academy of Engineering also sponsors engineering programs aimed at meeting national needs, encourages education and research, and recognizes the superior achievements of engineers. Dr. Charles M. Vest is president of the National Academy of Engineering.

The **Institute of Medicine** was established in 1970 by the National Academy of Sciences to secure the services of eminent members of appropriate professions in the examination of policy matters pertaining to the health of the public. The Institute acts under the responsibility given to the National Academy of Sciences by its congressional charter to be an adviser to the federal government and, upon its own initiative, to identify issues of medical care, research, and education. Dr. Harvey V. Fineberg is president of the Institute of Medicine.

The **National Research Council** was organized by the National Academy of Sciences in 1916 to associate the broad community of science and technology with the Academy's purposes of furthering knowledge and advising the federal government. Functioning in accordance with general policies determined by the Academy, the Council has become the principal operating agency of both the National Academy of Sciences and the National Academy of Engineering in providing services to the government, the public, and the scientific and engineering communities. The Council is administered jointly by both Academies and the Institute of Medicine. Dr. Ralph J. Cicerone and Dr. Charles M. Vest are chair and vice chair, respectively, of the National Research Council.

www.national-academies.org

Acknowledgments

THE PANEL to Review the Programs of the Bureau of Justice Statistics of the Committee on National Statistics (CNSTAT) is pleased to submit this final report on the programs and priorities of the Bureau of Justice Statistics (BJS). The work that leads to such a report always represents a collectivity—a devoted and talented staff of the National Research Council (NRC) and a set of volunteers, both panel members and those who met with the panel. Finally, the agency seeking advice from CNSTAT is key to the success of the endeavor.

The staff of BJS has been exceptionally receptive to our external review of the agency's programs. We benefited greatly from the energy and enthusiasm of BJS Director Jeffrey Sedgwick during the course of our study, particularly given his added responsibilities during the latter half of our work. In January 2008 he took on the duties of acting assistant attorney general for the Office of Justice Programs, the parent division of BJS, while retaining the BJS directorship. He was subsequently nominated as assistant attorney general in April 2008 and confirmed in October 2008, serving in that capacity until the end of the Bush administration. The other members of BJS's senior leadership were also unstinting in their support. Deputy Director Maureen Henneberg provided considerable assistance as the lead liaison between BJS and the panel, and BJS Senior Statistical Advisor Allen Beck gave greatly of his time and expertise in interacting with the panel and led a wide-ranging discussion of BJS data collections required by the Prison Rape Elimination Act. Patrick Campbell, special assistant to Director Sedgwick, also participated in the public sessions of the panel's meetings. Michael Rand, chief of victimization statistics, deserves particular credit for leading a thorough and extremely useful review of the National Crime Victimization Survey (NCVS) at the panel's first meeting. Program managers and BJS staff briefed the panel on their work and fielded numerous questions about their designated subject areas; for this service, we thank Thomas Cohen and William Sabol. More than just cooperation is notable; in its meetings with BJS staff, the panel observed clear devotion among BJS staff to the quality and efficiency of the agency's statistical activities and a common purpose of serving the country well through its activities.

Under a separate contract—but with the intent of supporting our panel's work—BJS commissioned the Council of Professional Associations on Federal Statistics (COPAFS) to host a dedicated one-day conference on BJS data user perspectives. Through its commissioned papers and its wide-ranging list of approximately 80 participants, the February 12, 2008, workshop provided the panel with a great deal of material to consider in this report. We appreciate the contributions of the authors and presenters of lead papers at the workshop: Lynn Addington, American University; Theodore Eisenberg, Cornell Law School; Brian Forst, American University; and Karen Heimer, University of Iowa. Foremost, though, we thank Edward Spar, COPAFS executive director, for his hard work in organizing and convening the session.

We greatly appreciate the work of the Justice Research and Statistics Association; its invitation to panel staff to attend the association's annual meeting in Denver in October 2006 was helpful in structuring the panel's work. Joan Weiss, executive director of the association, provided helpful comments and suggestions as the panel began its work, and research director Stan Orchowsky spoke to the panel about the sweep of the organization's work and its relationship with BJS.

Pursuant to the panel's charge, subgroups of the panel met with a variety of individuals and representatives of interested groups to elicit opinions on the range of BJS programs, the usefulness of BJS data series, and possible areas for change and refinement. To facilitate full and candid discussions, these subgroup sessions were done with the understanding that comments made therein were not intended for direct attribution. We thank those persons who gave generously of their time and expertise to speak to members of the panel: James Boden, chief, Justice Branch, U.S. Office of Management and Budget; Michael Crowley, policy analyst, Justice Branch, U.S. Office of Management and Budget; John Firman, research director, International Association of Chiefs of Police; Lawrence Greenfeld, senior statistical advisor, Office of the Inspector General, General Services Administration; Chris Koper, deputy director of research, Police Executive Research Forum; Bruce Kubu, senior research associate, Police Executive Research Forum; Shelly Martinez, statistician, U.S. Office of Management and Budget; Bruce Taylor, director of research, Police Executive Research Forum; and Katherine Wallman, chief statistician, U.S. Office of Management and Budget.

We gratefully acknowledge the other expert speakers who contributed to our plenary meetings: Kim English, research director, Division of Criminal Justice, Colorado Department of Public Safety; Mark Epley, senior counsel to the deputy attorney general, U.S. Department of Justice; Pat Flanagan, assistant division chief, Demographic Statistical Methods Division, U.S. Census Bureau; David Hagy, director, National Institute of Justice, U.S. Department of Justice; Paula Hannaford-Agor, principal court research consultant, National Center for State Courts; Douglas Hoffman, director, Center for Research, Evaluation, and Statistical Analysis, Pennsylvania Commission on Crime and Delinquency; Howard Hogan, associate director for demographic programs, U.S. Census Bureau; Krista Jansson, Home Office, United Kingdom; Ruth Ann Killion, chief, Demographic Statistical Methods Division, U.S. Census Bureau; Cheryl Landman, chief, Demographic Surveys Division, U.S. Census Bureau; Marilyn Monahan, chief of NCVS Branch, Demographic Surveys Division, U.S. Census Bureau; Richard Schauffler, director of research services and director of Court Statistics Project, National Center for State Courts; Jon Simmons, head of research analysis and statistics, Crime Reduction and Community Safety Group, Home Office, United Kingdom; and Philip Stevenson, Statistical Analysis Center director, Arizona Criminal Justice Commission.

The study director of the panel was Daniel Cork, whose ability to absorb reams of technical, administrative, and organizational information about BJS and the Department of Justice earned the admiration of all panel members. His wisdom in assembling and integrating the writing of panel members and in structuring and writing the report was notable. The panel's work is conducted in cooperation with the NRC's Committee on Law and Justice (CLAJ). As senior program officer to this panel, CLAJ Director Carol Petrie helped the panel integrate its work with prior studies and activities of the NRC concerning the Department of Justice. As is true with other CNSTAT studies, our panel benefited greatly from the regular and active participation and engagement in our meetings of Constance Citro, director of CNSTAT. Agnes Gaskin, the administrative assistant, made sure meetings were organized and conducted in the professional manner that CNSTAT always achieves.

The Survey Research Center of the Institute for Social Research at the University of Michigan hosted two deliberative sessions of the panel at Ann Arbor in June 2007 and August 2008. Deborah Serafin, Rose Myers, and Kelly Smid handled the arrangements for the panel with grace and efficiency. We also appreciate Chris Eskridge, University of Nebraska–Lincoln and executive director of the American Society of Criminology, for providing space for the panel to hold a writing and reviewing meeting at the end of the society's annual meetings in St. Louis, Missouri, in November 2008.

This report has been reviewed in draft form by individuals chosen for their diverse perspectives and technical expertise, in accordance with procedures approved by the Report Review Committee of the NRC. The purpose of this independent review is to provide candid and critical comments that will assist the institution in making the published report as sound as possible and to ensure that the report meets institutional standards for objectivity, evidence, and responsiveness to the study charge. The review comments and draft manuscript remain confidential to protect the integrity of the deliberative process.

We thank the following individuals for their participation in the review of this report: Eric Baumer, College of Criminology and Criminal Justice, Florida State University; David Cantor, Statistics Group, Westat, Inc.; Jan M. Chaiken, consultant, Criminal Justice Policy Research Institute, Hatfield School of Government, Portland State University; Kenneth Land, Department of Sociology, Duke University; Sharon Lohr, Department of Mathematics and Statistics, Arizona State University; Doris MacKenzie, Department of Criminology, University of Maryland; Pat Mayhew, consultant, London, England; Steven F. Messner, Department of Sociology, University at Albany; John V. Pepper, Department of Economics, University of Virginia; James M. Tien, College of Engineering, University of Miami; and Jeremy Wilson, Department of Criminal Justice, Michigan State University.

Although the reviewers listed above provided many constructive comments and suggestions, they were not asked to endorse the conclusions or recommendations nor did they see the final draft of the report before its release. The review of the report was overseen by Alfred Blumstein, School of Public Policy and Management, H. John Heinz III College, Carnegie Mellon University, and Philip J. Cook, Terry Sanford Institute of Public Policy, Duke University. Appointed by the NRC, they were responsible for making certain that an independent examination of the report was carried out in accordance with institutional procedures and that all review comments were carefully considered. Responsibility for the final content of this report rests entirely with the authoring panel and the institution.

Finally, a personal note of appreciation to my fellow panel members. NRC panels bring together unusually bright and accomplished scholars, deliberately chosen to have different areas of expertise and different perspectives on the focal topic. Our panel was no exception to this rule. The members were unusually active and contributed greatly to both the interim and final report. While never shy to voice their thoughts, they worked together with respect for each other's viewpoints and reached consensus on key recommendations with good will intact. I believe they all learned much from each other over the several meetings of the panel and formed intellectual bonds that will survive over time. I thank Bill Barron, Bill Clements, Janet Lauritsen, Colin Loftin, Jim Lynch, Ruth Peterson, "Raghu" Raghunathan, Steve Schlesinger, Wes Skogan, Bruce Spencer, and Bruce Western.

Robert M. Groves, *Chair*
Panel to Review the Programs of the Bureau of Justice Statistics

Contents

List of Figures

List of Tables

List of Boxes

Acronyms and Abbreviations

AAG	assistant attorney general (for the Office of Justice Programs)
ACASI	audio computer-assisted self-interviewing
ACS	American Community Survey
ADAM	Arrestee Drug Abuse Monitoring program (National Institute of Justice)
ADR	alternative dispute resolution
AO	Administrative Office of the U.S. Courts
ASJ	Annual Survey of Jails
BJA	Bureau of Justice Assistance (Office of Justice Programs, U.S. Department of Justice)
BJS	Bureau of Justice Statistics (Office of Justice Programs, U.S. Department of Justice)
BLS	Bureau of Labor Statistics
BOP	Federal Bureau of Prisons (U.S. Department of Justice)
CAPI	computer-assisted personal interviewing
CATI	computer-assisted telephone interviewing
CDC	U.S. Centers for Disease Control and Prevention (U.S. Department of Health and Human Services)
CES	Current Employment Statistics program (U.S. Bureau of Labor Statistics)
CFR	*Code of Federal Regulations*
CLAJ	Committee on Law and Justice (National Research Council)
CNSTAT	Committee on National Statistics (National Research Council)
COMPSTAT	"computerized statistics" or "comparative statistics" (term

	used to describe police information system developed in New York, Los Angeles, and other cities)
COP	community policing *or* community-oriented policing
COPAFS	Council of Professional Associations on Federal Statistics
COPS	Community Oriented Policing Services (as in the Office of Community Oriented Policing Services in the U.S. Department of Justice)
CSLLEA	Census of State and Local Law Enforcement Agencies
DAWN	Drug Abuse Warning Network (Substance Abuse and Mental Health Services Administration)
DEA	Drug Enforcement Administration (U.S. Department of Justice)
DOJ	U.S. Department of Justice
DUF	Drug Use Forecasting program (historical name for Arrestee Drug Abuse Monitoring [ADAM] program)
FAIR	Federal Activities Inventory Reform Act
FBI	Federal Bureau of Investigation (U.S. Department of Justice)
FJSRC	Federal Justice Statistics Research Center (Urban Institute)
GAO	U.S. Government Accountability Office (formerly, General Accounting Office)
IACP	International Association of Chiefs of Police
IADLEST	International Association of Directors of Law Enforcement Standards and Training
IAFIS	Integrated Automated Fingerprint Identification System
ICE	U.S. Immigration and Customs Enforcement (U.S. Department of Homeland Security)
ICPSR	Inter-university Consortium for Political and Social Research
ICR	Information Collection Review
III	Interstate Identification Index
JRSA	Justice Research and Statistics Association
LEAA	Law Enforcement Assistance Administration (predecessor to Office of Justice Programs)
LEMAS	Law Enforcement Management and Administrative Statistics
LEOKA	Law Enforcement Officers Killed and Assaulted
MFS	Monitoring the Future Survey
NACJD	National Archive of Criminal Justice Data (Inter-university Consortium for Political and Social Research, University of Michigan)
NASDA	National Association of State Departments of Agriculture

NASS	National Agricultural Statistics Service (U.S. Department of Agriculture)
NCANDS	National Child Abuse and Neglect Data System
NCES	National Center for Education Statistics (U.S. Department of Education)
NCHIP	National Criminal History Improvement Program
NCHS	National Center for Health Statistics (U.S. Department of Health and Human Services)
NCIC	National Crime Information Center (Federal Bureau of Investigation)
NCJISS	National Criminal Justice Information and Statistics Service (historical name for Bureau of Justice Statistics)
NCJJ	National Center for Juvenile Justice
NCJRS	National Criminal Justice Reference Service
NCRP	National Corrections Reporting Program
NCSC	National Center for State Courts
NCVS	National Crime Victimization Survey (formerly, National Crime Survey, NCS)
NIBRS	National Incident-Based Reporting System
NICS	National Instant Criminal Background Check System
NIJ	National Institute of Justice (Office of Justice Programs, U.S. Department of Justice)
NIS	National Inmate Surveys (data collection for the Prison Rape Elimination Act of 2003)
NJRP	National Judicial Reporting Program
NLSY97	National Longitudinal Survey of Youth 1997
NORC	National Opinion Research Center
NPS	National Prisoner Statistics
NRC	National Research Council
NSDUH	National Survey on Drug Use and Health (Substance Abuse and Mental Health Services Administration)
NSFG	National Survey of Family Growth (National Center for Health Statistics)
NW3C	National White Collar Crime Center
NYRBS	National Youth Risk Behavior Survey
OJARS	Office of Justice Assistance, Research, and Statistics (predecessor to Office of Justice Programs, U.S. Department of Justice)
OJJDP	Office of Juvenile Justice and Delinquency Prevention (Office of Justice Programs, U.S. Department of Justice)
OJP	Office of Justice Programs (U.S. Department of Justice)
OMB	U.S. Office of Management and Budget
ORI	Originating Agency Identifier

PART	Performance Assessment Rating Tool
PERF	Police Executive Research Forum
P.L.	Public Law
PPCS	Police-Public Contact Survey (periodic supplement to the National Crime Victimization Survey)
PREA	Prison Rape Elimination Act of 2003, P.L. 108-79
QCEW	Quarterly Census of Employment and Wages (U.S. Bureau of Labor Statistics)
rap	Record of Arrest and Prosecution ("rap sheet")
RFP	request for proposals
SAC	Statistical Analysis Center
SAMHSA	Substance Abuse and Mental Health Services Administration
SCPS	State Court Processing Statistics
SCS	School Crime Supplement (periodic supplement to the National Crime Victimization Survey)
SIFCF	Survey of Inmates in Federal Correctional Facilities
SILJ	Survey of Inmates in Local Jails
SISCF	Survey of Inmates in State Correctional Facilities
SISFCF	Survey of Inmates in State and Federal Correctional Facilities
SJS	State Justice Statistics
SRS	Summary Reporting System (of the Uniform Crime Reporting program)
SSV	Survey on Sexual Violence (data collection for the Prison Rape Elimination Act of 2003)
STRIDE	System to Retrieve Information from Drug Evidence (Drug Enforcement Administration)
SVORI	Serious and Violent Offender Reentry Initiative
T-CHRIP	Tribal Criminal History Record Improvement Program
Triple I	Interstate Identification Index; also, III
UCR	Uniform Crime Reporting program *or* Uniform Crime Reports
USC	U.S. Code

Abstract

THE BUREAU OF JUSTICE STATISTICS (BJS) of the U.S. Department of Justice (DOJ) is one of the smallest of the U.S. principal statistical agencies but shoulders one of the most expansive and detailed legal mandates among those agencies. BJS requested that this panel be convened to examine the full range of BJS programs and suggest priorities for data collection. We described the current methods of and future options for the National Crime Victimization Survey (NCVS) in an interim report (National Research Council, 2008b). This final report considers the balance of BJS's portfolio, its assistance to state and local authorities, and the functions of BJS as a whole.

We conclude that BJS's data collection portfolio is a solid body of work, well justified by public information needs or legal requirements and a commendable effort to meet its broad mandate given less-than-commensurate fiscal resources. We identify some major gaps in the substantive coverage of BJS data, such as white-collar crime, civil justice, juvenile justice, and contextual factors such as the interaction between drugs and crime. However, the methodological challenges involved in filling these major gaps preclude doing so under BJS's current funding; it would require increased and sustained support in terms of staff and fiscal resources.

BJS generally espouses the principles and practices of a federal statistical agency, but it has sustained major shocks to its position of independence as a national statistical resource in recent years. We suggest two strong organizational measures to reduce the likelihood that BJS and its officials are inappropriately treated in the future. Concluding that BJS's current administrative position within the Office of Justice Programs (OJP) is detrimental to the agency's function, we recommend that BJS be moved out of OJP. We further recommend that the position of BJS director be made a fixed-term presidential appointment with Senate confirmation.

BJS's independence as a statistical agency would be enhanced by fuller use of its flagship study. The NCVS has unique value in providing insight on the etiology as well as the characteristics of crime not reported to police. It is critically important for the NCVS to continue to provide annual estimates of levels and changes in criminal victimization—and be funded commensurately—but also that the NCVS's substantive reach grow through the use of topic supplements.

BJS's individual data series are of generally high quality but would benefit from attention to explicit conceptual frameworks on several levels. Most generally, the interrelationships of BJS's current set of collections are not always immediately clear; this is particularly so for BJS's law enforcement collections, the utility of which have been hurt by an overly restrictive focus on management and administration issues. Core-supplement frameworks should be implemented within BJS's major surveys, streamlining recurring basic content to a simplified "core" and adding structured topic supplements. In BJS's data series on adjudication, we urge a third type of framework—progression toward a more rigorous basis in probability sampling as computerized case management systems become more accessible.

The nation currently has two principal indicators of crime and justice: BJS's NCVS and the FBI's Uniform Crime Reporting (UCR) program, the latter of which covers crimes reported to the police. Both these series have unique strengths in studying crime but share the common problem of lengthy lag times between data collection and the release of the results. We suggest that BJS study the feasibility of compiling crime incident data already maintained in individual police departments' electronic systems. This new collection is not intended to duplicate the UCR, as it would not involve local police staff to record counts in a prescribed fashion; it is simply intended as a way to leverage the availability of existing local data and to produce a quick indicator of general national crime trends.

BJS data cover all the steps in the criminal justice process but, almost exclusively, this coverage is cross-sectional in nature. We see a longitudinal approach as essential to study the performance of the justice system as a whole. We recommend a variety of strategies for improving longitudinal structures, ranging from improving the linkage capacity of existing data to fielding panel surveys of crime victims or persons leaving incarceration.

Outreach and dissemination are areas in which BJS has made laudable strides. Its network of state Statistical Analysis Centers (SACs) stands as a strong example of federal-state cooperation. The network benefits BJS in terms of feedback and the inventiveness of research performed by the SACs, while the SACs benefit from technical assistance that would be cost-prohibitive to provide on their own; we urge continued strengthening of the BJS-SAC relationship. To further strengthen outreach, we suggest that BJS create a standing advisory committee and make continued use of ad hoc user and stakeholder workshops.

We have avoided ranking data collections for several reasons, among them that the current collections are necessary for coverage of events in the justice system; elimination of data series would make BJS appear more visibly to fail to fulfill its massive legal mandates. However, this report suggests a mix of short- and long-term ideas for improving the evidence with which crime and justice policy in the United States is developed. The strategic goals we suggest through this report provide BJS a set of principles against which the content of its data collection portfolio can be assessed. In its thirtieth year, BJS can look back on a solid body of accomplishment; our work in this report suggests further directions for improvement to give the nation the justice statistics—and the BJS—that it deserves.

Summary

THE BUREAU OF JUSTICE STATISTICS (BJS) of the U.S. Department of Justice (DOJ) requested that the Committee on National Statistics of the National Academies (in cooperation with the Committee on Law and Justice) convene this Panel to Review the Programs of the Bureau of Justice Statistics. The panel has a broad charge to:

> examine the full range of programs of the Bureau of Justice Statistics (BJS) in order to assess and make recommendations for BJS' priorities for data collection. The review will examine the ways in which BJS statistics are used by Congress, executive agencies, the courts, state and local agencies, and researchers in order to determine the impact of BJS programs and the means to enhance that impact. The review will assess the organization of BJS and its relationships with other data gathering entities in the Department of Justice, as well as with state and local governments, to determine ways to improve the relevance, quality, and cost-effectiveness of justice statistics. The review will consider priority uses for additional funding that may be obtained through budget initiatives or reallocation of resources within the agency. A focus of the panel's work will be to consider alternative options for conducting the National Crime Victimization Survey, which is the largest BJS program. The goal of the panel's work will be to assist BJS to refine its priorities and goals, as embodied in its strategic plan, both in the short and longer terms. The panel's recommendations will address ways to improve the impact and cost-effectiveness of the agency's statistics on crime and the criminal justice system.

This is the panel's second and final report. At BJS's request, the panel devoted the first phase of its work to considering broad options for conducting the National Crime Victimization Survey (NCVS), arguably the agency's flagship data collection and certainly its most expensive. We described such options in detail in our interim report, National Research Council (2008b), and the summary of that report is included in this volume as Appendix B. In this final report, we complement our work on the NCVS by considering the balance of BJS's data collection series, its work in providing assistance to state and local authorities, and the functions and priorities of the agency as a whole.

Established in its present form in 1979, BJS is one of the smallest of the principal agencies in the U.S. federal statistical system, yet it shoulders one of the most expansive and detailed legal mandates among the statistical agencies. Formally, BJS is tasked to provide information on crime and the operation of the justice system, across all levels of government. Like its administrative parent, the Office of Justice Programs (OJP), BJS was originally created as part of the former Law Enforcement Assistance Administration (LEAA); accordingly, BJS also inherits the LEAA's strong mandate to assist state and local authorities—both financially and technically—in developing justice information systems.

ASSESSMENT OF THE BJS DATA COLLECTION PORTFOLIO

In our review of BJS programs, we have used as a guide a model of the series of events in the justice system that was first developed by the President's Commission on Law Enforcement and Administration of Justice (1967). Graphically, this model resembles a funnel, in which large numbers of crimes and victimization incidents are processed—in decreasing numbers—through major parts of the system, including police work, prosecution and adjudication, and correctional supervision.

Through our review of existing BJS data collections and the manner in which they map to the funnel model of the justice system, we conclude that BJS's data collection portfolio is a solid body of work, generally well justified by public information needs or legal requirements. It represents a better-than-good-faith effort to marshal data relevant to an astoundingly large mandate, given that fiscal resources have typically been less than commensurate. Within its resources and the topics it has chosen to address, BJS has done well in the sense that nothing in its portfolio is obviously frivolous, wasteful, or inconsistent with its legal mandates.

Our basic observations about BJS's portfolio include:

- BJS's key *victimization* series, the NCVS, is its most expensive, most flexible, and most scrutinized collection. It is also, arguably, the agency's most underutilized collection, undercut by scarce resources, diminishing sample size, and—to a degree—a lack of innovation in analyzing and promoting the data.

- BJS's *corrections* data series is a good example of a well-designed and integrated system of collections, wisely using different methods (personal interviewing and facility or administrative records) at certain time intervals for a range of facilities (prisons, jails, support agencies). The set of collections is designed such that censuses build the frame for subsequent, more detailed surveys. Going forward, the challenge for BJS's efforts in the general area of corrections is to expand its coverage to include prisoner reentry issues, broadly, and to improve on its previous solid (but infrequent) studies of recidivism.

- BJS's work in *law enforcement* is hindered by an overly restrictive focus on management and administrative issues, with little direct connection to data on crime, much less providing the basis for assessing the effectiveness of police programs. It is also in the area of law enforcement—with the proliferation of numerous special-agency censuses and little semblance of a fixed schedule or interconnectedness of series—where the need for refining the conceptual framework for multiple data collections is most evident.

- Our critique of BJS's work in *adjudication* is more a reflection on the general difficulty of measurement in the justice system than a criticism of BJS. Information systems in state court systems—and, indeed, the structure and jurisdiction of those courts—vary strongly in their accessibility and sophistication. The dominant impression is of the agency (with its data collection partners) doing the best it can with what it has. That said, there are numerous areas where improvement is needed: bolstering the adjudication series' basis in statistical sampling and patching important gaps in statistical coverage of the justice system "funnel" (particularly declinations to prosecute and out-of-court settlements) would dramatically upgrade the relevance and utility of BJS's data series.

Through comparison to the funnel model, we identify four major substantive gaps in BJS's portfolio:

- White-collar crime, including various types of fraud, public corruption, and Internet crimes;

- Civil justice (as opposed to criminal justice) matters—ranging from prosecution of nonviolent crimes to property disputes to divorce and custody arrangements—which are currently covered by only a single BJS collection (the Civil Justice Survey of State Courts) that, by design, is not able to collect information on out-of-court settlements;

- Juvenile justice, authority for data collection of which is largely ceded to BJS's fellow OJP bureau, the Office of Juvenile Justice and Delinquency Prevention (OJJDP); and

- Contextual factors such as the interaction between drugs and crime that cut across various parts of the system and are thus difficult to measure completely and consistently.

These are important topics and ones on which the principal statistical agency of DOJ should be able to speak authoritatively. However, each of these areas involves major methodological complexities; to fill any of these gaps in BJS's portfolio would require increased and sustained support from Congress and the administration in terms of staff and fiscal resources. In the interim, we recommend (Recommendation 2.1) that BJS play a stronger "clearinghouse" role related to these and other substantive gaps, documenting and organizing those statistics that are available and pursuing research on what new statistics could be feasibly and usefully developed.

Our review of BJS programs leads us to suggest four broad strategic goals for the agency over the coming years.

A STRONG POSITION OF INDEPENDENCE

Strategic Goal 1: To establish and maintain a strong position of independence as a statistical agency; to serve as an independent and objective source of statistical information on crime and the administration of justice.

BJS generally espouses the expected principles and practices of a federal statistical agency, but it has sustained major challenges to its independence as a national statistical resource in recent years. These include:

- An attempt by Justice Department officials to alter the content and substantive emphases of a statistical press release announcing new estimates on police-public contact during traffic stops, an incident that led to the dismissal of a BJS director;

- The legal imposition, through the Prison Rape Elimination Act of 2003 (PREA), of reporting requirements on sexual violence in correctional facilities that oblige BJS to play a regulatory role, holding individual facilities up for sanction; and

- Legislative changes that have the effect of tethering BJS to the policy objectives of its immediate administrative parent, OJP, and DOJ generally.

BJS's uniquely precarious position—as a statistical agency nested within a program agency (OJP) that is dedicated to furthering policy objectives, nested in turn within the nation's principal law enforcement department— means that its independent function as a source of independent and objective information is under constant threat. For this reason, we identify the building of a position of independence as a statistical agency as BJS's chief strategic goal. The panel concludes that this goal requires strong corrective actions: changing BJS's administrative placement within DOJ and the term of service of the BJS director.

Organizational Changes

There exists no organizational arrangement that can completely shield a statistical agency from threats to its independence, guaranteeing complete freedom from political or structural interference or the appearance thereof. That said, the measures we recommend are strong and should reduce the likelihood that BJS and its officials are inappropriately treated.

> *Recommendation 5.3:* **BJS should be administratively moved out of the Office of Justice Programs, reporting to the attorney general or deputy attorney general.**

> *Recommendation 5.4:* **Congress and the administration should make the BJS director a fixed-term presidential appointee with the advice and consent of the Senate. To insulate the BJS director from political interference, the term of service should be no less than 4 years.**

By these recommendations and the suggested strategic goal statement, we also suggest that preserving a strong position of independence should be BJS's first criterion for undertaking new data collections or revising existing ones. Because of the experience with PREA reporting requirements, we recommend (Recommendation 5.1) that BJS should not provide or be required to provide individually identified data in support of functions that compromise BJS's role—functions that involve collecting and analyzing statistical data for policy-furthering, tactical, regulatory, and operational purposes. We further reinforce guidance issued by the U.S. Office of Management and Budget in the wake of the press release incident described above, recommending that any Justice Department review of BJS statistical products and related communications preserve the content, the release schedule, and the mode of dissemination planned by BJS (Recommendation 5.2).

BJS should position itself as a statistical *resource* to DOJ, not an "arm" for the furtherance of any policy objectives. To this end, reciprocal outreach and cooperation are necessary: The BJS director needs to reach out to other agencies within DOJ, forming partnerships to propose information collection initiatives (Recommendation 5.5), and DOJ needs to build provisions for BJS collection of statistical information—the raw material for monitoring and evaluation—into its program initiatives aimed at crime reduction (Recommendation 5.6).

Ensuring the Quality of the NCVS

Statistical agencies can safeguard their independence through the quality and visibility of their products. They are better suited to ward off threats to independence if their data are essential indicators of national well-being; this produces support by a vocal set of stakeholders who are committed to their objectivity and accuracy. In our assessment, the NCVS can and should be positioned as such a critical social indicator, with its unique ability to provide insight into the etiology of crime and the characteristics of crimes not reported to police. As such, the NCVS should have clear primacy in BJS's resource allocations. It is sufficiently core to BJS's legally mandated duties and its basic function as a statistical agency that it is difficult to imagine an effective BJS without a strong and continuing NCVS. Accordingly, we reiterate in this report two recommendations that we first offered in our interim report:

> *Recommendation 6.1:* **BJS must ensure that the nation has quality annual estimates of levels and changes in criminal victimization.**

> *Recommendation 6.2:* **Congress and the administration should ensure that BJS has a budget that is adequate to field a survey that satisfies the goal in Recommendation 6.1.**

We further recommend (Recommendation 6.4) that any additional resources made available for the NCVS should be used not only to increase the reliability of annual estimates (i.e., rebuild the NCVS sample size) but also to supplement the survey in ways that increase understanding of criminal victimization and keep the content of the survey fresh. This includes seeking ongoing topic supplements such as the School Crime Supplement, with the regular support and cooperation of other federal agencies. It also includes fuller use of the NCVS to measure citizen involvement with other parts of the system (e.g., experiences with adjudication procedures and law enforcement). Given the utility of NCVS information on the needs of victims and the compensation and assistance goals of the Justice Department's Office of Victims of Crime—we also suggest that Congress permit the use of funds obtained through the Victims of Crime Act to provide some funding for the collection and improvement of victimization data (Recommendation 6.3).

Continuing and Strengthening State Justice Statistics Program and SAC Partnerships

Through its State Justice Statistics program, BJS supports the operation of Statistical Analysis Centers (SACs) in states and several U.S. territories. The SAC network includes a mix of organizational arrangements—some

SACs function as independent state agencies, whereas others are affiliated with state justice or law enforcement departments or academic institutions.

In the panel's assessment, the BJS-state SAC network is a strong example of the federal-state cooperative systems that are important to statistical agency functioning. Support for the SAC network stands as a relatively low-cost activity on BJS's part with great dividends in terms of outreach and feedback, as well as dissemination of data and products to state policy makers. Moreover, the SAC network is mutually beneficial—states gain from the technical assistance that might otherwise be unavailable or cost-prohibitive to obtain and from national-level benchmarking that is useful for framing policy developments in the states, and BJS benefits from the inventiveness of research performed by its SAC partners. As a vital part of BJS's operations, we urge continued strengthening of the relationship between BJS and its SAC partners:

> **Recommendation 4.1:** **Through its Statistical Analysis Center and State Justice Statistics programs, BJS should continue to develop its ties with the states, and more fully exploit the potential for using states as partners in data collections.**

We further suggest that the experience of BJS's SAC partners be tapped as BJS pursues methodological improvements, including developments toward longitudinal and small-area measurement systems (Recommendation 4.2).

BUILDING STATISTICAL SYSTEMS AND CONCEPTUAL FRAMEWORKS

Strategic Goal 2: To build, maintain, and utilize statistical systems that describe the extent and characteristics of crime in our nation and the status and response of the justice system.

More than a basic statement of topic, the language of this goal suggests attention to "statistical systems" in two basic respects: first, ensuring that BJS's data collection portfolio has a greater sense of coherence and interconnectedness than it does at present and, second, making use of computerized record systems—including those supported by BJS's grants for system development—to develop new sources of data.

Conceptual Frameworks

Based on our review of past and current BJS data collections, several of our recommendations concern the refinement of conceptual frameworks for what BJS does, with respect not only to the whole structure of BJS's portfolio, but also to the design of BJS collections in major topic areas.

With respect to the whole portfolio, we recommend that BJS's strategic plan articulate a blueprint of interrelated data collection and product activities that could then be used to evaluate new opportunities (Recommendation 3.7). The BJS-developed mapping of data series to the criminal justice system funnel model is an important first step in this regard; our concern is that the unique characteristics, shared areas of substantive overlap, and interrelationships of BJS's current set of collections are not always immediately clear. We find this lack of an overall framework to be most striking in BJS's law enforcement data collections, where the agency's large number of special-agency censuses has an unmistakable scattershot appearance. In particular, we recommend that BJS develop an integrated conceptual plan for the periodicity of these law enforcement agency censuses and surveys, abandoning those with no meaningful prospect of being repeated (Recommendation 3.10). New one-time collections should have a clearly defined role within the broader portfolio.

Within BJS's major survey vehicles, we urge the implementation of a core-supplement framework, where the base content of the survey is reduced to a simplified common "core" of recurring content and coupled with recurring, structured topic supplements. We recommend such a structure for the NCVS in Recommendation 3.8 and for the Law Enforcement Management and Administrative Survey (LEMAS) in Recommendation 3.9. We think that such a structure is an important part of overcoming limitations in BJS's current holdings in the area of law enforcement generally, which are currently principally limited to the organizational and administrative focus of LEMAS. We suggest the use of LEMAS topic supplements to expand the substantive scope of that survey and the analysis of LEMAS-type data in combination with crime data (thus coming closer to informing assessment of police policy effectiveness). We further recommend (Recommendation 3.11) that the NCVS and its supplements be used more effectively as a tool for studying law enforcement, both in terms of the types of crime that are reported (and not reported) to police and the action that results from the reporting of a crime (such as has been done by the current Police-Public Contact Survey supplement).

As noted above, the panel's general impression of BJS's data collections in the area of adjudication is one of the agency doing the best it can with what it has—making good use of data collection partners but facing the basic problem of wide variation in the operational structure of state courts, in their levels of automation, and their use of computerized case management systems. Thus, it is more a statement on the condition of court information systems than on BJS's efforts that we conclude that BJS's adjudication series lack an effective basis in sampling. We recommend (Recommendation 3.12) that BJS work to implement more rigorous methods of probability sampling in its adjudication series as court records become more accessible through computerized case management systems; its recent steps to redesign the State Court Processing Statistics collection exemplify developments in this regard. We further recommend that BJS develop a research program to build representative samples of courts and to assess strategies for collection of case records for even such a small, but representative, sample (Recommendation 3.13). One major omission in BJS's current work in court statistics is information on declinations to prosecute and the use of alternative dispute resolution techniques; collecting these data in full detail would be a major enterprise, but we suggest that one low-cost way to get at least some information on this missing piece would be to add some basic questions on aggregate case processing to the National Prosecutors Survey, which currently has a purely administrative focus.

Making Use of Record Systems

Interpreting "statistical systems" more technically, this strategic goal statement also includes two recommendations regarding the use of computerized record systems.

First, it is part of BJS's explicit legal mandate to provide both technical and financial assistance to state and local authorities for purposes of developing information resources. Much of this work has been done under BJS's National Criminal History Improvement Program (NCHIP), under which BJS provides grants to local authorities to help their development of criminal history record databases. These local databases are used, in turn, to populate national-level record databases maintained by the Federal Bureau of Investigation (FBI) that are used, among other things, to conduct background checks for firearm purchases. From the purely statistical standpoint, BJS's NCHIP work is ripe for criticism because BJS's role is strictly a money-transfer operation; although it pays to develop the local databases, BJS has no access to the resulting data for research purposes. Other than generating rough summary statistics of firearm-purchase background checks, BJS has not been able to utilize the criminal history record data that its grant monies help to develop at the state and local levels. This failure has occurred despite a formal set-aside of funds in all of those grants for BJS evaluation purposes. Very recently, however, the FBI has begun to permit BJS the same access to criminal history records that is available to law enforcement agencies. Access to compiled history record data for research purposes opens very exciting avenues for study, which BJS intends to first explore in studies of prisoner reentry to the community and recidivism. We encourage BJS to pursue this work and expand it to include broader studies of criminal behavior:

> *Recommendation 4.3:* **BJS should actively utilize the NCHIP program to improve criminal history records necessary for longitudinal studies of crime.**

Second, we suggest that BJS study the feasibility of a records-based data collection that would be capable of providing a timely indicator of major crime trends. In the panel's assessment, the panel is well served by having multiple indicators of crime and justice in the United States; the FBI's Uniform Crime Reporting (UCR) program has the weight of being an official measure of crimes reported to police while BJS's NCVS has unique strengths in describing the etiology and contexts of crime and violence (including those incidents not reported to police). However, both series suffer in public perception from lengthy lag times between data collection and release of results. It has been argued that the UCR program would be better administered if it were transferred from the FBI to BJS; we do not recommend such a transfer at this time, arguing that the move would pick unnecessary turf fights and would incur a much more intensive—and costly—redesign than the simple organizational switch suggests at first glance. However, we suggest that BJS study the feasibility of an entirely different approach that is better in some respects, in order to determine whether crime incident data collected by local police departments on their own (and, in some cases, published on their websites) can be compiled on a probability sample basis. This new collection is not intended to be duplicative of the UCR, because it puts the burden of structuring the sample and compiling the data on BJS rather than requiring local departments to compile and record the counts in the UCR format; rather, it is simply intended to gain leverage from the availability of existing local data.

> **Recommendation 4.4:** To improve the timeliness of crime statistics, BJS should explore the development of a crime reporting system based on a probability sample of police administrative records. The goals of such a system would be national representativeness, high response, high data quality, timeliness and flexibility in terms of crime classification and analysis, and national statistics for the monitoring of crime trends.

IMPROVING COVERAGE OF THE JUSTICE SYSTEM

Strategic Goal 3: To provide comprehensive statistical coverage of all parts of the criminal justice system, including the longitudinal flow of persons and cases through the system and their return to the community.

Through this goal statement, we return to our analysis of the funnel model for the criminal justice system sequence of events and our conclusions from that analysis. Accepting the funnel model as a premise, this goal might be restated as saying that data collections should rise and fall in importance relative to their proximity to the funnel: The further a specific phenomenon is from the activities described by the funnel model, the less central it is to the mission of BJS. By its nature, the funnel model has strong implications for priority setting. First, it suggests that BJS should focus on activities that affect the most people and affect them most extensively. By definition, this means the incidence of crime and victimization, which the funnel graphically describes as the most prevalent phenomenon in the criminal justice system. Following this logic, and all other things being equal, describing crime, victimization, and the immediate consequences should be the principal focus of BJS. Second, however, the nature of a funnel also directs attention to the late stages of the process. As offenders make their way through the funnel, their numbers decrease but the consequences of justice system decisions increase, in terms of impacts on lives and public budgets. At the far end of the criminal justice system, in the correctional area, the effects of criminal justice decisions are so extensive that obtaining adequate data on these populations has importance well beyond their numbers. This principle for establishing the importance of any given activity need not correspond to the allocation of resources on that activity because some data can be collected more cheaply than others. Nonetheless, it should guide decisions as to where to put the intellectual energy of the agency if not in a one-to-one correspondence with its fiscal resources. A logical decision process may be to separately value activities by their proximity to the funnel, and then assess their relative cost; the final portfolio is a function of both importance and cost.

The word "longitudinal" in this goal statement is particularly critical. As the mapping of BJS data series to the funnel makes clear, BJS generally has good coverage of all the steps in the criminal justice process but, almost exclusively, this coverage is cross-sectional in nature. BJS's steps in true longitudinal measurement of persons and cases through multiple steps of the criminal justice system have been rare; in large part, this is due to the structural problem that there is no common identifier attached to a person (or case) that follows the person (or case) through all major steps (police, court processing, entry into corrections, and reentry into the community) to facilitate easy longitudinal measurement. We see a longitudinal approach as essential to study of the performance of the justice system as a whole, and we recommend a variety of strategies for improving longitudinal designs. Some of these are low-cost techniques that simply build on longitudinal structures within existing data resources such as building methods for record linkage into existing and emerging correctional data collections; one example is linking data from recidivism studies to the Census of Adult Correctional Facilities to understand how correctional facilities and programming affect recidivism (Recommendation 3.3). We also recommend that BJS work to emphasize transitions and flows in prison and jail populations, whereas BJS's current publications tend to emphasize current "stocks" or fixed-time counts of inmates. We suggest that BJS's regular correctional data products include both stock (Recommendation 3.1) and flow (Recommendation 3.2) information capable of disaggregation and cross-tabulation by state, offense category, and demographic group. We recognize that other suggestions are more costly, but would ultimately be beneficial in enhancing understanding of the justice system:

- Conduct research on the "common identifier" problem discussed above, determining the feasibility of measuring individuals' experiences in the justice system on a prospective, longitudinal basis, beginning as early as practicable in the process (arrest) and ending with their eventual exit (ranging from early dismissal of charge through completion of sentence) (Recommendation 3.4);

- Develop an approach to measure the victimization of experiences of *individuals* (as contrasted with the NCVS focus on households) on a longitudinal basis, beginning from a focal victimization and following the victim forward in time, measuring subsequent victimizations (Recommendation 3.5); and

- Develop a panel survey of people under correctional supervision to understand how individuals move between institutional and community settings, and to understand the social contexts of correctional supervision (Recommendation 3.6).

To extend this last point, BJS and the Justice Department should look for opportunities to leverage studies called for by the Second Chance Act of 2007 (enacted in April 2008) to build an active program on recidivism and reentry. Specifically, we recommend that BJS mount a feasibility study of the flow of individuals between correctional supervision and community settings. Repeated interviews of samples of about-to-be-released prisoners that track their successes and failures in reintegrating with the community would enhance understanding of this critical policy issue (Recommendation 3.14). We also suggest that BJS explore ways to reactivate its studies of probationers and parolees.

Under this general heading of justice system coverage, we further recommend that BJS continue to develop and study the measurement of emerging or hard-to-reach groups and develop more appropriate approaches to sampling and measurement of these populations (Recommendation 5.14). In addition, BJS (in conjunction with OJJDP) should develop juvenile victimization, crime, and justice statistical series suitable for describing the patterns of offending and victimization of youth, as well as studies of the longitudinal progression of youth through the juvenile and criminal justice systems (Recommendation 2.2).

FACILITATING ACCESS AND IMPROVING DISSEMINATION AND OUTREACH

Strategic Goal 4: To ensure access to statistics and data on crime and justice by the American public, the U.S. Congress, the U.S. Department of Justice and other executive agencies, and state and local government agencies.

Outreach and dissemination are areas that already play considerable roles in BJS operations and in which BJS has made laudable strides. We have already noted and endorsed BJS's formation of a strong communications and contact network with the states through its SACs. BJS also deserves credit for the extensive backfile of reports and summary tabulations that are available from the agency's website and the dissemination of its reports through the OJP-administered National Criminal Justice Reference Service. High-end users are well served by BJS's microdata holdings that are available for download through the National Archive of Criminal Justice Data. Through this goal statement, we encourage still further developments along these lines as BJS works to increase its public profile and its relevance in policy debates, while maintaining the high quality standards it has set for itself.

In the panel's assessment, there is value in BJS pursuing both formal and informal mechanisms for obtaining user input and feedback and for shaping possible new data collections. The law that created BJS in 1979 also created a formal advisory board for the agency, a provision that was later omitted in reauthorization language in 1984. We recommend that such an ongoing advisory group would be a useful vehicle for obtaining recommendations from diverse perspectives; specifically, we recommend that BJS establish an Advisory Group under the Federal Advisory Committee Act, the membership of which should include—at a minimum—leaders and practitioners from each of the major subject matters covered by BJS data, as well as those with statistical and other types of academic expertise (Recommendation 5.8). In the past, BJS has also made good use of ad hoc user and stakeholder workshops on targeted issues; the agency convened a very useful data user conference in February 2008 in partial support of this panel's activities. We recommend that BJS continue to hold such ad hoc stakeholder workshops to suggest areas of immediate needs or to get input on contemporaneous topics of interest (Recommendation 5.7).

The data extracts directly available from the BJS website tend to be selected tabulations and spreadsheets. As BJS works to enhance its website presence, it should explore methods for making more of its data available for direct tabulation, analysis, and—perhaps—mapping on the website. We also recommend that BJS take care in managing how some of its data collections (gathered by external data collection agents) are housed on non-BJS websites; BJS should articulate and describe the process by which links to external sites are allowed and used, and should also articulate and justify the use of its insignia—which carries with it important quality connotations—on external websites (Recommendation 5.12).

From the numerous references to BJS data and findings in legislative text and debate records, it is clear that BJS has a key and receptive audience in the U.S. Congress and its staff. However, from the types of legislative demands that are sometimes put on BJS (e.g., the PREA reporting requirements) and past difficulties BJS has experienced in securing funding, it is not quite as clear that the dialogue between BJS and Congress is effective. (At present, of course, a major reason for this problem in communication is that liaison with Congress is done principally by and through OJP; for instance, BJS directors do not directly testify to appropriations committees on agency needs.) Accordingly, we recommend that—regardless of the organizational structure of BJS and OJP—BJS should take a stronger role in cultivating a relationship with Congress:

> *Recommendation 5.9:* **DOJ should take steps to ensure that congressional staff are aware of BJS data that could be used in developing legislation; DOJ and BJS should learn from congressional staff how their data are needed to inform/support legislation so that they can improve the utility of their current data and so that they can develop new datasets that could enhance policy development.**

From the technical standpoint, an important means for BJS to increase its public relevance is to find ways to address long-standing concerns about the agency's data products: though they are generally held in high regard, they are frequently seen as lacking timeliness and, particularly for the NCVS, subnational geographic detail. One approach that we recommend to deal with this concern is that BJS evaluate each of its data programs to ascertain whether more timely estimates might be obtained by (a) making discrete data collections into more continuous operations and (b) issuing preliminary estimates, to be followed by final estimates (Recommendation 5.11).

More fundamentally, and consistent with the expected principles and practices of a federal statistical agency, we believe that the cultivation of an active and continuous research program is essential to an agency's progress. For BJS, such research should include work on statistical modeling to improve the temporal and spatial resolution of estimates (a program for which BJS could partner with other federal statistical agencies, most of which face similar pressures from their user constituencies). A vigorous research program could also suggest ways to better and more efficiently make use of state and local database resources, such as through steps to increase the automation and consistency of BJS's National Corrections Reporting Program data (Recommendation 5.10). The presence of a solid research program would also allow BJS to do something that sounds obvious—carefully study changes in the NCVS design before implementing them (Recommendation 5.13)—but that is difficult to accomplish in a climate of constrained resources. To carry out a research program and continually improve the agency's data collection holdings, we recommend that BJS improve the technical skill mix of its staff, including mathematical statisticians, computer scientists, survey methodologists, and criminologists (Recommendation 5.15).

PRIORITY SETTING AND CONSTRAINED RESOURCES

A possible criticism of this report is that it tends to suggest *adding* to BJS's inventory of collections rather than *subtracting*. The recommendations are generally geared to improvements within BJS's various existing data collections—for instance, ensuring a high-quality independent measure of crime in the NCVS, emphasizing conceptual frameworks in BJS's adjudication and law enforcement collections, and expanding its corrections series to study prisoner reentry into society. As a panel, we grappled with questions regarding this: How should these improvements be paid for? Should some existing series be cut to make ways for new ones? And, given the disproportionate share of BJS's current resources consumed by the NCVS, should some smaller collections be stopped to free up additional resources for the NCVS or should some NCVS resources be steered to other areas?

The reason why we rejected explicitly suggesting that some data series be cut in order to pay for others is certainly not that the current collections are perfect. The improvements we suggest in the recommendations are testimony to that. Rather, we believe that BJS cannot achieve its legislated goals by cutting programs. In our assessment, we think it can be stated as a fact: BJS has been given more responsibilities than can be achieved with current resources. The resources provided to BJS to carry out its work are not commensurate with the breadth—and importance—of the responsibilities assigned to the agency by its authorizing legislation.

Because of this, the agency has for some years walked a tight line of small cuts of sample or measurement, short delays of publications, and temporary hiring freezes—each of these tolerable in itself, but cumulating over the years such that core functions have broken down. On a routine basis, BJS must make decisions about addressing certain responsibilities and not others, trading off periodicity and completeness of data collections, and balancing continuity of existing data collections against the need to comply with directives from Congress and DOJ. Such decisions are hardly unique to BJS—at some point, all organizations must make such trade-offs—but BJS's mismatch between resources and responsibility makes the decisions particularly difficult.

Thus, in setting priorities, BJS directors have perforce had a short time horizon—responding to a certain set of demands even though those decisions may have negative long-term consequences for individual data collections and the health of the agency. Certainly, in the midst of year-to-year juggling of data series in order to keep production moving, longer-term investments in research and innovation become difficult or impossible to make. The most striking example of the consequences of this extremely tough climate is the current state of the NCVS: what was once, clearly, the best victimization survey in the world is now unable to satisfy its basic function of providing annual estimates of level and change in common-law crime. This decay happened gradually as BJS administrators were attempting to respond to immediate exigencies, aggravated by an overly broad mandate. Each single cut in sample size, or other cost-cutting measure, was justifiable given then-present alternatives. Cumulatively—as demonstrated most vividly by the declared "break in series" with the 2006 NCVS data—they lead to our conclusion in our interim report that "the [current] NCVS is not achieving and cannot achieve BJS's legislatively mandated goals" (National Research Council, 2008b:Finding 3.1).

Statistical infrastructure (data collections and record series) shares with physical infrastructure (such as roads, bridges, sewers, and cable) the fundamental problem that it lacks glamour and can be difficult to champion as a government spending priority. Yet it is undeniably important to the public knowledge and the public good; BJS's purview includes topics important to general welfare, spanning the measurement of interpersonal violence, the function and magnitude of law enforcement and corrections, and the operations of the judicial branch of government. The problem is not that there are parts of BJS's legal mandate that are unimportant or unworthy, but that its current resources do not permit it to cover its mandates as effectively as possible. Given this finding, we are loath to construct a list of series to terminate or reduce out of concern that the agency fail even more visibly to fulfill the charge given by its legislation.

Another reason for not considering a specific ranking of data collections in order to suggest possible cuts is consistency with the approach we took in our interim report on options for the NCVS. In that report, we presented an array of possible options and described how specific design choices corresponded to particular goals for the survey. However, we made a point of "not suggest[ing] one single path as the ideal for a redesigned NCVS" (National Research Council, 2008b:4). Different people and decision makers do not necessarily put the exact same weights on the goals and objectives of a program such as the NCVS, and we did not want to presume "that our preferred set of NCVS goals is correct to the exclusion of all others" (National Research Council, 2008b:4). The same logic applies to considering other collections in the BJS portfolio: deeming one collection more worthy than another involves complicated value judgments, not science. Specific constituencies for BJS data that are not represented on the panel or by groups who have spoken before the panel could make eloquent and compelling cases for their particular favored set of statistics; again, we do not wish to suggest that any weighting we could suggest is somehow paramount.

Although hoping that tight fiscal constraints may be alleviated somewhat in coming years, we cannot assume infinite resources either; we have not interpreted our task as assembling a "wish list" for everything that a justice statistics agency *could* do, but rather think that out suggestions are a mix of short- and long-term, low- and higher-cost ideas for improving the statistical evidence with which crime and justice policy in the United States is developed. Though we do not explicitly rank BJS's data collections, our suggested strategic goals provide BJS with a set of principles against which its data collection portfolio can be assessed. BJS, which marks its thirtieth anniversary in its present form this year, can rightly look back on a solid body of accomplishment. But our recommendations in this report suggest that there are many directions for continued improvement; much work remains to be done to ensure the quality, credibility, and relevance of statistics on justice in the United States—to achieve the BJS the country deserves.

– 1 –

Introduction

FINDING CONTINUAL EVIDENCE of "the inadequacy of the available data on crime and the criminal justice system," the President's Commission on Law Enforcement and Administration of Justice (1967:269) recommended that "a National Criminal Justice Statistics Center should be established in the Department of Justice." The commission anticipated that this center would:

> serve as a central focus for other statistics related to the crime problem, such as costs of crime, census data, and victim surveys. It would have to work in close coordination with the FBI's [Federal Bureau of Investigation's] Uniform Crime Reports Section, the Children's Bureau of the Department of Health, Education, and Welfare, the Federal Bureau of Prisons, and other existing agencies with continuing responsibility for collecting and reporting related statistics. It would combine their information into an integrated picture of crime and criminal justice.

The Omnibus Crime Control and Safe Streets Act of 1968 created the Law Enforcement Assistance Administration (LEAA) within the U.S. Department of Justice to provide financial and technical support to state and local law enforcement agencies. The act specifically authorized LEAA to "collect, evaluate, publish, and disseminate statistics and other information on the condition and progress of law enforcement in the several States" (P.L. 90-351 § 515(b); 82 Stat. 207). To meet this mandate, a National Criminal Justice Information and Statistics Service (NCJISS) was founded within the new LEAA.[1] The Justice Systems Improvement Act of 1979 (P.L. 96-157)

---

[1] The NCJISS originally bore the name "National Criminal Justice Statistics Center," drawing directly from the commission's recommendation, but the name was changed in short order.

renamed and authorized the NCJISS as the Bureau of Justice Statistics (BJS) within a new Office of Justice Assistance, Research, and Statistics. Subsequent reauthorization (the Justice Assistance Act of 1984, P.L. 98–473) completed the process of converting the former LEAA into the Office of Justice Programs (OJP), which remains BJS's administrative parent agency within the Justice Department.

From these beginnings, BJS has developed into one of the principal statistical agencies of the federal government. Armed with a broad mandate to provide statistical measures on the justice system, the agency maintains dozens of data collection series. Each year, it releases about 40 bulletins or reports summarizing its findings, disseminated through BJS's own website (http://www.ojp.usdoj.gov/bjs/) or through the OJP-administered National Criminal Justice Reference Service (http://www.ncjrs.gov/). BJS data are generally made available in processed spreadsheets on the BJS website or in microdata form at the National Archive of Criminal Justice Data hosted by the University of Michigan (http://www.icpsr.umich.edu/NACJD/). BJS also supports the maintenance of an online *Sourcebook of Criminal Justice Statistics* (http://www.albany.edu/sourcebook). BJS's report series and dissemination venues are described more completely in Box 1-1, and an illustrative front page of a BJS bulletin is shown in Figure 1-1.

BJS's signature data collection, and its most demanding in terms of budget resources, is the National Crime Victimization Survey (NCVS), an effort that—like BJS itself—is the direct result of a recommendation by the President's Commission on Law Enforcement and Administration of Justice (1967). The NCVS serves as a critical indicator of crime and violence in the United States because it includes crimes that are not reported to police as well as those that are. In this respect, it serves as an important counterpart to the Uniform Crime Reporting (UCR) program of the FBI—the nation's other key indicator of violent crime levels—because the UCR is strictly limited to incidents reported to law enforcement authorities. More significantly, the importance of the NCVS stems from its flexibility as a detailed survey measurement tool, permitting valuable insight into the nature and etiology of victimization incidents as well as public perceptions of and encounters with other parts of the justice system.

Although the NCVS represents a dominant share of BJS's budgetary resources, the balance of BJS's data collection portfolio covers an enormous range of phenomena. BJS is well known for its body of data series on populations under correctional supervision; these provide important information on the levels and dynamics of correctional populations, which are (like other institutionalized populations) commonly excluded from household surveys

---

Box 1-1 Bureau of Justice Statistics Publications and Data Dissemination Venues

Historically, BJS publications have followed a few fairly well-defined types, as described in the *Bureau of Justice Statistics Style Book* (Bureau of Justice Statistics, 1997a:5):

- *Bulletins* summarize findings from new data releases from BJS's more permanent data collections. They are meant to be relatively concise, having five to eight tables, and frequently have a detailed methodology section that describes the processes used to collect the data. The *Criminal Victimization* series based on the National Crime Victimization Survey (NCVS; e.g., Rand and Catalano, 2007) and the midyear count reports from the National Prisoner Statistics Program (e.g., Harrison and Beck, 2006; Sabol and Couture, 2008) are examples of BJS Bulletins.

- *Special Reports* cover "more restricted topics than do Bulletins, describing in 10 to 15 tables statistical relationships among findings from one or more datasets." They are meant to be more individual in nature than the Bulletins. Example Special Reports include analyses of victimization rates by level of urbanicity (Duhart, 2000), detailed concentration on violent felons using State Court Processing Statistics data (Reaves, 2006), tabulations of citizen complaints regarding police use of force from the Law Enforcement Management and Adminstrative Statistics survey (Hickman, 2006), and results from a supplemental survey of civil appeals (Cohen, 2006).

- *Data Briefs* are meant to cover a very limited set of findings and are generally a maximum of two formatted pages. For instance, a special NCVS tabulation of carjacking incidents covers three formatted pages (Klaus, 2004). However, depending on the topic, they can run longer; for instance, the analysis by Mumola (2007b) of cause-of-death information collected as part of the Deaths in Custody Reporting Program has a three-page narrative but includes eight pages of appendix tables. Because of space limitations, Data Briefs do not cover methodology in any depth but can contain references to other, more extensive reports. In recent years, very short reports have also been issued under the *Fact Sheet* label (see, e.g., Hughes, 2007, on analysis of National Center for Health Statistics data on unidentified human remains; and Hickman, 2003, on tribal law enforcement departments).

- *Selected Findings* "gathe[r] the most important facts, statistics, and conclusions from a number of data sources, usually separate statistical reporting programs within BJS." Examples of Selected Findings reports include summaries from both inmate and probationer data sets on prior physical or sexual abuse (Harlow, 1999) and analysis of punitive damage awards in selected large counties (Cohen, 2005b).

Other BJS publications have been labeled as BJS *Technical Reports*, including the results of pilot testing of computer security questions in the NCVS (Rantala, 2004) and results from a special NCVS data file for the New York, Los Angeles, and Chicago metropolitan statistical areas (Lauritsen and Schaum, 2005). Some longer publications are left "unbranded" by any of these labels, such as BJS's tabulation of statistics from various sources of American Indians and their experience of crime (Greenfeld and Smith, 1999; Perry, 2004) and the *Compendium of Federal Justice Statistics* (Bureau of Justice Statistics, 2006a).

(continued)

Box 1-1   (continued)

Established by BJS's parent Office of Justice Programs (OJP), the National Criminal Justice Reference Service (NCJRS; http://www.ncjrs.gov) maintains a library of OJP publications dating to the 1970s, many of which are downloadable online and others that can be ordered in hard copy. In addition to OJP and its component agencies, the NCJRS is also sponsored by the Office of National Drug Control Policy in the Executive Office of the President. "By referral from BJS," NCJRS also "handles major distributions as needed for White House and [Department of Justice] events and attends major conferences representing the statistical products available from BJS" (Bureau of Justice Statistics, 2005a:24).

The National Archive of Criminal Justice Data (NACJD; http://www.icpsr.umich.edu/NACJD/) is the designated official repository for data collections funded by three OJP bureaus: BJS, the National Institute of Justice, and the Office for Juvenile Justice and Delinquency Prevention. In addition to the OJP bureaus, NACJD also collects and disseminates data sets from research projects that are contributed by investigators; it also archives summary data files from the Federal Bureau of Investigation's Uniform Crime Reporting and National Incident-Based Reporting System collections (Heraux, 2007). (Section 4–C describes both collections in more detail.) The NACJD archive was created in 1978 and is hosted by the Inter-university Consortium for Political and Social Research (ICPSR), University of Michigan, and receives funding from the John D. and Catherine T. MacArthur Foundation in addition to the three OJP bureaus.

Arrangements with NACJD and ICPSR also give BJS an important venue for promoting accurate and effective use of BJS data. As part of ICPSR's regular Summer Training Program in Quantitative Methods, BJS sponsors an annual 4-week seminar program on the quantitative analysis of crime and criminal justice data; since 2000, the seminar has focused on key methodological and presentation issues rather than attempting a comprehensive review of all BJS data series. The seminar is open to researchers from academia, nonprofit organizations, and government agencies, but annual attendance is capped at 10 persons.

For fiscal year 2008, BJS's contribution to the maintenance of NCJRS and NACJD were estimated at $1,100,000 and $900,000, respectively.

---

administered by the government.[2] The "justice system" that BJS monitors with its data collections is sprawling, including not only correctional facilities and offices but the entire infrastructure of law enforcement: police departments at the state and local levels and their various support bureaus. It also includes the judiciary: the full array of federal and state courts, which vary strongly in organizational structure and information resources. BJS's

---

[2]An exception is the American Community Survey—the replacement for the traditional decennial census long-form sample—which does include prison populations in its coverage of the "group quarters" population. However, the BJS data series are likely better measures of correctional populations, particularly for shorter-term or mixed-term facilities such as local jails where the decennial census definition of "usual residence" may lead to poor counts; see Section 5–B.11. National Research Council (2006) provides additional discussion of correctional populations and the decennial census.

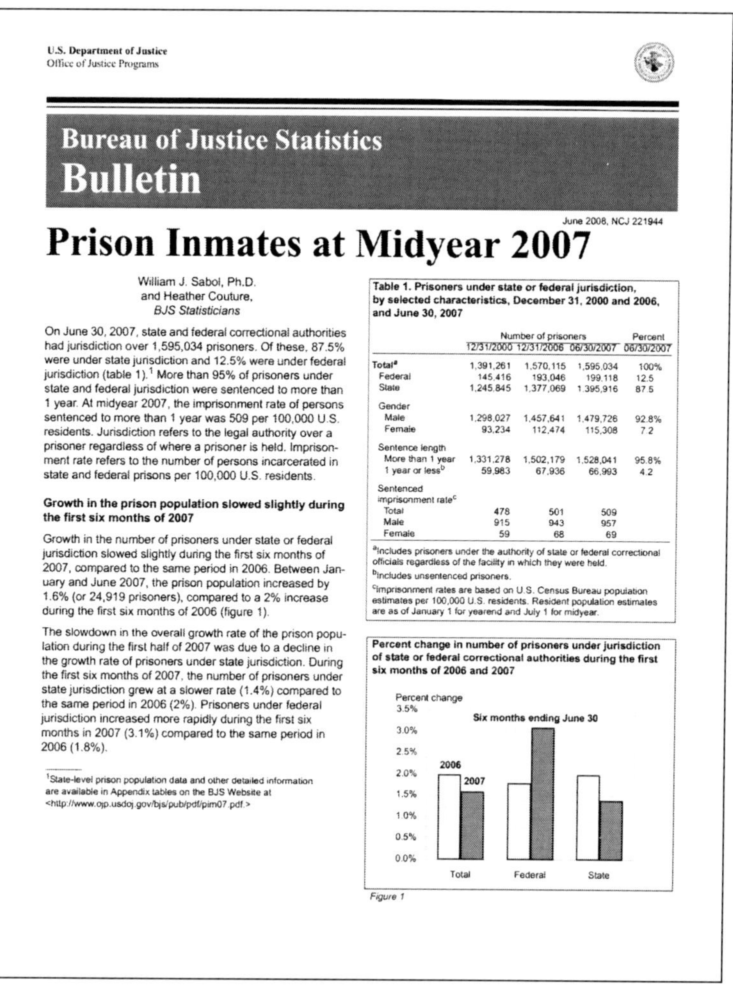

U.S. Department of Justice
Office of Justice Programs

Bureau of Justice Statistics
Bulletin

June 2008, NCJ 221944

Prison Inmates at Midyear 2007

William J. Sabol, Ph.D.
and Heather Couture,
BJS Statisticians

On June 30, 2007, state and federal correctional authorities had jurisdiction over 1,595,034 prisoners. Of these, 87.5% were under state jurisdiction and 12.5% were under federal jurisdiction (table 1).[1] More than 95% of prisoners under state and federal jurisdiction were sentenced to more than 1 year. At midyear 2007, the imprisonment rate of persons sentenced to more than 1 year was 509 per 100,000 U.S. residents. Jurisdiction refers to the legal authority over a prisoner regardless of where a prisoner is held. Imprisonment rate refers to the number of persons incarcerated in state and federal prisons per 100,000 U.S. residents.

Growth in the prison population slowed slightly during the first six months of 2007

Growth in the number of prisoners under state or federal jurisdiction slowed slightly during the first six months of 2007, compared to the same period in 2006. Between January and June 2007, the prison population increased by 1.6% (or 24,919 prisoners), compared to a 2% increase during the first six months of 2006 (figure 1).

The slowdown in the overall growth rate of the prison population during the first half of 2007 was due to a decline in the growth rate of prisoners under state jurisdiction. During the first six months of 2007, the number of prisoners under state jurisdiction grew at a slower rate (1.4%) compared to the same period in 2006 (2%). Prisoners under federal jurisdiction increased more rapidly during the first six months in 2007 (3.1%) compared to the same period in 2006 (1.8%).

[1]State-level prison population data and other detailed information are available in Appendix tables on the BJS Website at <http://www.ojp.usdoj.gov/bjs/pub/pdf/pim07.pdf.>

Table 1. Prisoners under state or federal jurisdiction, by selected characteristics, December 31, 2000 and 2006, and June 30, 2007

	Number of prisoners			Percent
	12/31/2000	12/31/2006	06/30/2007	06/30/2007
Total[a]	1,391,261	1,570,115	1,595,034	100%
Federal	145,416	193,046	199,118	12.5
State	1,245,845	1,377,069	1,395,916	87.5
Gender				
Male	1,298,027	1,457,641	1,479,726	92.8%
Female	93,234	112,474	115,308	7.2
Sentence length				
More than 1 year	1,331,278	1,502,179	1,528,041	95.8%
1 year or less[b]	59,983	67,936	66,993	4.2
Sentenced imprisonment rate[c]				
Total	478	501	509	
Male	915	943	957	
Female	59	68	69	

[a]Includes prisoners under the authority of state or federal correctional officials regardless of the facility in which they were held.

[b]Includes unsentenced prisoners.

[c]Imprisonment rates are based on U.S. Census Bureau population estimates per 100,000 U.S. residents. Resident population estimates are as of January 1 for yearend and July 1 for midyear.

Percent change in number of prisoners under jurisdiction of state or federal correctional authorities during the first six months of 2006 and 2007

Figure 1

Figure 1-1   First page of example from BJS "Bulletin" series of data releases

legal mandate is such that its work is not limited to the collection and analysis of data; it is also tasked with providing financial assistance to state and local governments for the development and maintenance of information resources such as background check databases. Some of BJS's data collections are performed on a regular basis whereas others are more sporadic (or one-time efforts), and some involve direct field data collection and interviewing whereas others are based on administrative and agency records.

Despite this wide scope, BJS has endured effectively flat funding for most of its existence. Figure 1-2 presents BJS's budget requests and total appropriated amounts for each fiscal year since 1981. Converting BJS's final budget allocations into real 2008 dollars, the funding has oscillated slightly but held relatively constant over the entire range. The figure also shows the amount of BJS's budget allocated to the NCVS for each fiscal year since 1990, in nominal dollars; that line suggests an upward creep in the basic costs of survey data collection. During fiscal years 2001 through 2007, the NCVS consumed at least 51.2 percent (in both 2001 and 2002) and as much as 64.0 percent (in 2004) of the total BJS appropriations. These fiscal constraints have led BJS to reduce the sample size of NCVS over time, along with other cost-cutting measures; what began as a survey of 72,000 households in 1972 reached only 38,000 households in 2006. In recent years, the diminishing NCVS sample size has combined with generally low and decreasing estimated overall victimization rates, with the result that only large percentage changes in violent crime victimization rates—at least 8 percent— would be a statistically significant year-to-year change.[3] As noted in the U.S. Office of Management and Budget (2007:8) annual review of statistical program funding, "cost cutting measures applied to the NCVS continue to have significant effects on the precision of the estimates—year-to-year change estimates are no longer feasible and have been replaced with two-year rolling averages" in BJS reports on victimization. In addition to decreased precision in the NCVS estimates, fiscal constraints and the large NCVS share of the overall budget have raised trade-off decisions in the portfolio of BJS programs: Should some resources currently devoted to the NCVS be applied to collections in other areas such as law enforcement or judicial processing (at the expense of NCVS accuracy), or do the variety and extent of non-NCVS collections detract from the accurate victimization-based measurement of crime and violence?

In 2005, BJS was subjected to review under the U.S. Office of Management and Budget's Performance Assessment Rating Tool (PART).[4] In general, PART found fault with only a few areas and BJS received high marks. However, one need that the evaluation suggested was an independent assessment of BJS's effectiveness:

> BJS would benefit from a comprehensive review that could provide specific evidence of BJS's impact overall. Major reviews of BJS statistical

---

[3]This figure is from a presentation by Michael Rand, BJS, at the panel's first meeting. For 2005, the violent crime victimization rate was estimated as 21.2 per 1,000 population, and the 95 percent confidence interval of this rate is ±8.1 percent. Since 2000, the estimated annual violent crime victimization rates have dipped from 27.9 to 21.2 per 1,000, and the 95 percent intervals have been in excess of ±7 percent for each of those annual estimates. These figures are discussed further in the panel's interim report (National Research Council, 2008b:App. C).

[4]The detailed report of the PART evaluation is accessible at http://www.whitehouse.gov/omb/expectmore/detail/10003805.2005.html.

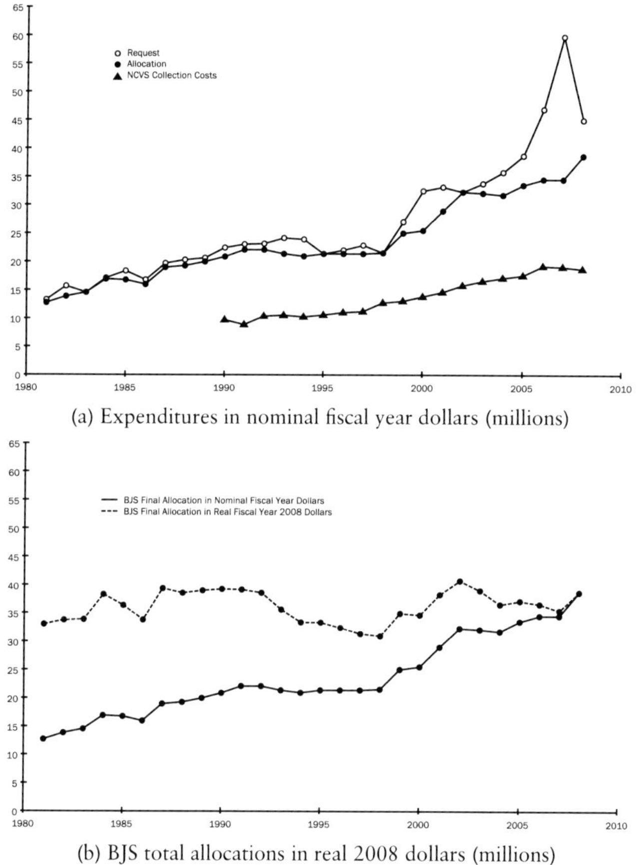

(a) Expenditures in nominal fiscal year dollars (millions)

(b) BJS total allocations in real 2008 dollars (millions)

Figure 1-2 Bureau of Justice Statistics budget requests to Congress (fiscal years 1981–2009), final total appropriations (1981–2008), and National Crime Victimization Survey (NCVS) data collection costs (1990–2008)

NOTES: Budget request and appropriations figures include both base and program costs. Final appropriations reflect amounts after any applicable budget recission or across-the-board cut. NCVS data collection costs exclude additional costs for developing and respecifying sample based on new decennial censuses, as well as costs associated with the automation (conversion to computer-based administration) of the survey. For fiscal year 2008, NCVS spending does not include an additional $3.9 million designated for redesign activities. Consistent with the approach used in National Research Council (2009:Table A-1), nominal dollars are converted to real 2008 dollars by the gross domestic product chain-type price indexes for federal government nondefense consumption expenditures (based on Table 3.10.4, line 34, at http://www.bea.gov/national/nipaweb/SelectTable.asp?Selected=N#S3).

SOURCE: Data provided by the Bureau of Justice Statistics to the panel at its February 2007, December 2007, and April 2008 meetings.

activities conducted by the Census Bureau, the American Statistical Association, the National Academy of Sciences, and other external groups have concluded that BJS adheres to standards of quality and practice that are consistent with the expectations for a national statistics agency. Such reviews are important for providing feedback and confirming success at an operational level. However, they do not provide any information on BJS's ultimate impact, which focuses on the production of national crime and justice statistics. Is BJS collecting the right kinds of statistics to meet the nation's needs? Are changes needed in the nation's system for collecting and producing crime and justice statistics? Would the nation be better served by an alternative organization (e.g., a consolidated statistical agency) for producing crime and justice statistics? These are the kinds of questions a more comprehensive review could address.

1–A CHARGE TO THE PANEL

Embracing this suggestion for external, independent review—and nearing a milestone of 30 years of operation in its present form—BJS requested that the National Research Council's Committee on National Statistics, in collaboration with the Committee on Law and Justice, establish a Panel to Review the Programs of the Bureau of Justice Statistics. The full charge to the panel is to:

> examine the full range of programs of the Bureau of Justice Statistics (BJS) in order to assess and make recommendations for BJS' priorities for data collection. The review will examine the ways in which BJS statistics are used by Congress, executive agencies, the courts, state and local agencies, and researchers in order to determine the impact of BJS programs and the means to enhance that impact. The review will assess the organization of BJS and its relationships with other data gathering entities in the Department of Justice, as well as with state and local governments, to determine ways to improve the relevance, quality, and cost-effectiveness of justice statistics. The review will consider priority uses for additional funding that may be obtained through budget initiatives or reallocation of resources within the agency. A focus of the panel's work will be to consider alternative options for conducting the National Crime Victimization Survey, which is the largest BJS program. The goal of the panel's work will be to assist BJS to refine its priorities and goals, as embodied in its strategic plan, both in the short and longer terms. The panel's recommendations will address ways to improve the impact and cost-effectiveness of the agency's statistics on crime and the criminal justice system.

Given the prominence of the NCVS in BJS operations—and its dominance of BJS budget resources—the panel was specifically asked to evaluate options for conducting the NCVS in our first year of work before turning to the agency's data collections in other areas. The panel's interim report, *Surveying Victims: Options for Conducting the National Crime Victimization Survey* (National Research Council, 2008b), focuses on that portion of the panel's charge. To give our work a unified presentation, we include the summary from the interim report as Appendix B. This summary includes all of the formal recommendations from that interim report and, to be clear, we stand by all of the guidance in that report—indeed, a few of the recommendations from the interim report are directly restated in this volume. That said, our intent in this final report is to complement the interim report rather than update it or incorporate it in full; the interim report delves into the methodology of the NCVS and the range of possible models for the conduct of the NCVS in much greater detail than the treatment of the NCVS in this report (as one part of the full suite of BJS programs) permits.

1–B   WHAT IS THE BUREAU OF JUSTICE STATISTICS?

It is impossible to properly evaluate the programs of BJS without discussing the agency itself—its functions and mandates, and its placement in both the U.S. Department of Justice and the overall federal statistical system. In this section, to get a sense of the responsibilities of and demands placed on BJS, we discuss various answers to the basic question, "What is the Bureau of Justice Statistics?" To be clear, we do not imply any hierarchy of importance through the ordering of this list, save that it makes sense to start with the two most basic definitions of the agency under the law.[5] In this section—and throughout this report—we refer frequently to the enumerated duties of BJS as they are presented in the agency's authorizing legislation; these are listed in Box 1-2.

1–B.1   Mission-Type Definitions

A first, basic definition of BJS is that it is a *gatherer of information on crime and on the justice system*. The first stated purpose of the section of legislation authorizing BJS is "to provide for and encourage the collection and analysis of statistical information concerning crime, juvenile delinquency, and the operation of the criminal justice system and related aspects of the civil justice system" (42 USC § 3731).

The second, dual purpose of BJS under its authorizing language is to serve as a *developer of justice information systems on all governmental levels*. Specifically, BJS is tasked to "support the development of information

---

[5]The portions of the Justice Systems Improvement Act of 1979 that created BJS (and subsequent revisions) are codified as Title 42, Chapter 46, Subchapter III of the U.S. Code (42 USC §§ 3741–3745). Functions of BJS and its role within OJP are also defined in Title 28, Part 0, Subpart P-1 of the Code of Federal Regulations (28 CFR §§ 0.90, 0.93).

Box 1-2 Statutory Functions of the Bureau of Justice Statistics

The Bureau is authorized to—

1. make grants to, or enter into cooperative agreements or contracts with public agencies, institutions of higher education, private organizations, or private individuals for purposes related to this subchapter; grants shall be made subject to continuing compliance with standards for gathering justice statistics set forth in rules and regulations promulgated by the Director;

2. collect and analyze information concerning criminal victimization, including crimes against the elderly, and civil disputes;

3. collect and analyze data that will serve as a continuous and comparable national social indication of the prevalence, incidence, rates, extent, distribution, and attributes of crime, juvenile delinquency, civil disputes, and other statistical factors related to crime, civil disputes, and juvenile delinquency, in support of national, State, and local justice policy and decisionmaking;

4. collect and analyze statistical information, concerning the operations of the criminal justice system at the Federal, State, and local levels;

5. collect and analyze statistical information concerning the prevalence, incidence, rates, extent, distribution, and attributes of crime, and juvenile delinquency, at the Federal, State, and local levels;

6. analyze the correlates of crime, civil disputes and juvenile delinquency, by the use of statistical information, about criminal and civil justice systems at the Federal, State, and local levels, and about the extent, distribution and attributes of crime, and juvenile delinquency, in the Nation and at the Federal, State, and local levels;

7. compile, collate, analyze, publish, and disseminate uniform national statistics concerning all aspects of criminal justice and related aspects of civil justice, crime, including crimes against the elderly, juvenile delinquency, criminal offenders, juvenile delinquents, and civil disputes in the various States;

8. recommend national standards for justice statistics and for insuring the reliability and validity of justice statistics supplied pursuant to this chapter;

9. maintain liaison with the judicial branches of the Federal and State Governments in matters relating to justice statistics, and cooperate with the judicial branch in assuring as much uniformity as feasible in statistical systems of the executive and judicial branches;

10. provide information to the President, the Congress, the judiciary, State and local governments, and the general public on justice statistics;

11. establish or assist in the establishment of a system to provide State and local governments with access to Federal informational resources useful in the planning, implementation, and evaluation of programs under this Act;

12. conduct or support research relating to methods of gathering or analyzing justice statistics;

13. provide for the development of justice information systems programs and assistance to the States and units of local government relating to collection, analysis, or dissemination of justice statistics *[Clause revised in 1984, replacing more general "financial and technical assistance" language]*;

14. develop and maintain a data processing capability to support the collection, aggregation, analysis and dissemination of information on the incidence of crime and the operation of the criminal justice system *[Clause added in 1984]*;

(continued)

Box 1-2   (continued)

15. collect, analyze and disseminate comprehensive Federal justice transaction statistics (including statistics on issues of Federal justice interest such as public fraud and high technology crime) and to provide technical assistance to and work jointly with other Federal agencies to improve the availability and quality of Federal justice data *[Clause added in 1984]*;

16. provide for the collection, compilation, analysis, publication and dissemination of information and statistics about the prevalence, incidence, rates, extent, distribution and attributes of drug offenses, drug related offenses and drug dependent offenders and further provide for the establishment of a national clearinghouse to maintain and update a comprehensive and timely data base on all criminal justice aspects of the drug crisis and to disseminate such information *[Clause added in 1988]*;

17. provide for the collection, analysis, dissemination and publication of statistics on the condition and progress of drug control activities at the Federal, State and local levels with particular attention to programs and intervention efforts demonstrated to be of value in the overall national anti-drug strategy and to provide for the establishment of a national clearinghouse for the gathering of data generated by Federal, State, and local criminal justice agencies on their drug enforcement activities *[Clause added in 1988]*;

18. provide for the development and enhancement of State and local criminal justice information systems, and the standardization of data reporting relating to the collection, analysis or dissemination of data and statistics about drug offenses, drug related offenses, or drug dependent offenders *[Clause added in 1988]*;

19. provide for improvements in the accuracy, quality, timeliness, immediate accessibility, and integration of State criminal history and related records, support the development and enhancement of national systems of criminal history and related records including the National Instant Criminal Background Check System, the National Incident-Based Reporting System, and the records of the National Crime Information Center, facilitate State participation in national records and information systems, and support statistical research for critical analysis of the improvement and utilization of criminal history records *[Clause added in 1988 and revised in 2006, including reference to specific program names]*;

20. maintain liaison with State and local governments and governments of other nations concerning justice statistics;

21. cooperate in and participate with national and international organizations in the development of uniform justice statistics;

22. ensure conformance with security and privacy requirement of section 3789g of this title and identify, analyze, and participate in the development and implementation of privacy, security and information policies which impact on Federal and State criminal justice operations and related statistical activities *[Clause added in 1984, replacing previous clause on privacy and security]*; and

23. exercise the powers and functions set out in subchapter VIII of this chapter. *[These include authority to issue rules and regulations, consult with the Bureau of Justice Assistance on grant evaluation programs, deny or terminate grants for failure to comply with regulations, convene hearings (including subpoena power) as necessary, and issue an annual report to Congress and the president.]*

SOURCE: Excerpt, 42 USC § 3732(c); italicized comments based on revision history and notes maintained by U.S. Code Service.

and statistical systems at the Federal, State, and local levels to improve the efforts of these levels of government to measure and understand the levels of crime, juvenile delinquency, and the operation of the criminal justice system and related aspects of the civil justice system" (42 USC § 3731). Federal regulation that describes the role of the BJS director takes a broader view of the audience for BJS support: The director "performs functions and administers programs, including provision of financial assistance, [to] provide a variety of statistical services for the criminal justice community" (28 CFR § 0.93).

These first two basic roles are embodied in the two clauses of BJS' formal mission statement, which BJS describes as an "operationalized" version of "the statutory statement of purpose" for the agency (Bureau of Justice Statistics, 2005a:10):

> It is the mission of BJS to collect, process, analyze, and disseminate accurate and timely information on crime and the administration of justice and to assist States and localities to improve criminal justice record-keeping.

1–B.2   Statistical System Definitions

A fundamental definition of BJS—which we examine in greater detail in Chapter 5—is that it is *a principal agency in the decentralized U.S. federal statistical system*. Relative to other countries, the U.S statistical system is exceptionally decentralized; rather than vest authority for the production of official statistics in a single entity, the U.S. model has been to add measurement programs (and agencies) over time as needs for information have evolved. Currently, about 80 federal government agencies each spent at least $500,000 on statistical activities, including data analysis and development of statistical models in addition to actual data collection efforts (U.S. Office of Management and Budget, 2007:Table 1). BJS is among the agencies that meet the definition of "federal statistical agency" expressed in the Committee on National Statistics' white paper, *Principles and Practices for a Federal Statistical Agency*: "a unit of the federal government whose principal function is the compilation and analysis of data and the dissemination of information for statistical purposes" (National Research Council, 2005b:2). More formally, BJS is considered a "principal" statistical agency because it holds a seat on the Interagency Council on Statistical Policy (ICSP), which is established as a coordinating body for federal statistics; see Box 1-3.

Consistent with BJS's role as a statistical agency, BJS's enabling legislation includes the provision that (42 USC § 3735):

> Data collected by [BJS] shall be used only for statistical or research purposes, and shall be gathered in a manner that precludes their use for law enforcement or any purpose relating to a private person or public agency other than statistical or research purposes.

Box 1-3 "Principal Statistical Agencies" and the Interagency Council on Statistical Policy

The 1995 reauthorization of the Paperwork Reduction Act vested authority to "coordinate the activities of the Federal statistical system"—to ensure "the efficiency and effectiveness of the system" and the "integrity, objectivity, impartiality, utility, and confidentiality of information collected for statistical purposes"—in the U.S. Office of Management and Budget (OMB; 44 USC § 3504(e)(1)). OMB's Statistical and Science Policy Office, headed by the Chief Statistician of the United States, is located in OMB's Office of Information and Regulatory Affairs.

The act also formalized the role of the Interagency Council on Statistical Policy, a group of heads of major statistical agencies that meets monthly and discusses common issues. The ICSP predates the 1995 amendments but functioned on a more informal basis prior to being written into law (44 USC § 3504(e)(8)). The ICSP is chaired by the Chief Statistician and those agencies with seats on the council are typically referred to as the "principal statistical agencies" of the U.S. federal government. Though the law mandates the creation of the ICSP, it does not explicitly list its membership, and so agencies have been added to the ICSP by OMB over time.

Currently, the ICSP has 14 members located in 12 cabinet departments or independent agencies. The major statistical agencies represented on the council are:
- from the Department of Agriculture, the Economic Research Service and the National Agricultural Statistics Service;
- from the Department of Commerce, the Bureau of Economic Analysis and the Census Bureau;
- from the Department of Education, the National Center for Education Statistics;
- from the Department of Energy, the Energy Information Administration;
- from the Department of Health and Human Services, the National Center for Health Statistics
- from the Department of Justice, the Bureau of Justice Statistics;
- from the Department of Labor, the Bureau of Labor Statistics; and
- from the Department of Transportation, the Bureau of Transportation Statistics.

Four other agencies or offices are also designated as principal statistical agencies—the Office of Environmental Information (Environmental Protection Agency), the Statistics of Income Division of the Internal Revenue Service (Treasury), the Science Resources Statistics Division (National Science Foundation), and the Office of Policy (Social Security Administration)—but are not included on some tabulations of major statistical agencies (e.g., U.S. Office of Management and Budget, 2007:App. B).

It is also fair to extend this definition in at least two other respects, relative to other units in the federal statistical system. First, BJS is *one of the smallest of the principal statistical agencies in the federal system.* BJS's current organizational structure, including counts of positions classified as statisticians, is shown in Figure 1-3. Of the 10 major agencies represented on the ICSP, BJS's actual fiscal year 2006 funding level of $50.2 million ranks ninth, ahead only of the Bureau of Transportation Statistics (U.S. Office of Management and Budget, 2007:Table 1).[6] Its full-time permanent staff of

---

[6]This $50.2 million does not match the funding level shown in Figure 1-2 because, as

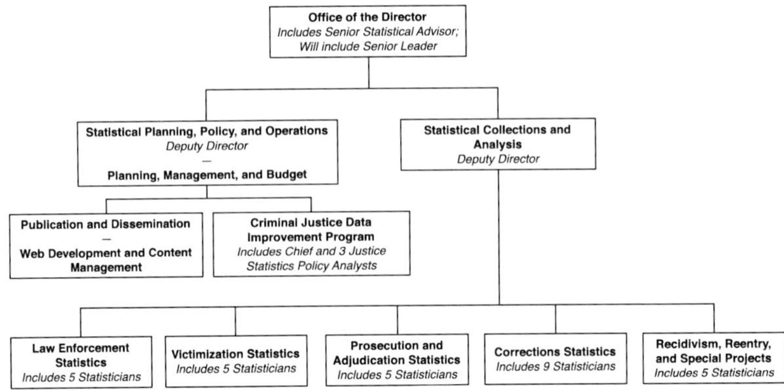

Figure 1-3 Bureau of Justice Statistics organizational structure, June 2008

SOURCE: Adapted from organizational chart provided by BJS to the panel, July 2008.

51 (again, as of actual numbers in fiscal year 2006) was the smallest of the 10 major agencies, the next-smallest being the National Center for Education Statistics at 91 (U.S. Office of Management and Budget, 2007:App. B). Introducing a hearing on the limitations of existing crime statistics, then-Representative Charles Schumer (D-New York) summarized the small resources of BJS relative to those of its peers (U.S. House of Representatives, Committee on the Judiciary, 1991:116):

> Polls show that crime and the economy are the two most important issues to most Americans, so let me draw a comparison between the two. With the economy, we have the Labor Department's Bureau of Labor Statistics. It has a staff of well over 2,000 people, an annual budget of $240 million, and it does an excellent job of measuring all sorts of economic indicators. When the Bureau of Labor Statistics says something, people know it is true. Markets go up and down, waiting for those statistics to be announced. For crime, the lead agency is the Bureau of Justice Statistics. It has a staff of 50 people, a budget of about $20 million, less than one-quarter of 1 percent of what we spend at the Federal level for the war on drugs and a miniscule proportion of what we spend as a nation on law enforcement in general. It is no wonder we have such an incomplete understanding of what is going on [in crime].

Referring to Box 1-2, it is useful to make another definitional point: BJS is *a statistical agency with one of the most elaborate and extensive lists of*

---

noted in U.S. Office of Management and Budget (2007:Table 1), "the amounts for BJS [include] estimated salaries and expenses that are not directly appropriated" and that "the FY 2006 amounts for BJS include carryover funds and any other prior year recoveries."

formal duties compared to those of its peers. By comparison, an entire title of the U.S. Code is dedicated to "Census" issues (Title 13) but the code articulates no formal list of duties for the Census Bureau or its director; rather, the code outlines duties and responsibilities of the Secretary of Commerce, and the Director of the Census is charged only to "perform such duties as may be imposed upon him by law, regulations, or orders of the Secretary" (13 USC § 21). The basic mandate of the Bureau of Labor Statistics is concise: "to acquire . . . useful information on subjects connected with labor, in the most general and comprehensive sense of that word, and especially upon its relation to capital, the hours of labor, the earnings of laboring men and women, and the means of promoting their material, social, intellectual, and moral prosperity" (29 USC § 1). The legal duties of some of the newer statistical agencies (such as BJS) tend to be more specific than those of the older agencies: enabling law describes nine specific data collection themes for the National Center for Health Statistics (42 USC § 242k(1)), while 9 specific duties and 13 data collection themes are outlined for the Bureau of Transportation Statistics (49 USC § 111(c)). Still, relative to its peers, the legal demands put on BJS are unusually detailed and wide-ranging, from specific data collection types (e.g., duty 16 on collection of information on drug-related crime) to administrative support functions for state and local agencies.

1–B.3 Organizational Definitions

A shorthand definition that BJS commonly uses to describe itself is that it is *"the statistical arm of the U.S. Department of Justice."* This phrasing is used in the agency's strategic plan (Bureau of Justice Statistics, 2005a:1) and in other materials (Greenfeld, 2004). The language used in the strategic plan as context for this definition strikes a few slightly different notes from language used in BJS's legally defined duties (see Box 1-2), saying that BJS is "responsible for the collection, analysis, publication, and dissemination of statistical information on crime, criminal offenders, victims of crime, and the operations of justice systems at all levels of government" (Bureau of Justice Statistics, 2005a:1). A semantic point, but an important one, is that BJS's self-description as a "statistical arm" of the Department of Justice (rather than "statistical agency" or "statistical office," or some such term) subtly connotes a dedication to the furtherance of Department of Justice policy objectives rather than the atmosphere of independence that statistical agencies should be permitted. Indeed, as we describe in more detail later in Section 5–A.2, BJS has endured incidents in which Justice Department policy practices have directly conflicted with statistical agency independence, culminating in a high-profile (and damaging) removal of a BJS director.

Setting aside the semantics of "statistical arm" for the moment, the formulation of BJS as "the" statistical arm of the Justice Department is odd. As it is presently operated, it is more correct to say that BJS is *a statistical arm of the U.S. Department of Justice.* Although BJS is responsible for a great deal of the statistical information collected by the Department of Justice (the organizational structure of which is described in Figure 1-4), it is far from the only data-gathering entity in the department.[7] Spanning the many program branches, policy branches, and federal law enforcement agencies in the department, an illustrative (but by no means exhaustive) list of major data-driven activities includes:

- Use of census and other data by the Civil Rights Division to study equity in legislative districting in certain states and the potential for vote dilution;

- Maintenance of extensive administrative (purchase and transfer) databases by the Bureau of Alcohol, Tobacco, Firearms, and Explosives, as well as operation of a national database of images of bullet and cartridge case evidence related to local crime scenes;

- Analysis of the effectiveness of "weed and seed" grants to localities by the Community Capacity Development Office;

- Compilation of data on arrests and seizures made by Drug Enforcement Administration officers;

- Population and inventory reports maintained by the Federal Bureau of Prisons; and

- Analyses of drug markets by the National Drug Intelligence Center, including fielding of a National Drug Threat Survey to law enforcement agencies.

The UCR program—the record of crimes reported to the police—is arguably the longest-standing and highest-profile of the department's statistical series. However, it is administered by the FBI, not BJS. Similarly, primary responsibility for data collections on juvenile offenders and victims is generally held by the Office of Juvenile Justice and Delinquency Prevention (OJJDP), one of BJS's sister agencies in OJP; this is the case even though, as a reading of the duties articulated in Box 1-2 suggests, BJS' authorizing language is replete in references to juvenile delinquency. (We discuss the relationships between BJS and the FBI and OJJDP, in particular, in Section 4–C and Section 2–C.3, respectively.)

---

[7]The fiscal year 2008 version of *Statistical Programs of the U.S. Government* (U.S. Office of Management and Budget, 2007:Table 1) lists four Department of Justice agencies in its basic table of agencies with direct funding for "statistical activities" of at least $500,000: BJS ($50.2 million in fiscal 2006), the Bureau of Prisons ($13.0 million), the FBI ($7.6 million), and the Drug Enforcement Administration ($2.2 million).

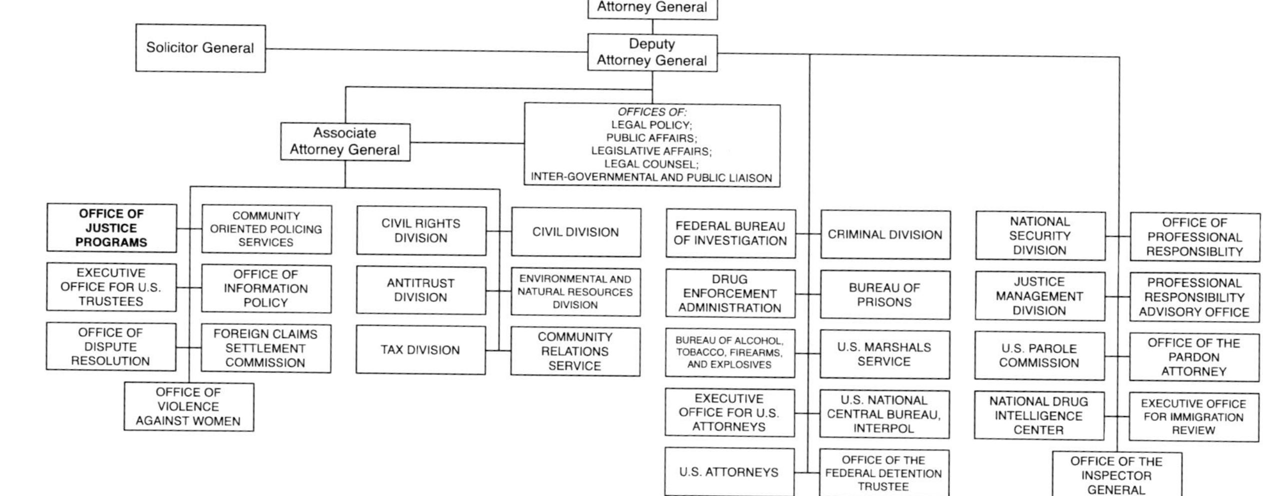

Figure 1-4 U.S. Department of Justice organizational structure, March 2009

NOTE: The Bureau of Justice Statistics (BJS) is a constituent agency of the Office of Justice Programs (OJP), indicated here by boldface type. Box 1-4 depicts the organizational structure within OJP, and Figure 1-3 shows the internal organization of BJS.

SOURCE: Adapted from http://www.usdoj.gov/dojorg.htm, organizational chart dated March 2, 2009.

Administratively, BJS is *an organ of OJP within the U.S. Department of Justice.* The current structure of OJP is described more fully in Box 1-4. This administrative placement has two fundamental ramifications for BJS and its work. First, it means that BJS is administratively nested within the Justice Department; BJS's director reports to the attorney general and the deputy attorney general through a designated assistant attorney general. This type of administrative layering and reporting structure is increasingly common for federal statistical agencies, although it is not ideal for a statistical agency's purpose as an independent broker of information. Second, the administrative tie to OJP means that, like other units in the office, BJS inherits a strong focus of attention on the needs of state and local law enforcement, since OJP is the legal successor of the previous Law Enforcement Assistance Administration. OJP is a program agency that takes as its general mission "provid[ing] federal leadership in developing the nation's capacity to prevent and control crime, administer justice, and assist crime victims" (Bureau of Justice Statistics, 2005a:10). The closing sentence of BJS's basic authorizing language explicitly directs that "[BJS] shall give primary emphasis to the problems of State and local justice systems" (42 USC § 3731).

1–B.4   Functional Definitions

In terms of BJS's functions, it may be said that BJS is principally *a paying sponsor of data collection efforts and a manager of grants.* Although BJS staff are actively engaged in the design of data collections and the analysis of resulting data, many of BJS's data series—including signature collections such as the NCVS—are not conducted in-house. Instead, they are done by contracting with an external data collection agent. Notably, BJS contracts with the Census Bureau for many of its data series, including the NCVS and its censuses of prisons and jails. It also issues contracts with, among others, the National Opinion Research Center, Westat, the Police Executive Research Forum, and the National Center for State Courts. For fiscal year 2008, budget estimates called for BJS to spend $45 million out of an expected $61.5 million in direct funding on purchasing statistical services (73 percent). Of that total, about half ($23.2 million) was budgeted to go to other federal agencies (e.g., the Census Bureau), a smaller share to private-sector organizations ($18.2 million), and a small share to state and local governments ($3.6 million) (U.S. Office of Management and Budget, 2007:Table 3). In addition to data collection–specific contracts, a major part of BJS's work is the provision of grants to state and local law enforcement units, consistent with the strong state and local focus described above. In particular, BJS administers the National Criminal History Improvement Program of grants to state and local governments, and has issued grants to try to encourage coop-

Box 1-4 Organizational Structure of the Office of Justice Programs

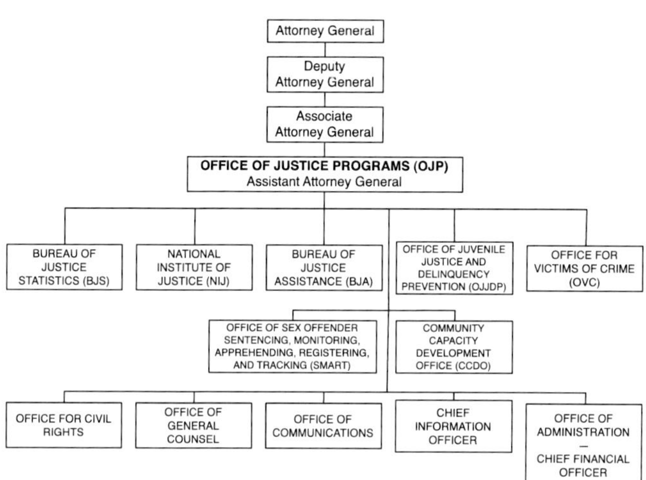

The Office of Justice Programs (OJP) was established in 1984, inheriting major functions from the former Law Enforcement Assistance Administration (LEAA). LEAA had been formed in 1968 to coordinate the flow of federal financial resources to state and local law enforcement efforts. OJP's mission is "to increase public safety and improve the fair administration of justice across America through innovative leadership programs," and it has adopted as its "vision" the following (U.S. Department of Justice, Office of Justice Programs, 2006:3):

> To be the premier resource for the justice community by providing and coordinating information, statistics, research and development, training, and support to help the justice community build the capacity it needs to meet its public safety goals; embracing local decision making and encouraging local innovation through strong and intelligent national policy leadership.

Current law defines powers of OJP as follows (42 USC § 3712(a)):

> The Assistant Attorney General [for OJP] shall—
>
> (1) publish and disseminate information on the conditions and progress of the criminal justice systems;
> (2) maintain liaison with the executive and judicial branches of the Federal and State governments in matters relating to criminal justice;
> (3) provide information to the President, the Congress, the judiciary, State and local governments, and the general public relating to criminal justice;
> (4) maintain liaison with public and private educational and research institutions, State and local governments, and governments of other nations relating to criminal justice;
> (5) provide staff support to coordinate the activities of the Office and the Bureau of Justice Assistance, the National Institute of Justice, the Bureau of Justice Statistics, and the Office of Juvenile Justice and Delinquency Prevention; and
> (6) exercise such other powers and functions [as may be defined or delegated].

(continued)

Box 1-4 (continued)

Section 5–A.2 describes the evolution of point (5) in this list in greater detail.

Aside from the Bureau of Justice Statistics (BJS), major units within OJP include:

- The **Bureau of Justice Assistance (BJA)** provides leadership in grant administration and policy development to support state, local, and tribal criminal justice strategies to achieve safe communities.

- The **Community Capacity Development Office (CCDO)** works with local communities to analyze public safety and criminal justice problems, develop solutions, and foster local-level leadership to implement and sustain these solutions. For example, it oversees the "Weed and Seed" initiative—community-based activities to promote the arrest and sanction of offenders ("weed") and crime prevention and community revitalization ("seed").

- The **National Institute of Justice (NIJ)** is the research, development, and evaluation agency of the U.S. Department of Justice, dedicated to researching crime-control and justice issues.

- The **Office of Juvenile Justice and Delinquency Prevention (OJJDP)** provides national leadership, coordination, and resources to prevent and respond to juvenile delinquency and victimization.

- The **Office for Victims of Crime (OVC)** is committed to enhancing the nation's capacity to assist crime victims and to providing leadership in changing attitudes, policies, and practices to promote justice and healing for all crime victims. It provides federal funds to support victim compensation and assistance and training programs.

- Most recently, the **Office of Sex Offender Sentencing, Monitoring, Apprehending, Registering, and Tracking** (or SMART Office) was created by law in 2006.

The assistant attorney general who oversees OJP operations is appointed by the president with the advice and consent of the Senate, as are the heads of BJA, BJS, NIJ, OJJDP, and OVC. The CCDO director is appointed by the attorney general while the SMART Office director is a presidential appointment without Senate confirmation.

SOURCES: Organizational chart adapted from http://www.ojp.usdoj.gov/about/bureaus.htm (8/1/08) and http://www.usdoj.gov/dojorg.htm (1/14/07). Capsule descriptions of agencies adapted from http://www.ojp.usdoj.gov/index.htm (1/14/07) and individual bureau websites.

eration by local police departments with the FBI's National Incident-Based Reporting System.

BJS also functions as a *synthesizer of information from other federal agencies and from state governments*. At the federal level, BJS acquires some of its data through partnerships with other agencies inside and outside of the Department of Justice, including the Bureau of Prisons and the Administrative Office of the U.S. Courts. It also acquires some of its data directly from state governments as, indeed, it is explicitly urged to do in its legal statement of purpose: "The Bureau shall utilize to the maximum extent feasible State governmental organizations and facilities responsible for the collection and analysis of criminal justice data and statistics" (42 USC § 3731).

In its relationship with the states, BJS has also cultivated a role as an *active broker for research and data dissemination within the states.* Through its State Justice Statistics program, BJS provides grant assistance to a strong network of state Statistical Analysis Centers (SACs), some of which are colocated with state government or law enforcement agencies and others of which are installed at academic institutions. These SACs act as a source of information to BJS as well as a conduit for dissemination of information among state policy makers (e.g., in fielding questions from state legislators). BJS funding also supports the Justice Research and Statistics Association, which provides coordination to the SACs and which operates a research journal.

BJS is also frequently called upon to play a role as a *"criminal justice investigator,"* typically when Congress mandates the collection of information on some aspect of the criminal justice system. Several of these are described in Box 3-3 and one major such request—for information on the occurrence of sexual violence in correctional facilities, as mandated by the Prison Rape Elimination Act of 2003—is described in detail in Section 5–A.1. On occasion, it is not Congress but the administration—the Justice Department—that creates an "investigatory" role for BJS, as with BJS's role in coordinating an 18-city tour to corroborate perceived increases in violent crime with accounts from local police officials (Rosenfeld, 2007). BJS has also been directed by legislation to produce and calculate the formulas used in grant programs administered by other OJP programs, such as the Bureau of Justice Assistance's Edward Byrne Memorial Justice Assistance Grant program (Hickman, 2005a). It has also assumed responsibility for producing certain compilations and summaries of state law, such as an overview series of state privacy and security laws that dates from the LEAA days (Bureau of Justice Statistics, 2003b) and, more recently, summaries of state laws and rules governing firearm sales and transfers (Regional Justice Information Service, 2006).

1–C LIMITATIONS OF THE STUDY

The preceding review of the varied roles of BJS and the expectations placed on the agency underscores the point that our panel's charter to review the full suite of BJS programs and suggest strategic priorities is a very broad one. Accordingly, although we believe our review to be as comprehensive as possible, it is important to make two caveats up front before proceeding to the substance of our analysis.

The first is that we intend our review to suggest directions for research, development, and study to further BJS's objectives and priorities for data collection; it is neither intended as a set of specific budgetary priorities to reduce BJS costs nor as an uncontrolled "wish list" of topics that only a BJS with a massive funding increase could hope to accomplish. On several occasions, BJS Director Jeffrey Sedgwick described the work of our panel in the context of a broader examination of BJS's mission and data series (Justice Research and Statistics Association, 2006a:12; see also Sedgwick, 2006):

> While we have routinely evaluated each of approximately four dozen data series (and received high marks for each), we have not evaluated whether we are doing the right things. . . . Put in econometric terms, we haven't asked whether the value of the least important data we collect exceeds the value of the most valuable data we don't collect.

Consonant with this direction, we have approached our study with an eye toward gaps in substantive coverage that should be filled as well as collections that should be suspended or dropped. However, to expand a note we struck in our interim report (National Research Council, 2008b:2), we recognize the fiscal constraints on BJS but do not intend to be either strictly limited by them (assuming relatively flat funding for the indefinite future) or completely indifferent to them. We approach our work as suggesting a map of several possible avenues for a more effective BJS but not necessarily a single, unique path with specific tasks to meet a certain budget constraint. In that regard, we identify research questions and development issues that necessarily precede a priority-setting agenda. Our panel is not constituted to perform a full financial audit of BJS—and we do not think such a task is envisioned in our charge—but a consequence is that our capacity to completely rank-order individual data collections and programs based on a cost-benefit analysis is necessarily limited. We return to these concepts, in particular, in Chapter 6 and our examination of priority setting in a climate of limited resources.

The second caveat is a simpler limitation of space: The wide scope of BJS programs and the magnitude of the task of assessing the whole portfolio preclude us from delving too deeply into the details of any single data series. Pursuant to BJS's requests, we provided a book-length treatment of one series—the NCVS—in our interim report. This report is not intended as a comprehensive sourcebook in which every BJS data series is granted the same level of review and analysis as we gave the NCVS; time is too short and the terrain is too broad for that level of detail on every series. Indeed, some of the measurement issues we touch on in our review have received extensive treatment from other National Research Council panels and workshops, including data needs in policing and prosecution (National Research Council, 2001b, 2004a) and measurement difficulties concerning the use of firearms and illegal drugs (National Research Council, 2001a, 2005a). In this report, we provide a brief overview of BJS data series, but speak more in terms of statistical coverage in major theme areas (e.g., law enforcement or adjudication, generally).

1–D REPORT CONTENTS

We begin our analysis in Chapter 2 by discussing general perspectives on the justice system on which BJS is tasked to provide statistical measures. Chapter 3 provides a brief inventory of BJS's data collections in the broad topic areas of victimization, law enforcement, corrections, and adjudication. BJS's partnership with the states through its SACs, and its grant programs to state and local governments are outlined in Chapter 4. We describe and assess BJS's programs and functions relative to expected principles and practices of a federal statistical agency in Chapter 5. Chapter 6 articulates a set of four strategic goals for BJS, and includes summaries of the mapping of our detailed recommendations and findings to each strategic goal.

– 2 –

____

Measurement in the Justice System

IRONICALLY, IT IS ONE OF THE SHORTER ENTRIES in the statutory duties of the Bureau of Justice Statistics (BJS) that assigns the agency its most gargantuan task: Point 4 in Box 1-2 obliges BJS to collect and analyze data "concerning the operations of the criminal justice system" at all levels of government. The complication, of course, is that the "criminal justice system" is not a simple construct.

Defining all the parts of a "criminal justice system" is difficult, much less organizing coherent techniques to measure aspects of said system, and different conceptual approaches lead to starkly different vantages of the system. An organizational theory perspective might focus on the interactions of government bodies within the system and on measures of throughput and effectiveness (e.g., police success rates in clearing cases and apprehending suspects and caseload processing rates by the courts); a person-level approach from the victim's perspectives might emphasize the availability and effectiveness of victim support and compensation programs whereas an offender-based approach would give higher prominence to physical and social conditions in correctional facilities as well as parole and prison reentry programs.

Measurement in the justice system is also complicated by the highly varied *units* of measurement that obtain throughout the process. The most basic of violent interpersonal crimes involves a triad of units—the victim, the offender, and the incident—each of which evinces a distinct geography, history, and set of circumstances and contexts, and the study of each of which may lead to different conclusions. However, crime has many types; "victims" and "offenders" need not be individual humans (they may, for instance, be businesses or corporations) and "incidents" need not be one-time acts of

47

violence. Police deal with "suspects" who may or may not be the actual perpetrators of crimes; courts deal with "cases" or "defendants," each of which involves one or more specific "charges," and the connection between these labels and the individual "prisoners" who serve correctional sentences may be lost when authority for them transfers from the courts to the corrections system. An immediate consequence of this unit-of-measurement problem, which we discuss in this chapter and elsewhere in the report, is that measuring the flow of individuals through the justice system is extremely complicated. As a general concept, though, it is important to bear in mind that an approach to the "justice system" that focuses on tracking the experience of individual persons as they move through the system will present a different picture of justice processing than inferences drawn from one that focuses on the progression of "cases" through the set of discrete operations that either move them forward toward resolution or divert them out of the system.

There is no single way, and certainly no uniquely correct way, to conceptualize the criminal justice system. Hence, BJS's task is to straddle a wide range of perspectives in selecting and defining its measurement programs, and to do so while ensuring collection at all (and widely varying, in themselves) levels of government.

In this chapter, we discuss the general challenge of measurement in the justice system. We begin by describing the basic model that we use as an orienting framework—the "funnel" model first developed by the President's Commission on Law Enforcement and Administration of Justice (1967) (Section 2–A)—and discuss how BJS's data collections roughly conform to that model (Section 2–B). Analysis of the funnel suggests some major gaps in U.S. justice statistics—and BJS's statistical coverage in particular—which we describe in Section 2–C; these include difficulty in addressing new and emerging types of crime and contextual factors that apply to a wide range of criminal activities.

2–A THE "FUNNEL" MODEL OF FLOWS IN THE CRIMINAL JUSTICE SYSTEM

Many date the emergence of criminal justice research on the public scene with the release of the 1967 report of the President's Commission on Law Enforcement and Administration of Justice, *The Challenge of Crime in a Free Society*. The report and its many companion volumes summarized what was known about criminal justice and called in virtually every section for more information about how the system operated.

The first chapter of the main report introduced as its organizing metaphor a flowchart presenting "a simple yet comprehensive view of the movement of cases through the criminal justice system" (President's Commission on Law Enforcement and Administration of Justice, 1967:8–9). The chart was meant to emphasize that "a study of the system must begin by examining it as a whole," that "the criminal process . . . is not a hodgepodge of random actions" but rather "a continuum—an orderly progression of events" (President's Commission on Law Enforcement and Administration of Justice, 1967:7). BJS continues to publish the chart—essentially identical in content to the 1967 version—because the model remains popular and useful for studying flows in the criminal justice system; a current version of BJS's publication of the chart is shown in Figure 2-1.

Moving from "crime" through the many paths by which accused persons might eventually progress "out of system," this chart illustrated some of the report's main points. One was that the major institutions that make up the criminal justice system—the police, the courts, and corrections—are interdependent (President's Commission on Law Enforcement and Administration of Justice, 1967:7)

> What each one does and how it does it has a direct effect on the work of the others. The courts must deal, and can only deal, with those whom the police arrest; the business of corrections is with those delivered to it by the courts. How successfully corrections reforms convicts determines whether they will once again become police business and influences the sentences the judges pass; police activities are subject to court scrutiny and are often determined by court decisions. And so reforming or reorganizing any part or procedure of the system changes other parts or procedures.

Another was the complexity of the system, which the report contrasted to "the popular, or even the lawbook, theory of everyday criminal process" (President's Commission on Law Enforcement and Administration of Justice, 1967:7). This complexity was illustrated by the many forks in the road along which individuals might travel, some of which led to a quick exit whereas others promised to involve them for years to come. Finally, the report expressed concern about the fairness as well as the effectiveness of the system that was on view. It noted that, "throughout the system the importance of individual judgment and discretion, as distinguished from stated rules and procedures, has increased." It concluded that "a consideration of the changes needed to make it more effective and fair must focus on the extent to which invisible, administrative procedures depart from visible, traditional ones" (President's Commission on Law Enforcement and Administration of Justice, 1967:10).

Practically, the chart also served as a starting point for the commission to discuss what was not known then—and, in some cases, what is still not known—from data on the justice system. As the chart's legend notes, "the differing weights of line [in the chart] indicate the relative volume of cases disposed of at various points in the system, but this is only suggestive since

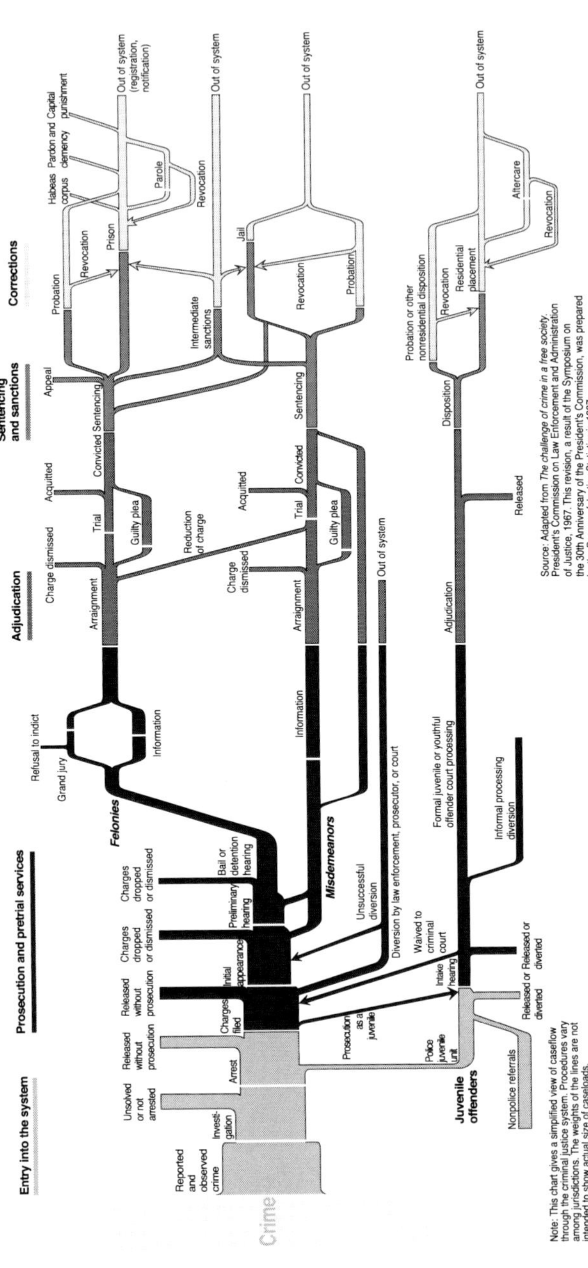

Figure 2-1 Sequence of events in the criminal justice system

SOURCES: Adapted from President's Commission on Law Enforcement and Administration of Justice (1967:8–9) in 1997; posted at http://www.ojp.usdoj.gov/bjs/flowchart.htm.

no nationwide data of this sort exists" (President's Commission on Law Enforcement and Administration of Justice, 1967:8).

Our review of BJS's statistical programs generally follows the flow of this historic chart, for many of them are also organized around the institutions and organizations that make up the justice system. The review also reflects many of the commission's original concerns. Like the commission's report, we emphasize the importance of understanding the interface between the parts of the system as well as the decisions that structure the flow of individuals within each of its components and their eventual exit from the system. Reflecting continued concern about the fairness issues raised by the commission, we examine the utility of the data for assessing the distribution of the outcomes of these decisions, decisions that constitute "justice" for those who are subject to them.

2–B   BJS DATA COLLECTIONS IN THE CONTEXT OF THE CRIMINAL JUSTICE SYSTEM

For the panel's benefit, BJS staff provided a listing of its varied data collections, indicating their approximate coverage of various steps relative to a stylized version of the crime sequence model (Figure 2-1) that more graphically resembles a "funnel." The resulting diagram is shown in Figure 2-2. The diagram shows individual BJS data series and so does not explicitly mention BJS's grantmaking functions, which would show up more directly in a listing of BJS programs. For instance, "Firearm Inquiry Statistics" are a basic summary of use of the National Instant Criminal Background Check System, which BJS does not directly administer but which it provides grants to local agencies to improve.

The "ranges" of coverage indicated on the diagram are approximate in nature. This particular schematic underplays the potential range of the National Crime Victimization Survey (NCVS) and its supplements. Because its primary focus is on victimization incidents, it makes most sense to place the NCVS at the leftmost point on the diagram, corresponding to the occurrence of crime. However, the NCVS includes victimization incidents that are not reported to police and hence do not start the criminal process, as suggested by the diagram; part of the NCVS's unique value is its ability to provide information on the characteristics of these incidents that "leak" out of the system in the earliest stages. Moreover, the NCVS is flexible enough to provide information on incidents of interpersonal violence that may not formally be "crime," and it can also speak to individuals' experiences with other, later parts of the system, such as contacts with police that do not result in arrest or experiences with court proceedings.

Likewise, the coverage range may overstate the scope of the *Justice Sys-*

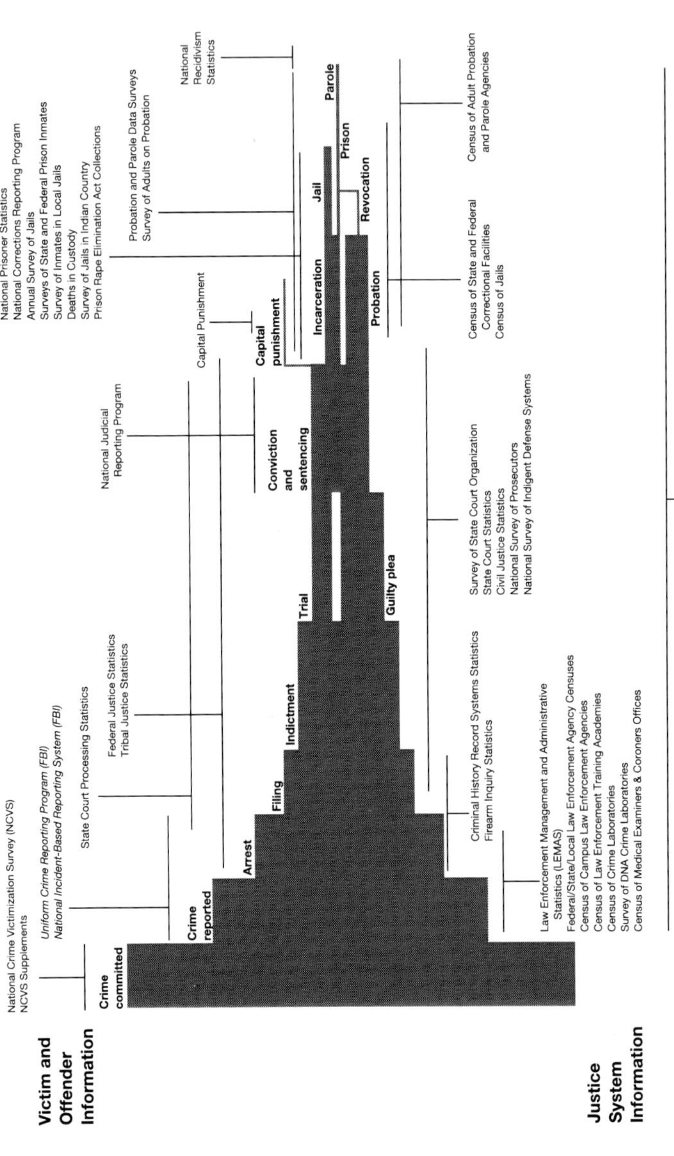

Figure 2-2 Mapping of Bureau of Justice Statistics data series to sequence of events in the criminal justice system

SOURCE: Adapted from chart provided by the Bureau of Justice Statistics to the panel for its first meeting, February 2, 2007.

tem Expenditure and Employment Extracts, which appear to be the widest-ranging and, seemingly, most comprehensive of the data series. This measure of local governments' reported spending on justice-related activities dates back to a special study in the late 1960s in which the Census Bureau compiled spending totals on police, corrections, and judicial processing for some of the large jurisdictions in their annual finance and employment surveys. Based on this initial work, a supplemental mail survey of local governments specifically on justice expenditures was first developed and fielded by the Census Bureau in 1969 and repeated until 1979. Though sample size was boosted in 1979 (with the passage of the Justice System Improvement Act that created BJS in its present form), it was canceled 1 year later for budgetary purposes. An expenditure survey has subsequently been conducted on a sporadic basis (albeit not since 1997). For regular collection of expenditure data, BJS has since adopted the original strategy for the collection: The annual Justice System Expenditure and Employment Extracts are derived from data collected by the Census Bureau's Governments Division.[1] The extracts are comprehensive relative to the system because they attempt to capture expenditures for the system's major components; however, they suffer from being less substantive (comporting with the categories in the Census Bureau's finance surveys) and less geographically detailed than might be possible with a survey-based collection (U.S. House of Representatives, Committee on the Judiciary, 1991:160–161). Tabulations from the Justice System Expenditure and Employment Extracts are currently released in electronic-only format as a set of spreadsheet files (e.g., Bureau of Justice Statistics, 2007a); the most recent report summarized fiscal year 2003 data and was released in 2006 (Hughes, 2006).

2–C   GAPS IN THE COVERAGE OF CRIME AND JUSTICE STATISTICS

Study and comparison of the "funnel" model of flows through the criminal justice system (Figure 2-1) and the illustration of BJS data series that correspond to those flows (Figure 2-2) suggest a few fundamental issues in BJS's statistical coverage.

One such issue is that the stylized BJS mapping (Figure 2-2) emphasizes a forward flow through the system from stage to stage. However, in doing so, it understates a key element that the fuller conceptual model of Figure 2-1 depicts: *There are significant "leaks" in the funnel structure.* These leaks, or diversions from the forward flow of the process, include matters that are of

---

[1]Specifically, the data are derived from the Census Bureau's Annual Government Finance Survey and Annual Survey of Public Employment. The Governments Division is part of the Census Bureau's economic directorate, and the Census Bureau's government programs were recently reviewed in National Research Council (2007).

keen interest for understanding the system as a whole: for instance, acts of violence that never enter the system because they are not reported to police, dismissals of charges in the early stages of prosecution, alternative court resolutions that prevent trials from following the usual channel, and plea agreements that circumvent parts of the process. These leaks are inherently difficult to measure because, by their nature, they may not show up in the extant data resources at any particular stage in the process.

Another issue with the forward-flow sieve model is that it tends to cast the late stages of the process (e.g., corrections) as terminal or ending states. Accordingly, the model *understates issues of reentry into the community from corrections, as well as recidivism.* Figure 2-1 generally shows the last stage in the process as an individual being "out of system"; it does not explicitly show the trajectory that would place individuals back in custody (e.g., a "feedback loop" all the way to the start of the process and the possibility of either committing a new offense or being victimized again).

Any search for gaps in BJS's program coverage—types of crime that are not described well or at all in current collections or additional frontiers in the understanding of crime and justice on which statistical information would be valuable—will inevitably generate a considerable list. The topic area of crime and justice is sufficiently broad that is relatively easy to rattle off long lists of important and interesting topics that a better funded BJS could take on; the perennial problem is reconciling that "ideal" list of desired knowledge with realistic resource assignments. Singling out "gaps"—major or minor— in BJS's coverage is not meant as a criticism of BJS in any way, but rather a reflection of practical realities and a suggestion of possibilities. Indeed, BJS itself goes through this exercise, recognizing gaps in its coverage. In a 2007 meeting with the executive committee of the Justice Research and Statistics Association, BJS staff articulated a "top 20" list of data needs and information gaps that it had developed as part of the process of preparing its budget requests (Sedgwick and Ramker, 2007:3):

1. Statistics on the extent and usage of private security services;

2. Expanded analysis of elderly victims of crime, including data on the prosecution of elder abuse and mistreatment;

3. Data on crimes of human trafficking;

4. Development of a statistical system to study outcomes of ex-offender employment programs, and prisoner reentry in general;

5. Standardization of the Record of Arrest and Prosecution ("rap sheet") criminal histories maintained by law enforcement and accelerated adoption of new standards for rap sheets;

6. Information on the characteristics of "frequent fliers" who are incarcerated in local jails on a repeated basis;

7. Development of an infrastructure to study recidivism using criminal history record databases;

8. National data on the use of lethal force by police;

9. Data on the nature and extent of citizen complaints concerning behavior of local law enforcement;

10. Information on the sentencing of felony identity theft offenders;

11. Inventory of law enforcement "cold case" forensic units;

12. Systematic collection of data on security threats and conditions in state and local courthouses;

13. Fuller collection of contextual information on weapon usage, gang involvement, and drug involvement in incidents included in the National Incident-Based Reporting System (NIBRS);

14. Data on predation and exploitation of children via the Internet;

15. Data on juveniles processed as felony defendants in (adult) criminal courts;

16. Creation of an establishment survey of U.S. businesses on computer security;

17. Expanding participation by law enforcement agencies in common information-sharing databases;

18. Fuller data on law enforcement use of excessive (but nonlethal) force;

19. Fuller data on electronic crime and identify theft; and

20. Articulation of particular challenges to law enforcement (including staffing and resources) in a post-9/11, heightened security environment.

In the balance of this section, we discuss four major "gaps" in BJS's data collection portfolio—things that are conspicuously absent in Figure 2-2. Two of these gaps are related because, as noted earlier, the "funnel" model of the justice system is perhaps too easily equated with the processing of violent crime; other classes of crime, such as white-collar offenses, and civil judicial proceedings are not well captured by current systems. A third gap is that contextual factors associated with crime are inherently difficult to describe— and even characterize consistently—at all steps in the criminal justice system; we describe the relation of drugs and crime as an example. A fourth major gap that is plainly apparent from a comparison of Figure 2-2 with the general framework of Figure 2-1 is the processing of juvenile offenders and victims. In discussing this topic, we also consider the relationship between BJS and one of its sister data-gathering units in the Department of Justice, the Office of Juvenile Justice and Delinquency Prevention (OJJDP). Our assessment of what should be done concerning each of these gaps is similar, and so we close the section and chapter with a common discussion.

We defer our commentary on a fifth—and possibly most severe—"gap" or flaw in BJS's existing coverage until the next chapter; this gap is the concept that the justice system is more than the sum of its parts, and that longitudinal flows throughout the system as a whole are not well measured at present. However, this topic is best considered after we have discussed more of the content of BJS's portfolio in Chapter 3, and so we defer the discussion of longitudinal structures until Section 3–F.1.

2–C.1   White-Collar Crime

The degree to which the measurement of "crime" is primarily focused on violent or street crime and major property crimes such as theft and arson is due, at least in part, to the long-standing definition of "Type I crimes" in the Federal Bureau of Investigation's (FBI's) Uniform Crime Reporting (UCR) program (which we describe in more detail in Section 4–C.1). The major definitions and conceptions of crime from the UCR were carried over to BJS's NCVS, and other programs. Yet both the UCR and the NCVS— the nation's two principal indicators of crime—have been critiqued for being slow to catch up to new crime types. A report of the FBI's Criminal Justice Information Services Division (Barnett, 2000:2) conceded that "it is well documented that the major limitation of the [UCR] Summary Reporting System is its failure to keep up with the changing face of crime and criminal activity." Likewise, our panel's interim report discussed the challenges involved in achieving the NCVS's full flexibility in studying new types of victimization (National Research Council, 2008b:Sec. 3–C.1).

A focus on certain forms of violent and property crime does not account for all important types of crime, or crime types that emerge with the introduction and maturation of new technologies. A particular gap is in the measurement of many forms of what could loosely be labeled as white-collar crime. The term "white-collar crime" has been in currency since sociologist and criminologist Edwin Sutherland introduced it in 1939. Some interpret the term narrowly (e.g., Sutherland's early focus on crimes committed by a person of high responsibility in the pursuit of his or her occupation) whereas others interpret it more broadly. For instance, the Federal Bureau of Investigation (1989:3) has defined white-collar crime as:

> those illegal acts which are characterized by deceit, concealment, or violation of trust and which are not dependent upon the application or threat of physical force or violence. Individuals and organizations commit these acts to obtain money, property, or services; to avoid the payment or loss of money or services; or to secure personal or business advantage.

We use the term more broadly to refer to crimes such as corporate fraud, health care fraud, financial institution fraud, money laundering, government fraud, consumer fraud, public corruption, and Internet crimes.

Recent survey evidence suggests that the public views white-collar crime at least as seriously as traditional forms of crime, and offenses committed by organizations or by higher-status persons as more serious than those committed by individuals or lower-status persons. With sponsorship from the Bureau of Justice Assistance, the National White Collar Crime Center (NW3C) conducted a "National Public Survey on White Collar Crime" in 2005. Based on the survey, the NW3C concluded that the general public views white-collar crime as seriously as traditional crime types. In addition, the survey also yielded the finding that (Kane and Wall, 2006:3):

> Crimes involving physical harm are seen as significantly more serious than those crimes that incur a monetary loss only; organizational offenses are viewed more harshly than those committed by individual offenders; and crimes committed by high-status offenders (those in a position of trust) are seen as more severe than those crimes committed by [non-high-status] persons.

There is also evidence that white-collar crime is a concern to law enforcement. Amidst its recent major change in strategic priorities to the deterrence of terrorism, the Federal Bureau of Investigation (2004a) prominently included a white-collar crime plank in its strategic goals for 2004–2009:

IIH.1 Reduce levels of corporate fraud by targeting those groups or individuals engaged in major corporate fraud schemes that significantly impact the investing public and financial markets.

IIH.2 Reduce the incidence of large scale health care frauds, involving both government-sponsored and private insurer programs.

IIH.3 Reduce fraud perpetrated by criminal enterprises targeting financial institutions.

IIH.4 Disrupt and dismantle the most significant money laundering institutions and facilities.

IIH.5 Reduce the impact of telemarketing, insurance, and investment fraud on businesses and individuals, particularly schemes originating from outside the United States.

IIH.6 Address those investigative matters which represent the most significant economic losses within federally-funded procurement, contract, and entitlement programs, environmental crimes, bankruptcy fraud, and anti-trust offenses.

In its strategic plan, the FBI further argues that "the ability of the U.S. Government and industry to function effectively [is] threatened by complex frauds" and that continuance of its "successful efforts in the white collar crime arena" is important to "ensure the integrity of government expenditures of taxpayer funds [and] protect individuals and businesses from catastrophic economic loss."

However, the FBI strategic plan leaves it unclear how the FBI and the Justice Department plan to assess its success toward achieving these goals. Indeed, it is unclear how such an assessment plan could be developed because of the lack of data. Responding to a questionnaire for a United Nations intergovernmental expert group on fraud, the U.S. Department of Justice, Criminal Division, Fraud Section (2006:6–7) had to acknowledge that "there is no single government agency or private-sector entity that compiles statistical data on the principal types of fraud that occur within or affect the United States." The Fraud Section followed that disclosure by outlining a lengthy list of what it described as "some of the more frequently reported types of fraud" committed in the United States—including advance-fee fraud,[2] telemarketing fraud, and identity document (including passport and visa) fraud—though it could provide no quantitative evidence on the actual frequency of these crimes.

When BJS was founded in 1979, white-collar crime was originally intended to be a part of the agency's portfolio; introductory text declared that the purpose of the law creating the agency was to "provide for and encourage the collection and analysis of statistical information concerning crime (including white-collar crime and public corruption)" (93 Stat. 1176). However, this parenthetical was stricken from law in 1984 (98 Stat. 2079). Though not part of the agency's formal charge, BJS is not completely silent on issues of white-collar crime. As we discuss in Section 3–A, BJS's supplement to the NCVS on identity theft has developed into a regular feature of the survey; the NCVS cybercrime supplement also touched on experiences with Internet fraud. However, the information that is available on this class of crime is not well organized or displayed on the BJS website. There is no front-page link to white-collar crime as a topic (the BJS home page is displayed in Figure 5-5). Using its "BJS only" search engine box, a query for white-collar crime provides only a few links to old reports available only in paper form and stray links to the Survey on Prosecutors in State Courts and the Compendium of Federal Justice Statistics. Broadening the search to include "OJP and NCJRS"—BJS's parent Office of Justice Programs (OJP) and the OJP–sponsored National Criminal Justice Reference Service (NCJRS), respectively, produces some more useful links. Chief among these is a designated "topic page" for white-collar crime on NCJRS, but the presentation on that page (http://www.ncjrs.gov/App/Topics/Topic.aspx?topicid=73, accessed 11/6/2008) lacks hard information, with a link to BJS's report on the NCVS identity theft data (Baum, 2007) being the only one that suggests obvious empirical data. Other links on the page are to National Institute of

---

[2]"Advance-fee fraud [can] encompass any type of fraud scheme in which victims are induced to pay money to criminals for nonexistent "taxes," "fees," or "customs duties" before the criminals are expected to provide whatever goods or services (e.g., offshore tax shelters) they have promised to victims.

Justice–sponsored research, including the connections between white-collar crime and terrorism and a legal analysis of the global scope of intellectual property law.

The NCJRS page does provide a link to the nonprofit, private-sector NW3C, which is supported by funds from BJS's sister organization in OJP, the Bureau of Justice Assistance. The NW3C specializes in white-collar crime law enforcement training and houses an academic research group; in a joint venture with the FBI, the NW3C operates the Internet Crime Complaint Center. Though the NW3C has fielded surveys on public perceptions of white-collar crime, it is not clear whether any kind of formal arrangement exists (or has been broached) between BJS and NW3C, such as BJS maintains with the National Center for State Courts and the Urban Institute's Federal Justice Statistics Resource Center.

In fairness to BJS, there are at least three major reasons that partially explain the dearth of information on white-collar crime:

- The definitional problems involved in even conceptualizing a data collection program in white-collar crime are immense, and are more than a small agency with limited resources—already grappling with massive and ill-defined data collection areas such as adjudication—can reasonably take on. As noted above, white-collar crime can be defined both narrowly and expansively, and selecting a set of activities to study is an initial and formidable hurdle.

- An even more significant conceptual hurdle is that opening the door to studies of white-collar crime would involve a major shift in focus and style for BJS: it would require a fuller examination of the concept of businesses and corporations as actors in the justice system, as the victims or perpetrators of crime. BJS's forays in this area, using businesses as the unit of analysis, have been relatively rare. The National Crime Survey program (now the NCVS) originally included a commercial victimization component, but that part of the survey was short-lived. BJS's work with the Federal Justice Statistics Resource Center uses "defendant-cases" as the unit of analysis as it tracks flows through the federal court processing system, and this does permit the notion of a business as a defendant. And, most recently, BJS has fielded a survey of businesses on their experiences with cybercrime, on a pilot basis in 2001 (Rantala, 2004) and as a full-scale survey of about 8,000 establishments in 2005 (Rantala, 2008). But—those exceptions aside—BJS has tended to use individual persons or agencies (e.g., corrections departments or law enforcement agencies) as the unit of analysis.

- By its nature, fuller collection of information on white-collar crime would necessitate the collection of financial and monetary data, in terms both of estimated losses and of award amounts. Accurate measurement (and disclosure) of financial data is a long-standing challenge for statistical agencies.

That said, we think it fair to say that BJS's data collections (and the information available on its website) have historically been dominated by violent crime and "street crime." This produces—or at least contributes to—a partial and perhaps misleading image of crime and criminals in America. Knowledge about the nature of corporate and white-collar crime—its levels and trends, its handling in the court system, and its impact on criminal justice system operations—would be as useful and informative as data on "street crime." If BJS is to be positioned as the primary data-gathering agency for crime and criminal justice in the United States, it would be most sensible for BJS to take the lead to organize the available statistics on white-collar crime and other new forms of crime, as we discuss in Section 2–D.

2–C.2 Civil Justice

A second major gap in BJS's overall portfolio is similar to white-collar crime in that it arises from the principal focus on violent or street crime, and on crime against the person. The construct described by both Figures 2-1 and 2-2 is commonly referred to as the "justice system" or the "criminal justice system," and the degree to which this nomenclature is used interchangeably severely underplays the scope of *civil justice* matters. Civil proceedings account for a major portion of the activity of the nation's court systems—ranging from prosecution of nonviolent crimes to property disputes to divorce and custody arrangements—yet are covered by only one basic data collection in BJS's portfolio.

On one level, the question of whether BJS belongs in the business of studying civil justice, as opposed to criminal justice, is clearly answered: collection and analysis of information on "civil disputes" and "civil justice" are explicitly mentioned in points 3, 6, and 7 of BJS's list of legal duties (Box 1-2). On the other hand, BJS is also bound by a clause in OJP's enabling legislation that exemplifies the tension between "civil" and "criminal" justice (42 USC § 3789n):

> Authority of any entity established under this chapter shall extend to civil justice matters only to the extent that such civil justice matters bear directly and substantially upon criminal justice matters or are inextricably intertwined with criminal justice matters.

The Civil Justice Survey of State Courts, administered every 4–5 years since 1992, has most recently been conducted by the National Center for State Courts (NCSC) on BJS's behalf. The sample is meant to be representative of large counties but is constructed in a somewhat unusual manner; in 2001, the survey represented a sample of civil bench and jury trials in "46 jurisdictions chosen to represent the 75 most populous counties," where 75 was chosen "based on cost and practicality" (Bureau of Justice Statistics, 2004b:1). Generally, these jurisdictions are counties, chosen from the 75 using strata defined by their level of civil dispositions recorded in 1990 and their population (Bureau of Justice Statistics, 2004b:3); however, only part of the caseload in Los Angeles County (the central district of the Los Angeles County Superior Court) is included in the sample, whereas some of the "judicial districts" sampled in New England may span multiple counties. The number of counties actually included in the sample has been stable over the years, at 46, 45, and 46 in 1992, 1996, and 2001, respectively.

For sampled jurisdictions, the NCSC reviewed tort, contract, and real property rights cases disposed of by trial during calendar year 2001; either NCSC or court staff coded details from cases onto a standardized form. For general civil cases, NCSC coded as many cases as possible but used sampling "based on 'take rates' generated by WESTAT" in jurisdictions where the number of such general trials became unworkable. However, "every medical malpractice or product liability case was included to over sample these case types" (Bureau of Justice Statistics, 2004b:3). Ultimately, the 2001 survey included "data on 6,215 civil jury trial cases, 1,958 civil bench cases, and 138 other civil trial cases that met the study criteria" (Bureau of Justice Statistics, 2004b:4).

It is very much to BJS's credit that, by sponsoring the Civil Justice Survey of State Courts, it takes a first step toward measuring and monitoring a major part of the nation's judicial workload. BJS staff also deserve credit for, arguably, making the civil justice survey one of the agency's most scrutinized and reported data series; BJS has published numerous analyses characterizing verdicts from the data series (Cohen, 2004, 2005a), the amount of trial awards (Cohen, 2005b), and the nature of appeals (Cohen, 2006). BJS has also worked with state partners to collect and analyze related data on specific case types, distinct from the civil justice survey; for instance, Cohen and Hughes (2007) analyze data from closed medical malpractice insurance claims in seven states.

However, BJS's current approach to measurement of civil justice suffers from the key limitation that it is based on only cases that end in a bench or jury trial and not other civil dispositions including settlements, summary judgments, and default judgments. As Cohen and Smith (2004:80) summarize, disposition through trial—whether a jury trial or a bench trial presided over only by a judge—"represent[s] the pinnacle event in the civil justice process," yet "both jury and bench trials are relatively rare." Civil lawsuits result in trial only "when at least one party refuses to settle" pretrial, out of court. These settlements constitute "the vast majority" of activity in civil proceedings; by their nature, they are also generally private and hence difficult to measure accurately or systematically. To be sure, the fact that case-based measurement systems miss the major activity of settlement does not mean that they lack value: documentation of trial outcomes—and the "perceived trial outcomes and potential award amounts" that are derived from analysis of completed cases—influences the propensity of parties to pursue or settle civil litigation.

Negotiated pretrial, out-of-court, settlements are part of a broader class of alternative dispute resolution (ADR) techniques. In adjudication generally, but particularly in civil matters, the nature and extent to which ADR techniques are applied are important but, to date, largely unanswered questions. ADRs include referral, by mutual consent of the parties, to an independent facilitator, mediator, or arbitrator (the difference between the specific techniques being the formality of the proceedings—an arbitrator's decision is imposed on the parties while a mediator facilitates agreement on a resolution but does not directly impose rulings). As a measure of the growing acceptance of ADR techniques, several state court systems maintain specific offices to monitor and suggest ADR avenues; as of 2004, 15 state court administrative offices completely fund and staff ADR offices and another 16 states partially fund such offices (Rottman and Strickland, 2006:Table 21). However, aside from general perceptions that ADR techniques are being increasingly used to divert cases from entering the court system, hard statistical information on ADRs is scant, including the types of cases to which they are applied, the specific techniques used, the nature of the agreements, and the number (and success or failure of) appeals of ADR settlements to the courts. In an Internet "frequently asked questions" page, the NCSC notes that "[ADR] programs and rules vary widely from state to state, and even from court to court within a single state. As a result, national data are nearly impossible to find" (http://www.ncsconline.org/WC/CourTopics/FAQs.asp?topic=ADRMed). The Federal Mediation and Conciliation Service, an independent federal government agency, provides summary statistical information in its annual reports; however, the agency's primary focus is on resolving disputes between labor unions and management, which is but one part of the larger phenomenon.

It should be noted that BJS has produced reports on one particular subset of civil justice cases, making use of data other than the Civil Justice Survey of State Courts. Using the Civil Master File of the Administrative Office of the U.S. Courts—which in turn combines case filing data from U.S. federal district courts—BJS has regularly generated tabulations of civil rights complaints, including disposition (jury trial, bench trial, or directed verdict) and award amount. Kyckelhahn and Cohen (2008) describe trends in civil rights filings between 1990 and 2006.

2–C.3 Juvenile Justice System

Arguably, the most striking difference between the justice system "funnel" model (Figure 2-1) and BJS's statistical coverage of that model (Figure 2-2) is that the former contains a parallel track that has developed over the past several decades. The separate stream of processing of juveniles shown in Figure 2-1, including the operation of separate courts and correctional facilities for juveniles, is essentially absent from the BJS coverage funnel.[3] It is true that the two tracks—and the experiences of juveniles and adults in crime and justice—are not as entirely separate as the stylized funnel model suggests. That is, some juveniles are tried as adults in criminal proceedings, some are confined in adult correctional facilities or transition into the adult justice system, and—of course—both juveniles and adults may commit or be victims of violence. Hence, it is possible for juveniles to show up in BJS's data collections that are principally intended to cover the adult population; for instance, DeFrances and Strom (1997) summarize information from the National Survey of Prosecutors on juveniles processed in state courts, Strom (2000) studies characteristics of inmates of state prisons under age 18, and Strom et al. (1998) and Rainville and Smith (2003) describe results from the State Court Processing Statistics project on juveniles charged with felonies in adult courts. Further, in at least one instance, BJS was explicitly mandated by Congress to conduct a study within juvenile correctional facilities: the data collections requested by the Prison Rape Elimination Act of 2003 (see Section 5–A.1). BJS has also studied decade-long trends (1993–2003) in juvenile victimization and offending, utlizing both NCVS and UCR (Supplementary Homicide Reports) data (Baum, 2005).

However, with those exceptions—and despite BJS's legal list of duties (Box 1-2) being replete with references to data gathering on juvenile delinquency—it is one of BJS's sister agencies in OJP that has assumed principal responsibility for collection and organization of information on juvenile justice. Under its establishing legislation, OJJDP has the authority to (42 USC § 5661(b)(2)):

> undertake statistical work in juvenile justice matters, for the purpose of providing for the collection, analysis, and dissemination of statistical data and information relating to juvenile delinquency and serious crimes committed by juveniles, to the juvenile justice system, to juvenile violence, and to other purposes consistent with [OJJDP's charter].

A separate provision of law obligates the directors of both BJS and the National Institute of Justice to "work closely with the Administrator of the

---

[3]In this section, we make a simplification and follow a common practice by defining a juvenile as someone under age 18; note, however, that 18 is not the legal age of maturity for various purposes (including treatment in the adult justice system rather than the juvenile justice system) in all states.

[OJJDP] in developing and implementing programs in the juvenile justice and delinquency field" (42 USC § 3789i)

In comparing OJJDP's statistical collections on juveniles with BJS's series for adults, it is important to keep in mind that the two agencies have quite different missions. BJS's duties under the law (Box 1-2) are primarily oriented toward data collection, with some provisions for assistance to state and local authorities. By comparison, the stated purposes in OJJDP's enabling legislation (42 USC § 5602) are exclusively concerned with programmatic support for state and local assistance:

> The purposes of this section of [this section of law] are—
>
> (1) to support State and local programs that prevent juvenile involvement in delinquent behavior;
>
> (2) to assist State and local governments in promoting public safety by encouraging accountability for acts of juvenile delinquency; and
>
> (3) to assist State and local governments in addressing juvenile crime through the provision of technical assistance, research, training, evaluation, and the dissemination of information on effective programs for combating juvenile delinquency.

The agency's formal mission statement is likewise geared toward influencing policy at the state and local levels: OJJDP "provides national leadership, coordination, and resources to prevent and respond to juvenile delinquency and victimization" and "supports states and communities in their efforts to develop and implement effective and coordinated prevention and intervention programs" (http://ojjdp.ncjrs.org/about/missionstatement.html).

That said, in carrying out its mission, OJJDP develops and sponsors a series of data collection programs—in some cases developing operations parallel to BJS work for adult facilities and processes, and in others making use of BJS data. Its efforts in data gathering and coordination are routinely made available in an online "Statistical Briefing Book" (http://ojjdp.ncjrs.gov/ojstatbb/index.html) that is maintained by the Pittsburgh-based National Center for Juvenile Justice (NCJJ) through a grant from OJJDP. OJJDP also regularly produces a "National Report" in its *Juvenile Offenders and Victims* series (Snyder and Sickmund, 2006).

Regarding original data, OJJDP sponsors the collection of primary data in three areas: (1) juvenile facilities and juveniles in residential placement; (2) juvenile court statistics; and, to a more limited extent (3) juvenile victimization and offending. On the first of these, corrections and residential placement, OJJDP has sponsored a person-based Census of Juveniles in Residential Placement (CJRP) in odd-numbered years and a facility-based Juvenile Residential Facility Census (JRFC) in even-numbered years since the late 1990s.[4] In their structure, OJJDP's collections mirror the strategy used by

---

[4]Typically, the reference date for the CJRP is in October of an odd-numbered year; "how-

BJS in its series on adult prisons and jails, alternating between inmate-based and facility-based data; they also bear similarities because OJJDP employs the Census Bureau as its data collection agent as is true of the adult series. However, the OJJDP person-based series differ from their adult counterparts in that they rely entirely on facility records and reporting rather than direct personal interviewing. Collectively, the CJRP and JRFC provide information on facilities (type, physical layout, counts, use of locked doors/gates) and individual juveniles held in residential facilities (demographic characteristics, placement authority, most serious offense, adjudication status, security status, etc.). Because they use juvenile-specific facilities as their frame, the OJJDP collections do not capture those juveniles held in adult prisons or jails; mental health and drug treatment facilities are also not a part of the CJRP or JRFC frames, and so juveniles housed there are not determined either. Although the OJJDP Statistical Briefing Book provides a tool for generating national and state summaries from the CJRP,[5] more detailed data from the CJRP are available to researchers only on a case-by-case basis.

Through NCJJ, OJJDP makes available juvenile court statistics with information on the activities of juvenile courts in the United States and the cases disposed of in these courts. Similar to the FBI's UCR program, data in this archive seem to be based upon voluntary submissions by juvenile courts. Thus, the number of participating courts varies each year; in addition, some courts provide detailed information for each case whereas others provide only aggregate counts. These data are used to provide national portraits of juvenile offenders and juvenile court activity, including information on court caseloads, variation in delinquency cases by demographic characteristics of youth involved, detention, disposition of cases, and the flow of cases through the juvenile justice system. Status offense cases are also considered, but to a lesser degree, and juveniles waived to adult courts are not followed beyond the waiver decision.

Finally, with regard to juvenile victimization and offending, OJJDP does not routinely sponsor supplements to BJS's NCVS. However, it sponsored a module on crime, delinquency, and arrest for inclusion in the National

---

ever, a set of unforseen circumstances prevented the 2005 mailout from taking place in October of that year." The mailout was pushed to February 2006, hence an apparent deviation in OJJDP's annual collection strategy (http://www.ojjdp.ncjrs.gov/ojstatbb/cjrp/asp/methods.asp). The CJRP and JRFC continue and extend an earlier data collection fielded by the Justice Department since the early 1970s: the Census of Public and Private Juvenile Detention, Correctional, and Shelter Facilities, better known as the Children in Custody census. The innovation in creating the two newer, separate series was the addition of individual-level characteristics in the CJRP rather than summary-level counts.

[5] As part of its strategy to protect confidentiality and prevent individual juveniles from being identified from such tabulations, "OJJDP has adopted a policy that requires all published table cells be rounded to the nearest multiple of three" (http://www.ojjdp.ncjrs.gov/ojstatbb/cjrp/asp/methods.asp).

Longitudinal Survey of Youth 1997 (NLSY97), conducted by the Bureau of Labor Statistics.[6] The NLSY97 is based on a national sample of youth who were 12–16 years old as of December 31, 1996; they are interviewed annually, with the tenth round in 2006 reaching 7,559 of the original respondents (http://www.nlsinfo.org/nlsy97/nlsdocs/nlsy97/97sample/introsample. html). Through the OJJDP module, and related questions on general risk behaviors among youth, the NLSY97 includes questions on substance use, delinquency and deviance (e.g., status offenses, gang membership, arrests, property offenses, carrying guns), as well as incidents of criminal victimization.

In addition to its own sponsored data sets, OJJDP's Statistical Briefing Book compiles information from a wide variety of agencies and organizations to permit assessments of juvenile victimization, offending, and law enforcement experiences. Specifically, the briefing book includes results from, among others:

- The NCVS;

- The FBI's UCR and, for participating jurisdictions, the National Incident-Based Reporting System;

- The Department of Health and Human Services' National Child Abuse and Neglect Data System Child File (NCANDS);

- The National Institute of Drug Abuse's Monitoring the Future Survey; and

- The Centers for Disease Control and Prevention's National Youth Risk Behavior Survey (NYRBS).

These other data sources provide more richness in detail for particular incident types than the NCVS can provide alone. For example, the NCANDS allows assessments of victim, caretaker, and perpetrator characteristics, and responses to abused or neglected children, and so is useful for studying specific kinds of victimization of youth. The NYRBS is a school-based sample (9th to 12th graders) that allows for additional information on victimization and participation in additional types of crimes and errant behaviors such as fights, suicide attempts, and alcohol and drug use.

Considering both OJJDP and BJS's portfolios, there remain important gaps in the coverage of youths' activities. A major one is that most of the available data collections—particularly the NCVS and the high school–based NYRBS—provide reasonable coverage of adolescents but not the complete juvenile population. Through its focus on caretakers and perpetrators, the NCANDS is unusual in its potential coverage of all ages. Further, whatever the age range covered by the data series, the data are typically aggregated

---

[6]See   http://www.nlsinfo.org/nlsy97/nlsdocs/nlsy97/97sample/introsample.html   on   the credit for OJJDP's sponsorship of the questions.

and reported for broad age categories (e.g., under age 12, 12–15, over 15, or simply under 18). Thus, it is not always possible to assess activities and situations for young people across the youthful life course. Importantly, too, some youthful populations are simply not considered. For example, OJJDP reports do not typically consider youth in adult courts or custodial facilities, beyond documenting who is waived. BJS gives attention to this group, but these data are not linked with the records from juvenile courts and facilities. Consequently, meaningful assessments of which types of youth are handled in the juvenile justice versus the criminal justice system are not readily available. Nor is there much information on how youth fare who transition out of the juvenile court to the adult courts. Ideally, record and data-reporting systems regarding juveniles under age 18 and adults over 18 would not be quite so hard and fast because the division at 18 does not represent hard and fast differences in maturation of young people.

2–C.4 Drugs and Crime

Amidst growing expenditures in the federal government's "war on drugs" policies, reauthorization legislation enacted in 1988 added three specific clauses on data collection on drugs and crime to BJS's statutory mandates, as shown in items 16–18 in Box 1-2. The new clauses referred to the "drug crisis" and the "overall national anti-drug strategy," and tasked BJS with establishing a "a national clearinghouse for the gathering of data generated by Federal, State, and local criminal justice agencies on their drug enforcement activities," particularly data that could be used to demonstrate the efficacy of programs and intervention efforts. By this language, Congress and the administration—not for the first time, and not for the last—assigned BJS a task that was difficult under the best of circumstances and virtually impossible given tight fiscal resources.

Contextual factors associated with crime pose particular difficulties for measurement. They can be difficult to define in general and can be particularly difficult to define consistently across multiple data sources or throughout the different steps in justice system processing. In addition, different data collection types—for instance, personal survey interviewing versus coding from written police reports—may be especially strong at capturing some factors but weak at others. The kinds of contextual factors we refer to here include what Cook (1991) has described as "the technology of personal violence"—the use of weaponry, particularly firearms, in crime—which can affect the probability of success of the crime, the consequences to the victim, the responses of law enforcement, and the implications for punishment in the court system. The geography of crime—more than just latitude and longitude, including social and physical conditions and community resources in an area—is another crucial and challenging contextual factor that has grown into a particularly vibrant area of criminological research in recent decades. In this section, we focus on another important crosscutting contextual factor—the interplay between drugs and crime—as an example principally because of the explicit references in BJS's enabling legislation.

Drug crimes—offenses involving the possession and sale of controlled substances—are a major area of criminal activity that resists an easy fit with the conceptual model of the crime funnel. Drug offenses account for about 30 percent of admissions to state prisons, but arrests for drug offenses tend not to follow the initial sequence of events in the funnel where a crime (or victimization) leads to a report or complaint to police and, in turn, to an arrest. Rather, the pattern of drug-specific arrests is more closely linked to targeted enforcement efforts which vary greatly across jurisdictions. In addition, within the court processing steps of the funnel, drug cases resist simple categorization because laws regulating drugs vary across states and within states, and change over time. Thus possession of a small quantity of drugs, for example, may be only a minor violation in some jurisdictions, but a felony in others. Over and above the difficulties involved in mapping drugs and drug crime to the funnel framework, drug-related activities are generally difficult to measure. The sensitive nature of inquiries on drug use make it a behavior that can be challenging to measure accurately through self-report techniques such as survey interviewing.

Those data resources that do cover aspects of drug use and drug-related criminal and violent behavior include both probability surveys and records-based series, and each has unique strengths and weaknesses. A partial list of these resources—past and present, conveying the range of data collections that have touched on drug-related issues—includes the following:

- The *National Survey on Drug Use and Health* (NSDUH; formerly the National Household Survey on Drug Abuse) is sponsored by the Substance Abuse and Mental Health Services Administration (SAMHSA). It has been in operation since 1971 and has utilized RTI International, Inc., as its data collection agent since 1988; it is perhaps the main source of statistical information on the use of illegal drugs by the U.S. population. The *Monitoring the Future Survey* has been conducted by the Institute for Social Research at the University of Michigan since 1975, asking secondary school students about drug use as well as other risk behaviors. Both of these surveys share a public health orientation, asking about drug use and dependency, and offer relatively little information to distinguish criminal drug activities.

- The *Arrestee Drug Abuse Monitoring* (ADAM) program (formerly the Drug Use Forecasting [DUF] program) was a data collection sponsored by the National Institute of Justice that produced data between 1987 and 2003. In select, participating sites, new arrestees were interviewed within 48 hours and asked a battery of questions on arrest history, drug use patterns, drug acquisition, and prior participation in treatment programs. Survey results were combined with—and could be compared against—the results of urinalysis to detect the presence of 10 drugs (but focusing in particular on cocaine, marijuana, methamphetamine, opiates, and phencyclidine [PCP]). A wide-ranging redesign in 1999 gave ADAM a sounder basis in probability sampling, started a "calendaring" routine in the questionnaire to cue arrestees to document drug use patterns over longer period of times, and positioned ADAM to have relatively easy "crosswalk" connections for linkage to other data collections such as NSDUH (National Institute of Justice, 2003:4, 13). DUF and ADAM data were subject to criticism over their representativeness and because they could function only as an indirect indicator of drug market activity (Caulkins, 2000:397–398); ultimately, the major flaw of the program is that it became unsustainable in light of constrained budgetary resources. Data collection was suspended in 2004 and has not since been reactivated.

- Like the ADAM program, the *Drug Abuse Warning Network* (DAWN) avoids the use of self-reporting in measuring drug use. Instead, this SAMHSA-funded surveillance system collects data from hospital emergency departments (on drug-related visits) and medical examiner offices (on drug-related deaths). DAWN suffered from well-documented problems and key limitations (e.g., Caulkins et al., 1995), prompting a major redesign effort between 1997 and 2003 (Substance Abuse and Mental Health Services Administration, 2002). As of November 2008, links to publications based on "New DAWN" data on the program's website (http://dawninfo.samhsa.gov), from 2006 onward, all include a "caution" note that "SAMHSA is currently reviewing the estimates in this report and expects to publish revised estimates at a future date," suggesting potential instability in the redesigned program's estimation routines.

- The Drug Enforcement Administration's (DEA) *System to Retrieve Information from Drug Evidence* (STRIDE) compiles information from drug sales to undercover federal agents. By their nature, STRIDE data are uniquely positioned to provide information on the prices paid for drugs in those undercover transactions and the quality (purity) of the purchased drugs. However, the reliability of these data for economic and policy analyses—for instance, how closely the prices paid in the transactions logged by undercover federal agents track with prices in the broader illegal drug market, and the degree to which they represent federal (DEA) interdiction priorities rather than local-level activities— was ruled to be inadequate by the National Research Council (2001a).

That previous National Research Council (2001a) panel reviewed the extant data sources on drugs, drug markets, and the connection between drugs and crimes, and provides fuller descriptions of these and other data programs.

In policy analysis, public conversation, and research, measures of drug crime tend to be proxied by measures of enforcement, though the empirical relationship between drug crime and drug arrests is poorly understood. Drug arrest rates grew substantially through the 1980s, for example, while self-reported drug use among high school seniors was falling. Still, patterns such as these have multiple interpretations. Enforcement efforts may have been moving in the opposite direction of trends in use. Drug enforcement may have been reducing drug use. Increased enforcement may also have reduced survey respondents' propensity to report drug use.

The summary of the state of quantitative knowledge of drugs and crime by Caulkins (2000:394, 395) remains apt:

> In the drug policy arena we have an abundance of numbers, but the glass of insight is at best half full. . . . We know quite a bit about drug offenders within the criminal justice system but much less about their activities on the street. We know quite a bit about how many drug users there are but little about why there are so many. In contrast, we understand why people sell drugs but know little about how many upper level dealers there are, let alone how they operate. . . . More generally, existing data systems are reasonably adequate for *describing* patterns and trends but generally are incapable of *explaining* them, in part because opportunistic instead of random samples and the absence of control groups makes it difficult to tease out causal relationships.

Though BJS refers to some of the data sources listed above on its website, it generally attempts no analytical work based on those data. "Drugs and crime" is a top-level link on the BJS home page (http://www.ojp.usdoj. gov/bjs/drugs.htm), the principal link on which is to the electronic BJS publication *Drugs & Crime Facts* (Dorsey et al., 2004); that report does briefly attempt to pull together a series of findings about drugs from a number of BJS and non-BJS data sets. A number of BJS analyses have also summarized findings of drug-related questions in BJS's standard data series; for instance, Wilson (2000) summarizes both the Annual Survey of Jails' information on drug services provided by facilities and the Survey of Inmates in Local Jails' queries on arrests for drug offenses and prior drug use (these two collections are summarized in Section 3–B.2). Still, in a section of BJS's strategic plan that briefly itemizes data collections under major section headings, the section on drug crime statistics notes only that "many ongoing BJS statistical series collect and analyze data on drugs and crime" (Bureau of Justice Statistics, 2005a:Fig. 1); in the absence of a coherent overview of what is and is not known from existing data, the interactions of drugs and crime must be considered a gap in the coverage of BJS's overall portfolio.

2–D ASSESSMENT: FILLING THE GAPS

In the panel's assessment, the four topics we have described in Section 2–C constitute clear gaps in BJS's statistical coverage of the events and interactions inherent in the criminal justice funnel model of Figure 2-1:

> **Finding 2.1:** The data on crime currently collected by BJS are primarily focused on street crime. This focus on certain forms of violent and property crime does not account for important or emerging types of crime—notably, many forms of white-collar crime such as corporate fraud, health care fraud, financial institution fraud, money laundering, government fraud, consumer fraud, public corruption, and Internet crimes. The broad area of civil justice proceedings—distinct from criminal justice—is represented by one principal data series in BJS's extensive portfolio, and is limited by its construction to cover only completed court cases (and not out-of-court settlements). BJS's slate of cross-sectional series also does not readily provide for comprehensive analyses of contextual factors such as drugs and their impact on crime and violence.

> **Finding 2.2:** Responsibility within the U.S. Department of Justice for coordinating and organizing data collections on juveniles is generally assumed by the Office of Juvenile Justice and Delinquency Prevention (OJJDP), instead of BJS. Though BJS's series do cover some segments of the juvenile population (e.g., juveniles housed in adult correctional facilities), the results of BJS and OJJDP studies are not well integrated. Within both BJS's and OJJDP's statistical coverage, there remain substantial gaps in data for juvenile offenders and victims with respect to their processing through the justice system "funnel."

Clear though these gaps are, it is equally clear that considerable care and caution are in order when suggesting what to do about them. The four topics we have profiled share the basic quality that they are massive and complex, and that crafting full and effective data collection strategies for them would require major innovations in BJS's current concepts and protocols.

- A full focus on white-collar crime would require BJS to shift from its historical norm of using either people or a relatively limited number of establishments (e.g., correctional agencies) as their unit of analysis and grapple with the unique problems of businesses as a unit of study. BJS coverage of white-collar crime would ultimately benefit the large number of agencies that are already involved to varying degrees in the monitoring of such crimes, but would also require extensive coordination with those agencies; this would likely involve the need for an ongoing interagency advisory or coordinating board.[7]

- Civil trials are a large part of the overall justice system, posing harms to both persons and businesses, and so fuller knowledge of civil justice would be highly beneficial. However, civil justice is also replete with serious definitional issues (hard-to-define concepts, exacerbated by extensive state-by-state variation in legal standards); even defining the range of possible ADRs and determining their applicability in various states is difficult, much less generating reliable counts of their use. Further, attempts to make the measurement problem more tractable by focusing on filed cases that enter the system miss the large fraction of potential "cases" that are resolved in private. As with white-collar crime, effective expansion of civil justice data collection would necessitate involvement and coordination with a number of other actors, including the Administrative Office of the U.S. Courts, the individual state court systems and NCSC, and the American Bar Foundation.

- It is complicated, or impossible, to reach the youngest end of the juvenile population through traditional self-interviewing methods, because of their age and cognitive development. The NCVS limits itself to respondents ages 12 and older and other existing data series similarly focus on adolescents and teens; although it is possible that age levels might feasibly be pushed lower (e.g., to age 10 or 11), we know of no evidence of the possible effect of such a switch on the accuracy of self-reports. Collection and release of data on juveniles are also subject to a wide array of legal and ethical restrictions, and studies of juvenile justice raise special sensitivities and heighten the involvement of intensely interested interest groups. Measuring the entry of juveniles into supervision or residential placement is somewhat complicated because referrals come from a variety of sources other than the police, including family members or guardians and state child welfare bureaus. OJJDP's person-level measure of juveniles in correctional facilities relies on indirect responses, through reference to facility records and contact with administrators. A fuller assessment of the quality and coverage of data that may be available in school or juvenile facility records would have to accompany expanded data collections on the juvenile population.

To be clear, these conceptual and operational complexities are only one part of the difficulty in suggesting that new data collections be developed to

---

[7]These agencies include, at a minimum, the Federal Trade Commission, the Federal Deposit Insurance Corporation, the Federal Reserve Board, the Office of the Comptroller of Currency, the Office of Thrift Supervision, the National Credit Union Administration, the Mortgage Bankers Association, the U.S. Department of the Treasury, the National Center for Missing and Exploited Children, the FBI, the U.S. Department of Justice Criminal Division and Tax Division, the U.S. Postal Inspection Service, and the Securities and Exchange Commission.

fill these gaps. The other—and more acute—difficulty is that none of these gaps can be filled without extensive planning, new financial resources, and additional personnel. We think that it is clear that these areas of study are things with which the principal statistical agency of the U.S. Department of Justice *should* be concerned, but it is equally clear that it is unreasonable to expect that major progress could be made on any of these gaps within what is effectively BJS's current flat-funding situation. Addressing any of these gaps would require commitment and a sense of high priority from the Justice Department, the administration, and Congress.

Given these complexities, our basic suggestion on how to proceed borrows from two sources: the "clearinghouse" role defined by law for BJS on data on drugs and crime (point 17, Box 1-2) and OJJDP's detailed online Statistical Briefing Book. In any of these gap areas, the necessary first step is a structured accounting of what is and is not known from existing data resources, both internal and external to BJS. BJS's *Drugs & Crime Facts* is a first step toward such an accounting, as is the compilation of data from a wider variety of sources in the online *Sourcebook of Criminal Justice Statistics*. Still, these gap areas would benefit from a more analytical approach and more complete exploration of existing data sets (and their limitations); BJS's website (and reports) should more completely catalog external data sources and research, particularly for subject areas where BJS's own collections are limited. Through such a mapping of problem areas—and the more refined list of specific information needs that the mapping would suggest—BJS would have a more useful template for soliciting input on new data collections, should commitment of resources be secured.

> *Recommendation 2.1:* **Consistent with its legal mandate to collect, analyze, and disseminate statistical information on all aspects of the justice system, BJS should (a) document and organize the available statistics on forms of crime not covered by the NCVS, the FBI's UCR and NIBRS data systems, and other major data series maintained by other statistical agencies, (b) pursue research on what new statistics could be feasibly and usefully developed, and (c) propose such new data collections as the research suggests to be both feasible and useful. BJS should strive to function as a clearinghouse of justice-related statistical information, including reference to data not directly collected by BJS.**

Given our panel's charge to consider BJS's relationship to other data-gathering entities within the Department of Justice, we think that the gap of coverage of the juvenile population warrants its own specific recommendation. Having concluded that the funnel model of the justice system (Figure 2-1) is a useful and sound one, and given the numerous references to juvenile delinquency in BJS's legal mandate, we think it odd that BJS's ceding of a complete branch of the funnel to another Justice Department entity is as complete as it is. In short, the measurement of juvenile justice is something in which the principal statistical agency of the U.S. Department of Justice should be fully engaged:

> *Recommendation 2.2:* **In line with its original charge and to better document and understand the contribution of juveniles to street crime and violence, the victimization of youth, and the consequences for youth and society of their victimization and offending, BJS should develop juvenile victimization, crime, and justice statistical series suitable for describing the patterns of offending and victimization of youth, longitudinal progression of youth through the juvenile and criminal justice systems, and reentry into the community and criminal system. Taking on this responsibility would require additional resources.**

We hasten to add, however, that this recommendation should not be construed as saying that BJS should necessarily usurp (or "reclaim") data collection functions from OJJDP. Like BJS, OJJDP has invested considerable time and effort in developing its relationships with its data collection providers, and upending those relationships should not be taken lightly. What we do envision through this recommendation is BJS-OJJDP collaboration on research on the full juvenile population, including, at a minimum, fuller study of juveniles processed by adult courts and correctional facilities.

– 3 –

Overview of Bureau of Justice Statistics Data Series

SOME OF THE DATA COLLECTIONS maintained by the Bureau of Justice Statistics (BJS) are very recent innovations whereas others were developed as the agency took its current form over the course of the 1970s. By comparison, at least one of BJS's collections can trace its origin to the turn of the 20th century (and, by extension, to the 1850 decennial census). Its portfolio includes those that have been successfully repeated in subsequent years (and hence have developed series continuity), but it also includes one-shot efforts that could have been repeated but were not because of budget or other constraints. In their content and structure, they range from extensively developed population surveys to targeted administrative questionnaires filled out by institution managers to hand-coded summaries of court docket folders—and, accordingly, range considerably in their associated level of expense. As illustration of the varying costs of BJS's data series, Table 3-1 summarizes BJS's expected spending in fiscal year 2008.

This chapter provides a brief overview of BJS's major data collection efforts, divided into four major topic areas: victimization (Section 3–A), corrections (3–B), law enforcement (3–C), and adjudications (3–D). (In recent years, BJS has undertaken a series of data collections specifically focusing on justice issues on American Indian reservations and tribal lands; these collections slightly overlap the major topic areas and are separately described in Box 3-1.) Within each of these sections, a table illustrates the years of collection for the various series under that topic heading. As noted in Chapter 1, these summaries are not intended to be full dossiers on the collections, their

Table 3-1 Estimated Funding for Bureau of Justice Statistics Criminal Justice Statistics Program, Fiscal Year 2008

Program	Estimated Funding (thousands of dollars)
Victimization	
National Crime Victimization Survey—Collection	18,700
National Crime Victimization Survey—Redesign Effort	3,900
Law Enforcement	
Census of State and Local Law Enforcement Agencies	905
Prosecution and Adjudication Statistics	
Civil Trial Court Cases	340
Court Statistics Project	415
National Judicial Reporting Program (Year 1)	340
State Court Processing Statistics	300
Survey Development, Two New Collections	300
Corrections	
Annual Probation and Parole Statistics	175
Annual Survey of Jails	230
Capital Punishment Statistics	260
Census of Probation and Parole Agencies	250
Deaths in Custody Reporting Program	553
National Corrections Reporting Program	647
National Prisoner Statistics	130
State Prison Expenditures	300
Federal Justice Statistics Program	800
Criminal Justice Employment and Expenditures	222
Firearm Background Check Statistics	360
Tribal Statistics	
Tribal Criminal History Improvement Program	704
State Tribal Crime Reports	145
State Justice Statistics Program	
State Statistical Analysis Centers	2,300
Technical Assistance to SACs/Multi-State Projects	1,400
Publication and Dissemination	2,872
Management, Administration, and Joint Federal Statistics Efforts	2,131
Total	38,679

NOTE: Expenditures for National Criminal History Improvement Program grants and for data collections pursuant to the Prison Rape Elimination Act of 2003 are funded through separate lines in the BJS budget.

SOURCE: Adapted from table provided by BJS.

uses, and their associated methodological challenges, but rather a general orientation. In particular, our interim report (National Research Council, 2008b) describes the development and protocols of the National Crime Victimization Survey (NCVS) in considerably more detail than we attempt in this more limited treatment.

We close the chapter in Section 3–F with general assessments of BJS's portfolio and the structure of BJS collections within that portfolio.

3–A VICTIMIZATION

The President's Commission on Law Enforcement and Administration of Justice (1967) that developed the justice system "funnel" model we use as a framework in this report (Section 2–A) and recommended the creation of what would become BJS also pioneered a new approach to studying crime. The commission sponsored the National Opinion Research Center (NORC) to survey the members of 10,000 households on their experiences as victims of crime and violence, and the commission then compared the results to the reported-to-police estimates from the Uniform Crime Reporting (UCR) program. This prototype National Survey of Criminal Victims demonstrated to the commission that "for the Nation as a whole there is far more crime than ever is reported" (President's Commission on Law Enforcement and Administration of Justice, 1967:v):

> Burglaries occur about three times more often than they are reported to police. Aggravated assaults and larcenies over $50 occur twice as often as they are reported. There are 50 percent more robberies than reported. In some areas, only one-tenth of the total number of certain kinds of crimes are reported to the police.

As a consequence of the commission's report, the Omnibus Crime Control and Safe Streets Act of 1968 specifically mandated the new Law Enforcement Assistance Administration (LEAA) to "collect, evaluate, publish, and disseminate statistics and other information on the condition and progress of law enforcement in the several States" (P.L. 93-83 § 515(b); see also U.S. Census Bureau, 2003:A1-5). This data-gathering authority was invoked to begin pilot work and implementation of what would become the National Crime Survey; later, at the culmination of an extensive redesign process, this survey was renamed the National Crime Victimization Survey. Drawn from an original mandate to study the progress of law enforcement, the NCVS was designed to do so by asking respondents about victimization incidents generally, whether or not they were reported to authorities. Accordingly, the NCVS presented the unique ability to shed light on the "dark figure of crime"—the phrase coined by Biderman and Reiss (1967) to describe criminal incidents that are not reported to police.

Table 3-2 shows the years of collection of the NCVS and, more specifically, the topic supplements to the NCVS. As we will discuss, there are some instances in which content from a supplement was subsequently integrated into the core NCVS; hence, the table's suggestion that a supplement was not repeated in later years does not necessarily mean that the topic was dropped.

Box 3-1 Bureau of Justice Statistics Collections on Tribal Justice

Tribal justice agencies are included in several of BJS's principal data series such as the Census of State and Local Law Enforcement Agencies (Hickman, 2003). However, on tribal lands, "criminal jurisdiction . . . is divided among the Federal, State, and tribal governments" depending on the "nature of the offense, whether the offender or victim was a tribal member, and the State in which the crime occurred" (Perry, 2005:1). In recent years, BJS has developed one-shot and continuing collections on the justice authorities that operate in American Indian tribal lands. BJS typically has worked with the U.S. Department of the Interior's Bureau of Indian Affairs on developing address and contact lists; the U.S. Department of Justice also maintains an Office of Tribal Justice that facilitates interactions between the Justice Department and the tribes, and the Federal Bureau of Investigation (FBI) shares law enforcement authority on tribal lands.

Conducted on a one-shot basis to date, in 2002, the Census of Tribal Justice Agencies (Perry, 2005) was the U.S. Department of Justice's first comprehensive effort to document tribal justice agencies as systems in their own right. The major difference between this collection and BJS's standard series is that it sought to articulate the use of "indigenous forum" arrangements (e.g., councils of elders or "sentencing circles") that are distinct from court proceedings, and that may combine adjudication and law enforcement functions on reservation lands. For those tribes that have developed specific law enforcement agencies, the census sought staffing and policy information (including the level of interface and cross-deputization with nontribal authorities). Questions on the census were also intended to provide information on tribes' criminal history record-keeping and ability to provide crime reports to federal efforts such as the FBI's National Crime Information Center and National Sex Offender Registry. The census achieved participation from 314 of 341 federally recognized tribes; however, "participation by Alaska Native tribes or villages was not extensive enough to enable their inclusion" in the results of the census (Perry, 2005:iii). BJS contracted with Falmouth Institute and Policy Studies, Inc., as the data collection agent for the census.

In 1998, BJS began specifically collecting information on tribal jails as a component of the Annual Survey of Jails; the collection has since developed into a separate Survey of Jails in Indian Country that continues on an annual basis. This collection mirrors the content of the Survey of Inmates of Local Jails (Section 3–B.2) but concentrates on those facilities and detention centers operated by the federal Bureau of Indian Affairs or by individual tribal authorities. The Survey of Jails in Indian Country also includes questions on facility programs and services, such as health care and counseling, but these questions typically are asked only in selected years. See, e.g., Minton (2006, 2008) for reports of survey results.

BJS's National Criminal History Improvement Program of grants to assist local law enforcement agencies develop criminal history and other information databases has recently expanded to include a tribal justice–specific component. Perry (2007) summarizes work in the first few years of the program, 2004–2006. Dubbed the Tribal Criminal History Record Improvement Program, the program's grant solicitation for 2008 puts particular priority on "enhancing automated identification systems, records of protective orders involving domestic violence and stalking, sex offender records, [and] DWI/DUI conviction information," as well as developing tribal interfaces to federal background check data.

The experiences of American Indians living on and off reservation lands in the justice system have been analyzed by BJS staff using the National Crime Victimization Survey and other resources; see, e.g., Greenfeld and Smith (1999) and Perry (2004).

Table 3-2 Bureau of Justice Statistics Data Collection History and Schedule, Victimization, 1981–2009

Series	81	82	83	84	85	86	87	88	89	90	91	92	93	94	95	96	97	98	99	00	01	02	03	04	05	06	07	08	09	Total
National Crime Victimization Survey	●	●	●	●	●	●	●	●	●	●	●	●	●	●	●	●	●	●	●	●	●	●	●	●	●	●	●	●	●	29
NCVS Supplements or New Content Areas																														
Crimes Against Disabled																											●	●	●	3
Cybercrime																				●	●	●	●							4
Hate Crimes																			●	●	●	●	●	●	●	●	●	●	●	11
Identity Theft																								●	●	●	●	●	●	6
Police-Public Contact Survey/Police Use of Force																●			●			●			●			●		5
School Crime Supplement									●						●				●		●		●		●		●		●	8
Supplemental Victimization Survey (stalking)																										●				1
Vandalism											●	●	●	●	●	●	●	●	●	●	●	●	●	●	●	●				16
Workplace Violence Supplement																						●								1
City-Level Victimization Surveys (12)																		●												1

NOTES: ●, data collected. ·, data not collected. Preliminary versions of NCVS questions on crimes against the disabled were asked as early as 2000, but only those used starting in 2007 were meant for or suitable for producing regular estimates.

SOURCE: Bureau of Justice Statistics.

In its role as an indicator of crime levels in the United States, the NCVS is often compared to data from the UCR program maintained by the Federal Bureau of Investigation (FBI); we discuss the UCR and related programs in more detail in Section 4–C.

3–A.1   National Crime Victimization Survey

Since the outset, BJS has commissioned the U.S. Census Bureau as the data collection agent for the NCVS; the NCVS is one of the major federal surveys administered by the Census Bureau's Demographic Surveys Division. The NCVS is a household survey using a rotating panel sample of addresses, meaning that addresses are chosen to be eligible for interviewing for a certain number of interviews over a fixed period of time. Currently, contacts are made for interviews at sample addresses for 3 years, seven interviews at 6-month intervals.[1] As sample addresses complete their time in sample, they are replaced with new ones. When the NCVS began in 1972, the NCVS was administered to 72,000 households—a large sample, meant to produce reliable estimates of year-to-year change in victimization as well as information on relatively rare crime types. However, sample size reductions (for purposes of cost savings) since the 1980s have reduced the sample size of the NCVS by almost half—in 2005, the NCVS was administered to about 38,600 households or 67,000 people. The sample sizes and response rates for the NCVS between 1996 and 2006 are shown in Table 3-3.

Although the current sample size qualifies the NCVS as a large data collection program, occurrences of victimization are essentially a rare event relative to the whole population: many respondents to the survey do not have incidents to report when they are contacted by the survey. Consequently, as we described in Chapter 1, the reduced sample size (combined with generally low and decreasing estimated overall victimization rates) is such that only a large percentage change in violent crime victimization rates—at least 8 percent—is a statistically significant year-to-year change. Indeed, as noted in the U.S. Office of Management and Budget (2007:8) annual review of statistical program funding, "cost cutting measures applied to the NCVS continue to have significant effects on the precision of the estimates—year-to-year change estimates are no longer feasible and have been replaced with two-year rolling averages" in BJS reports on victimization.

---

[1]The sample is further divided into six "rotation groups," and each of these into six "panels." One panel from each of the rotation groups is designated for interviewing each month, hence the "rotating panel" nomenclature. In large part, the use of this rotating panel structure derives directly from the choice and retention of the Census Bureau as data collector for the NCVS; to achieve some efficiencies in collection, such as sharing a pool of interviewers, some design features of the NCVS were chosen to emulate those of the Census Bureau's Current Population Survey (CPS), then the Census Bureau's largest intercensal survey (National Research Council, 2008b:123; see also Cantor and Lynch, 2000:107).

Table 3-3 Number of Households and Persons Interviewed by Year, 1996–2006

	Households		Persons	
Year	Sample Size	Response Rate	Sample Size	Response Rate
1996	45,000	93	85,330	91
1997	42,910	95	79,470	90
1998	43,000	94	78,900	89
1999	43,000	93	77,750	89
2000	43,000	93	79,710	90
2001	44,000	93	79,950	89
2002	42,000	92	76,050	87
2003	42,000	92	74,520	86
2004	42,000	91	74,500	86
2005	38,600	91	67,000	84
2006	38,000	91	67,650	86

NOTE: These sample sizes correspond to the number of separate households and persons designated for contact in a particular year. Participation rates for a particular year would be roughly double these, accounting for two interviews with sample addresses in the same year.

SOURCE: Bureau of Justice Statistics (2006b, 2008c).

A multiyear redesign effort, culminating in 1992 with the first collection using all of the new procedures (and the renaming of the survey to NCVS), focused principally on implementing a screening procedure. The first part of an NCVS interview is the screening questionnaire, which uses a series of carefully constructed questions to elicit counts—but not yet full information—about crime victimization incidents in the past 6 months. After this screener has been completed, the NCVS interviewer guides the respondent through the completion of a detailed incident report on the circumstances of each incident counted by the screener. This interviewing process is repeated for each person age 12 or older in the household at the sample address (although only the first respondent is asked about general characteristics of the household). Because the number of victimizations experienced by respondents varies—and, with it, the number of incident reports that must be completed—the length of time that it can take to complete an NCVS interview can also vary. BJS estimates that the average face-to-face interview lasts 26 minutes (Bureau of Justice Statistics, 2008d:10).

Over time, the mode of NCVS interviewing and the use of interviews to calculate estimates has shifted. BJS still insists that the first interview with a sample household be conducted in person by a Census Bureau enumerator. Beginning in 1980, BJS began to permit every other interview (after the first contact) to be conducted via phone; by 2003, NCVS interviews were being advised to complete their interviews by phone "whenever possible" (U.S. Census Bureau, 2003:A1-11). In the late 1990s and early 2000s, BJS also invested in the use of survey automation procedures, so that the NCVS and its supplements are fully electronic; face-to-face interviews are done by computer-assisted personal interviewing (CAPI) with the interviewer using a laptop computer, and telephone interviews can be completed using the same computer interface.[2] In terms of how NCVS interviews result in final estimates, BJS currently produces "collection year" estimates from the NCVS, accumulating the data from all interviews completed in a particular year t. Because of the 6-month reference period of the survey, this means that a year t estimate from the NCVS includes events that may have occurred in the last half of the year $t - 1$ and does not include all incidents that actually occurred in year t (since late year t incidents would only be picked up in interviews in year $t + 1$).

Until recently, a key feature of the NCVS was that the first interview with a sample household was not included in the estimates. Instead, it was withheld and used as a bounding interview: counts and incidents reported in the second interview could be checked against the bounding interview to correct for the same incidents being reported multiple times. As one of a bundle of cost-cutting measures, BJS and the Census Bureau began to include these "bounding interviews" in the production of estimates, beginning with estimates from the 2006 administration of the survey. As we discuss in Section 3–A.3, these changes produced anomalous results that led BJS to declare a "break in series" that prevents comparison with previous years' data.

NCVS results are annually described in two BJS report series, *Criminal Victimization* (e.g., Rand and Catalano, 2007) and *Crime and the Nation's Households* (e.g., Klaus, 2007). BJS staff have issued a number of "Special Reports" dedicated to analysis of particular content from the NCVS, such as the involvement of weapons and firearms in victimization incidents (Perkins, 2003), violence in the workplace (Duhart, 2001), reporting of rape or attempted rape to the police (Rennison, 2002b),[3] and specific reports on victimization experiences by racial and ethnic groups (e.g. Rennison, 2002a; Harrell, 2007). A complete list of references to publications that have used and analyzed NCVS data is beyond the scope of this report, but some recent references suggest the breadth of application of the data. NCVS data have been probed to study the issue of violence and police response to incidents in disadvantaged areas and neighborhoods (Baumer, 2002; Baumer et al.,

---

[2]As mentioned in the next section, BJS and the Census Bureau attempted to do many of the NCVS telephone interviews from centralized Census Bureau call centers, but this practice was eliminated in the most recent set of cost-cutting measures.

[3]Rennison (2002b) is officially designated a "Selected Findings" report rather than a "Special Report."

2003); the NCVS has also been used in lines of research on the reporting of crime by women, particularly of rape, to authorities (Baumer et al., 2003; Felson et al., 2005; Xie et al., 2006; Addington and Rennison, 2008) and the effect of victimization on residential turnover (Dugan, 1999; Xie and McDowall, 2008). Longer-term analyses have considered the comparability of NCVS with other sources of data on crime such as the UCR (Lynch and Addington, 2007) and the National Violence Against Women Survey (Rand and Rennison, 2005), and differential trends in violence by gender (Lauritsen and Heimer, 2008).

3–A.2 NCVS Supplements

Over time, NCVS's flexibility as a survey vehicle has been exploited to gather occasional or one-time data through survey supplements. These supplements have been supported by contributions from partner agencies or grants from other organizations; supplements have also been directly developed to respond to new mandates from Congress.

A battery of questions on crime against the disabled was developed in direct response to P.L. 105-301, the Crime Victims with Disabilities Awareness Act of 1998, which directed that the NCVS be used to measure "the nature of crimes against individuals with developmental disabilities" and "the specific characteristics of the victims of those crimes." The module of questions is meant to assess whether victims of crime were in poor health, had any physical or mental impairments, or had disabilities that affected their everyday life. They are also asked to judge if any of these provided an opportunity for their victimization.

Similarly, since 1999, the NCVS has collected data on hate crimes, a content area motivated by the Hate Crime Statistics Act of 1990 (P.L. 101-275). The hate crime questions were not a standalone module but rather a section added directly to the NCVS questionnaire.

BJS collaborated with its fellow Office of Justice Programs agency, the National Institute of Justice, to include a module of questions on crime in schools in the 1989 version of the survey (Bureau of Justice Statistics, 2008e:5). After a repeat administration in the mid-1990s, the National Center for Education Statistics (NCES) of the U.S. Department of Education has paid for BJS to include the School Crime Supplement (SCS) to the NCVS on a biennial basis. As described by the U.S. Census Bureau, Demographic Surveys Division (2007:51):

> The supplement contains questions on preventative measures employed by the school to deter crime; students' participation in extracurricular activities; transportation to and from school; students' perception of rules and equality in school; bullying and hate crime in school; the presence of street gangs in school; availability of drugs and alcohol in

the school; attitudinal questions relating to the fear of victimization in school; access to firearms; and student characteristics such as grades received in school and postgraduate plans.

The school crime questions were administered to all individual respondents in NCVS sample households who were between ages 12 and 18, "who were enrolled in primary or secondary education programs leading to a high school diploma, and who were enrolled in school sometime during the six months prior to the interview" (U.S. Census Bureau, Demographic Surveys Division, 2007:51).[4] The SCS is administered to NCVS respondents (maintaining the age-12-and-older restriction of the main survey) who attend schools. In the 2005 administration of the SCS (running from January through June), 61.7 percent of the 11,525 eligible SCS respondents completed the questions (Bureau of Justice Statistics, 2008e:6). SCS data are distinct from, and provide a different vantage on crimes in school from, the Schools Survey on Crime and Justice that NCES contracts directly with the Census Bureau to construct; that survey is essentially an establishment survey, meant to be completed by principals of a nationally representative sample of public elementary and secondary schools. Results from the various administrations of the SCS, and related data resources, are described by Dinkes et al. (2007), and the SCS data have been used in analyses by, for example, Addington (2003).

Questions on experiences with identity theft were first added to the NCVS in July 2004. Whereas other NCVS supplements such as the SCS function as a true supplement to the NCVS interviewing experience—a standalone questionnaire that is meant to be answered by everyone meeting certain eligibility requirements—the identity theft "supplement" was built into the NCVS screening interview. The version of the identity theft questions used in 2004 is illustrated in Figures 3-1 and 3-2. As noted in Section 2–C.1, this set of identity theft questions is one of few available quantitative measures of the levels of some types of fraud. With additional sponsorship from the Federal Trade Commission and other bureaus within the Office of Justice Programs, BJS planned to field a more comprehensive identity theft supplement to the NCVS beginning in 2008 (Baum, 2007:4).

Another important supplement, the Police-Public Contact Survey (PPCS), stems from a brief provision in the Violent Crime Control and Law Enforcement Act of 1994 (P.L. 103-322 § 210402): "the Attorney General shall, through appropriate means, acquire data about the use of excessive force by law enforcement officers." As a direct result of this mandate, BJS developed the PPCS to measure the extent of all types of interactions between the police and members of the public (of which those involving

---

[4] However, "students who were home schooled were not included past the screening questions since it was determined that many of the questions in the SCS were not relevant to their situation" (Bureau of Justice Statistics, 2008e:5).

HOUSEHOLD RESPONDENT'S IDENTITY THEFT QUESTIONS

FIELD REPRESENTATIVE – *Read introduction.*

INTRO 1: **The next few questions are related to identity theft. They refer to episodes of identity theft discovered by you or anyone in your household during the last 6 months.**

45c. Since ______________ , 20 _____ , have you or anyone in your household discovered that someone –

   (a) **Used or attempted to use any existing credit cards or credit card numbers without permission to place charges on an account?** . . . .

     | 107 | 1 ☐ Yes   2 ☐ No   3 ☐ Don't know

   (b) **Used or attempted to use any existing accounts other than a credit card account – for example, a wireless telephone account, bank account or debit/check cards – without the account holder's permission to run up charges or to take money from accounts?** . . . .

     | 108 | 1 ☐ Yes   2 ☐ No   3 ☐ Don't know

   (c) **Used or attempted to use personal information without permission to obtain NEW credit cards or loans, run up debts, open other accounts, or otherwise commit theft, fraud, or some other crime?** . . . .

     | 109 | 1 ☐ Yes   2 ☐ No   3 ☐ Don't know

CHECK ITEM C1   Look at 45c. How many times is box 1 (Yes) marked in 45c?

   1 ☐ None (no entries of Yes) – **SKIP** *to Check Item D*
   2 ☐ One or more times – *Ask 45d*

45d. Was the misuse of – (the credit card account(s)/any existing account(s) other than credit cards/personal information or new account(s)) one episode or more than one episode of identity theft?

   | 110 | 1 ☐ One – **SKIP** *to 45g*
              2 ☐ More than one

45e. Did these episodes of identify theft occur separately or at the same time?

   | 111 | 1 ☐ Separately
              2 ☐ At the same time – **SKIP** *to 45g*

45f. Which episode of identity theft was most recently discovered?

   | 112 | 1 ☐ Existing credit cards
              2 ☐ Existing accounts other than a credit card
              3 ☐ Personal information to obtain new accounts

---

INTRO 2: **The following questions refer only to the most recent discovery of identity theft by you or anyone in your household.**

45g. How did you become aware of the identity theft?

   Mark (X) all that apply.

   | 113 | 1 ☐ Block was placed on a credit card or other existing account
              2 ☐ Money missing from account or charges placed on an account
              3 ☐ Contacted by a credit bureau, collection agency, credit card company or other company about late/unpaid bills
              4 ☐ Contacted by a bank
              5 ☐ Noticed that a credit card, check book, etc. was missing
              6 ☐ Notified by a law enforcement agency
              7 ☐ Denied credit or a loan
              8 ☐ Noticed an error in a credit report
              9 ☐ Other – *Specify* ⬏

45h. What was the total dollar amount of the credit, loans, cash, services, and anything else the person obtained while misusing (the credit card account(s)/any existing accounts other than credit cards/personal information or new account(s)?

   | 114 | $ _________________ .00   Amount taken
              X ☐ Don't know
              0 ☐ None

45i. Has the misuse of – (the credit card account(s)/any existing accounts other than credit cards/personal information or new account(s)) stopped (e.g. you or a household member closed a checking account)?

   | 115 | 1 ☐ Yes
              2 ☐ No
              3 ☐ Don't know

45j. Is the misuse of – (the credit card account(s)/any existing accounts other than credit cards/personal information or new account(s)) still causing problems for you or any other household member? For example, are you still spending time clearing up credit accounts or your credit report?

   | 116 | 1 ☐ Yes – **SKIP** *to 45l*
              2 ☐ No
              3 ☐ Don't know

Figure 3-1   Module of questions on identity theft in the 2004 National Crime Victimization Survey (part 1)

NOTE: The questions are part of NCVS-1, the screening questionnaire part of the NCVS interview.

HOUSEHOLD RESPONDENT'S IDENTITY THEFT QUESTIONS

45k. How much time did it take to resolve ALL PROBLEMS associated with the misuse of – {the credit card account(s)/any existing account(s) other than credit cards/personal information or new account(s)} after the misuse was discovered?

Less than one month:

117 ______________ Days

OR

1 – 6 Months:

118 ______________ Months

x ☐ Don't know days or months

45l. As a result of (any of) the misuse of {the credit card account(s)/any existing account(s) other than credit cards/personal information or new account(s)} discovered in the last 6 months, have you or anyone in your household . . .

(Read answer categories 1–10)

Mark (X) all that apply

119
1 ☐ Been turned down for a loan?
2 ☐ Had banking problems?
3 ☐ Had problems with credit card accounts?
4 ☐ Had phone or utilities cut off or been denied new service?
5 ☐ Had to pay higher interest rates on credit cards, loans, etc.
6 ☐ Been turned down for insurance or had to pay higher rates?
7 ☐ Been contacted by a debt collector or creditor?
8 ☐ Been the subject of a civil suit or judgment?
9 ☐ Been the subject of a criminal investigation, warrant, proceeding, or conviction?
10 ☐ Had some other problems?– *Specify*

11 ☐ No problems

CHECK ITEM C2 Briefly summarize the identity theft that occurred against the respondent or another household member.

Figure 3-2 Module of questions on identity theft in the 2004 National Crime Victimization Survey (part 2)

NOTE: See Figure 3-1.

"excessive force" is logically a subset). The survey was first conducted on a pilot basis in 1996 (Greenfeld et al., 1997); after refinement, it was fully fielded as an NCVS supplement in 1999 and has become a continuing occasional supplement. The survey gathers detailed information about the nature of police-citizen contacts, respondent reports of police use of force and their assessments of that force, and self-reports of provocative actions that respondents may have themselves initiated during the encounter. We return to discussion of the PPCS in Section 5–A.2 because of events that transpired in the release of data from the 2002 administration of the supplement; those events notwithstanding, PPCS data have driven useful analyses of racial profiling by, for example, Engel and Calnon (2004) and Engel (2005).

BJS fielded a Supplemental Victimization Survey (SVS) from January through June 2006; the report of findings from the supplement was released in January 2009 (Baum et al., 2009). The SVS was funded by the Justice Department's Office on Violence Against Women (OVW) and focused principally on the measurement of victimization by stalking or harrassing behavior. The SVS content was determined on the basis a 1-day expert workshop convened by OVW and BJS and a subsequent year-long working group of OVW, BJS, and Census Bureau staff (Baum et al., 2009:10). An interesting feature of the SVS is that it deliberately did not use the term "stalked" (or variants thereof) until the final question. NCVS respondents were routed into the SVS based on their response to a screening question listing a number of behaviors (e.g., "leaving unwanted items, presents, or flowers") that "frightened, concerned, angered or annoyed" the respondent. Only in the final question were SVS respondents asked whether the behaviors they had just described constituted "stalking" (Baum et al., 2009:11–12).

The display of NCVS supplements in Table 3-2 favors supplements that have been mounted in the past 15 years, and misses some older one-shot supplements. Other NCVS supplements over the years have included:

- *Crime seriousness:* An early supplement gathered national data on the perceived seriousness of crime, information that has been used to differentially weight incidents to reflect their impact on the public. On a one-shot basis, BJS collaborated with the Office of Community Oriented Policing Services in conducting community safety surveys by telephone in 12 cities, wholly distinct from the NCVS (Smith et al., 1999).

- *Attitudes and lifestyles:* Another supplement gathered extensive data on the attitudes of individuals and the relationship between crime and how they conduct their lives (Murphy, 1976; Cowan et al., 1984).

- *Workplace risk:* A module of questions on nonfatal violence in the workplace was sponsored by the National Institute for Occupational Safety and Health (Centers for Disease Control and Prevention) in 2002.

- *Harassment and stalking:* As described above, the SVS was conducted from January through June 2006 on, as yet, a one-shot basis with sponsorship from OVW. Questions focused primarily on perceived experiences with harassment and stalking.

It is important to note two aspects of the existing set of NCVS supplements. The first is that they have been topic-based supplements—modules of additional questions asked to some or all NCVS respondents interviewed at a particular time. "Supplement," interpreted more broadly, could connote the addition of persons or households to the sample, such as adding sample in particular geographic areas to support subnational estimates or "targeting" additional sample units in particular age, race, or gender groups. The second noteworthy aspect of the current NCVS supplements is that the depiction in Table 3-2 may suggest more of a structure to the existing supplements than actually exists. Though some supplements have come to follow a regular pattern of inclusion in the NCVS, supplements are generally done on an as-available basis, requiring both funding from any external sponsoring agency and time to develop and test questionnaires. As we discuss further in Section 3–F.2, there is currently no regular, rigorous schedule of ongoing NCVS supplements.

3–A.3   The NCVS "Break in Series"

We describe the methodology of the NCVS in greater detail in our interim report (National Research Council, 2008b), the executive summary of which is reprinted as Appendix B of this report. Hence, our description of the NCVS in this report is meant to be a brief synopsis rather than a comprehensive overview. However, in this report's description of BJS data series, we believe it is important to mention a complication concerning the 2006 data from the NCVS that was encountered as our interim report was in the end stages of production.

On December 12, 2007, BJS released its first NCVS estimates for data collected in 2006—doing so with a strong warning that these newest estimates were fundamentally different from, and incomparable to, previous years' estimates. In processing 2006 NCVS results, BJS and the Census Bureau detected "variation in the amount and rate of crime [that] was too extreme to be attributed to actual year-to-year changes." After consulting with individual external experts, BJS concluded that these differences were sufficiently large as to declare that that "there was a break in series between 2006 and previous years that prevent[s] annual comparison of criminal victimization at the national level" (Rand and Catalano, 2007:1).

A technical note in the support documentation for the 2006 NCVS data, as logged in the National Archive of Criminal Justice Data (NACJD), indicates BJS's conclusion that (Bureau of Justice Statistics, 2008d):

> [The break] was mainly the result of three major changes in the survey methodology:
>
> 1. introducing a new sample beginning in January 2006, based on the 2000 Decennial Census to account for shifts in population and location of households that occur over time
>
> 2. incorporating responses from households that were in the survey for the first time (called "bounding interviews") in the production of survey estimates
>
> 3. replacing paper and pencil interviewing (PAPI) with computer-assisted personal interviewing (CAPI).

On the first point, the Census Bureau "redesigns" samples for the household surveys that it performs under contract to other federal agencies following the completion of a new decennial census. The use of results from the 2000 decennial census to update survey samples (and derive the population "controls" used to weight sample survey data to reflect the whole population) was delayed for several reasons. Key among these reasons was final determination of exactly which census results—whether the initial census totals or figures that had been statistically adjusted for nonresponse—should be applied. Though a 1999 U.S. Supreme Court decision precluded the use of adjusted census numbers for purposes of congressional apportionment, it left open the possibility of adjustment for data used in legislative redistricting, in deriving survey controls, or for other purposes. In a series of recommendations, the Census Bureau ultimately decided against adjustment of 2000 census results for any purpose, but said determination required 2 years of additional research and analysis (National Research Council, 2004b).

The change to a sample based on the 2000—and not the 1990—census began for the NCVS in January 2006. In their analyses, BJS and the Census Bureau concluded that the shift to the new sample had contributed to severely anomalous results (Bureau of Justice Statistics, 2008d):

> Of the new areas included in the 2006 sample, about two-thirds were in areas designated as rural areas. . . . Introduction of the new sample in rural areas showed that the rate of violent victimization increased by 62% between 2005 and 2006. However, there was very little change in rates of violent victimization for urban and suburban sample areas, and violent victimization rates for 2006 continuing areas (urban, suburban, and rural) were not significantly different from 2005.

BJS and the Census Bureau concluded that the addition of the new sample might also lead to a more subtle effect on the estimates: "during every sample redesign, the selection and integration of a new sample requires hiring and training interviewers to administer the survey in new areas." Accordingly, the introduction of the new sample might have led to a different mix of new versus experienced interviewers and, accordingly, differences in the effectiveness in eliciting victimization incidents from respondents.

The second cause for the "break in series"—the inclusion of first, bounding interviews—was one cost-cutting measure introduced in 2006 in order to remain within BJS's budgetary resources. Another cost-cutting measure implemented at the same time was an across-the-board 14 percent cut in the NCVS sample size. The technical documentation note indicates (Bureau of Justice Statistics, 2008d):

> Because of telescoping and panel bias (sometimes called respondent fatigue), respondents tend to report more incidents of crime during the first interviews than in subsequent interviews. A weighting adjustment factor was applied to mitigate over-counting of crime.

Despite these weighting adjustments, though, BJS and the Census Bureau were unable to fully parse the effects of including the first interviews.

Finally, BJS and the Census Bureau concluded that at least part of the anomalous findings might be attributable to mode effects: differences in survey response due to the medium through which questions are posed, such as face-to-face interviewing, self-response on paper forms, self-response by telephone, or self-response via the Internet. For several preceding years, BJS had invested in converting the NCVS from a paper-based survey to a fully automated collection, with interviewers reading from and entering responses into an electronic version of the questionnaire on a laptop computer. (Since the first interview with a sample household must be completed in person, but later interviewers may be completed by telephone, the fully-automated NCVS is an example of both CAPI [first and any subsequent face-to-face interview] and computer-assisted telephone interviewing [CATI].) This automation work happened to be completed at exactly the same time as the other methodological changes (Bureau of Justice Statistics, 2008d):

> In July 2006, NCVS was converted to a fully automated data collection. . . . Previous research suggested that computer-assisted telephone interviewing (CATI) enhances data accuracy and produces higher and more accurate estimates because the computer-based interviewing process ensures that correct skip patterns are followed so that respondents answer all relevant questions. [However,] limited time and financial resources prohibited the Census Bureau and BJS staff from fully assessing the effects of CAPI on the 2006 estimates.

In releasing its first report on 2007 NCVS data (Rand, 2008), BJS substantially softened the rhetoric suggesting an irrevocable break in series. Instead, the report tentatively characterizes the 2006 results as "a temporary anomaly in the data" and expresses "a high degree of confidence that survey estimates for 2007 are consistent with and comparable to those for 2005 and previous years." Generally, the report characterizes the still-not-fully-understood changes from 2005 to 2006 and again from 2006 to 2007 as "substantial fluctuations" that "do not appear to be due to actual changes in crime" (Rand, 2008:1, 2). In technical notes, the report summarizes evaluative work done by the Census Bureau on BJS's behalf that concludes that the effects of sample size reduction and inclusion of bounding interviews had little effect on NCVS estimates; it nudges toward concluding that the "hiring and training of new interviewers to administer the survey" as part of the redesign sample in 2006 was a major factor in the spike in victimization rates observed in that year (Rand, 2008:10). The report notes that "BJS continues to work with the U.S. Census Bureau to better understand the impact of these [methodological] changes upon survey estimates" and indicates that adjusted estimates for 2006 and 2007 may be issued at a future date (Rand, 2008:2). In the interim, in a footnote (Rand, 2008:Note 2), BJS encourages "users . . . to focus on the comparison between 2005 and 2007 victimization rates until the changes to the NCVS in 2006 are better understood."

BJS's December 2007 announcement of the "break in series" in the NCVS was made as our panel's interim report on the NCVS was in the very late stages of review. In that report, we could only acknowledge the BJS announcement in a footnote (National Research Council, 2008b:86).

Limited though it was, the brief footnote on the NCVS "break in series" was made as a comment to a pair of sentences that provide a good starting point for fuller discussion:

> The decision to include unbounded, first interviews in NCVS estimates was made as our panel was being established and assembled, and so we do not think it proper to second-guess it; we understand the fiscal constraints under which the decision was made. However, it serves as an example of a seemingly short-term fix with major ramifications, and it would have benefited from further study prior to implementation.

We return to a discussion of the NCVS break in series at various points in Chapter 5, particularly Section 5–B.8.

3–B CORRECTIONS

The maintenance of statistics on persons under correctional supervision in the United States dates back to the 1850 decennial census, giving corrections data the longest lineage of BJS data series. As discussed by National Research Council (2006:Sec. 3–D), the 1850 census was the first to give enumerators formal rules for determining residence. One of these was the specific direction to treat jailors and other superintendents of institutions as heads of "families," counting prisoners under their supervision as members of the family; as the term was used in censuses of the period, "family" had no direct tie to blood relations. This practice continued in the next several censuses, with the 1880 and 1890 censuses introducing a special form for enumerators to record information on individual prisoners. In 1904, the newly permanent Census Bureau began the annual publication "Prisoners in State and Federal Institutions," beginning to tally commitments to the institution in a calendar year rather than a single reference date (as in the decennial census). In a 1923 count, the Census Bureau began to count discharges from prison or jail, along with information on time served (Cahalan, 1986:1–2; see also Beattie, 1959). In 1950, authority for this annual collection was transferred to the Federal Bureau of Prisons (BOP), which in turn was transferred to the LEAA. Contracting with the Census Bureau as data collector, BJS has conducted the collection as the still-continuing National Prisoner Statistics (NPS) series since 1973.

Data on the correctional population has grown in importance and meaning, given the massive growth in that population since the 1970s; see Table 3-4. Counts of the prison population draw particular concern—tripling between 1980 and 1995 after decades of remarkable stability (Blumstein and Beck, 1999) before settling into slower rates of annual growth—as state governments have struggled to keep pace and develop facility capacity. In doing so, new and ever more varied styles of incarceration have developed, including use of privately built and operated facilities and community correctional facilities; some states have also relied on establishing contracts to house their prisoners in other states where capacity still exists. Significant though the prison population has become, the much larger number of people under probation supervision also rapidly escalated over the course of the 1980s. A major challenge for corrections systems (and corresponding challenge of measurement) concerns the experience of the formerly incarcerated when sentences are completed or parole is granted, and prisoners reenter the community.

BJS's data collections in the area of correctional supervision (see Table 3-5) include censuses and surveys of prisons and jails, intended to monitor the stocks and flows of inmates within these facilities. They also include periodic surveys of the inmate population, which provide an opportunity to study criminogenic factors in their backgrounds. Recently, BJS has conducted a special Survey on Sexual Violence among inmates of prisons, jails, and juvenile correctional facilities pursuant to the Prison Rape Elimination Act of 2003; those collections, and the legislative act, are described in Section 5–A.1 rather than this chapter. BJS's survey-based methods are supplemented by collections of administrative data providing counts of inmates, probationers, and parolees. BJS also occasionally conducts specialized studies of correctional populations; for instance, its studies of recidivism have provided valuable information about patterns of rearrest and reincarceration among state prisoners released in 1983 and 1994.

3–B.1 Prisons

National Corrections Reporting Program and National Prisoner Statistics

As described in the beginning of the chapter, the NPS series continues annual collection of the numbers of prisoners in both state and federal prisons that were begun by the Census Bureau in 1926. The Census Bureau discontinued the publication of the series in 1946 (but continued some data collection), and authority for the series was shifted to BOP in 1950 (Cahalan, 1986:6). When the BJS predecessor, National Criminal Justice Information and Statistics Service, was formed in 1971, it took responsibility for the series, engaging the Census Bureau as its data collection agent. In 1983, data collection for the NPS was combined with a parallel collection on parole—the Uniform Parole Reports (see Section 3–B.6)—and the resulting program was renamed the National Corrections Reporting Program (NCRP). In 1984, the Census Bureau began to collect data from federal prisons as well as state prisons, in addition to the California Youth Authority (Bureau of Justice Statistics, 2007d:3).

NCRP collection is based on facility or administrative records, as reported by correctional authorities, and data are compiled on both an annual

Table 3-4　Estimated Number of Adults Under Correctional Supervision in the United States, 1980–2006

Year	Prison	Jail	Probation	Parole	Total
1980	319,598	182,288	1,118,097	220,438	1,840,400
1981	360,029	195,085	1,225,934	225,539	2,006,600
1982	402,914	207,853	1,357,264	224,604	2,192,600
1983	423,898	221,815	1,582,947	246,440	2,475,100
1984	448,264	233,018	1,740,948	266,992	2,689,200
1985	487,593	254,986	1,968,712	300,203	3,011,500
1986	526,436	272,735	2,114,621	325,638	3,239,400
1987	562,814	294,092	2,247,158	355,505	3,459,600
1988	607,766	341,893	2,356,483	407,977	3,714,100
1989	683,367	393,303	2,522,125	456,803	4,055,600
1990	743,382	405,320	2,670,234	531,407	4,350,300
1991	792,535	424,129	2,728,472	590,442	4,535,600
1992	850,566	441,781	2,811,611	658,601	4,762,600
1993	909,381	455,500	2,903,061	676,100	4,944,000
1994	990,147	479,800	2,981,022	690,371	5,141,300
1995	1,078,542	507,044	3,077,861	679,421	5,342,900
1996	1,127,528	518,492	3,164,996	679,733	5,490,700
1997	1,176,564	567,079	3,296,513	694,787	5,734,900
1998	1,224,469	592,462	3,670,441	696,385	6,134,200
1999	1,287,172	605,943	3,779,922	714,457	6,340,800
2000	1,316,333	621,149	3,826,209	723,898	6,445,100
2001	1,330,007	631,240	3,931,731	732,333	6,581,700
2002	1,367,547	665,475	4,024,067	750,934	6,758,800
2003	1,390,279	691,301	4,073,987	774,588	6,883,200
2003	1,390,279	691,301	4,120,012	769,925	6,924,500
2004	1,421,345	713,990	4,143,792	771,852	6,995,100
2005	1,448,344	747,529	4,166,757	780,616	7,051,900
2006	1,492,973	766,010	4,237,023	798,202	7,211,400

NOTE: Entries in "Total" column are rounded to the nearest 100 "because a small number of individuals may have multiple correctional statuses." Counts for probation, prison, and parole populations are for December 31 of each year; jail population counts are for June 30 of each year.

SOURCE: Table 6.1.2006, *Sourcebook of Criminal Justice Statistics* Online (http://www.albany.edu/sourcebook/wk1/t612006.wk1).

Table 3-5 Bureau of Justice Statistics Data Collection History and Schedule, Corrections, 1981–2009

Series	81	82	83	84	85	86	87	88	89	90	91	92	93	94	95	96	97	98	99	00	01	02	03	04	05	06	07	08	09	Total
Prisons																														
Census of State and Federal Prison Facilities	·	·	·	•	·	·	·	·	·	•	·	·	·	·	•	·	·	·	·	•	·	·	·	·	•	·	·	·	·	5
National Corrections Reporting Program	·	·	•	•	•	•	•	•	•	•	•	•	•	•	•	•	•	•	•	•	•	•	•	•	•	•	•	•	•	27
National Prisoner Statistics	•	•	•	•	•	•	•	•	•	•	•	•	•	•	•	•	•	•	•	•	•	•	•	•	•	•	•	•	•	29
State Prison Expenditures	·	·	·	·	·	·	·	·	·	·	·	·	·	·	·	·	·	•	·	·	·	·	·	•	·	·	·	·	•	3
Survey of Inmates in Federal Correctional Facilities	·	·	·	·	·	·	·	·	·	·	•	·	·	·	·	·	•	·	·	·	·	·	·	•	·	·	·	·	·	3
Survey of Inmates in State Correctional Facilities	·	·	·	·	·	•	·	·	·	·	•	·	·	·	·	·	•	·	·	·	·	·	·	•	·	·	·	·	·	4
Jails																														
Annual Jail Survey	·	•	·	•	•	•	•	•	·	•	•	•	•	·	•	•	•	•	•	•	•	•	•	•	•	•	•	•	•	25
Annual Jail Survey of Indian Country	·	·	·	·	·	·	·	·	·	·	·	·	·	·	·	·	·	•	•	•	•	•	•	•	•	•	•	•	•	12
Census of Local Jails	·	·	•	·	·	·	·	•	·	·	·	·	•	·	·	·	·	·	•	·	·	·	·	·	•	·	·	·	·	5
Survey of Large Jails	·	·	·	·	·	·	·	·	·	·	·	·	·	·	·	·	·	·	·	·	·	·	·	•	·	·	·	·	·	1
Survey of Lcoal Jail Inmates	·	·	•	·	·	·	·	·	•	·	·	·	·	·	·	•	·	·	·	·	·	•	·	·	·	·	·	·	·	4
Probation, Parole, and Recidivism																														
National Census of Parole Agencies	·	·	·	·	·	·	·	·	·	·	•	·	·	·	·	·	·	·	·	·	·	·	·	·	·	•	·	·	·	2
National Parole Statistics	•	•	•	•	•	•	•	•	•	•	•	•	•	•	•	•	•	•	•	•	•	•	•	•	•	•	•	•	•	29
National Probation Statistics	•	•	•	•	•	•	•	•	•	•	•	•	•	•	•	•	•	•	•	•	•	•	•	•	•	•	•	•	•	29
National Survey of Adult Probationers	·	·	·	·	·	·	·	·	·	·	·	·	·	·	•	·	·	·	·	·	·	·	·	·	·	·	·	·	·	1
Recidivism Studies	·	·	•	·	·	·	·	·	•	·	·	·	·	•	·	·	·	·	·	·	·	•	·	·	·	·	·	•	•	6
Prison Rape Elimination Act of 2003 Collections																														
Survey of Sexual Violence—Adult Facilities	·	·	·	·	·	·	·	·	·	·	·	·	·	·	·	·	·	·	·	·	·	·	·	•	•	•	•	•	•	6
Survey of Sexual Violence—Juvenile Facilities	·	·	·	·	·	·	·	·	·	·	·	·	·	·	·	·	·	·	·	·	·	·	·	·	•	•	•	•	•	5
National Inmate Surveys																														
Prisons and Jails	·	·	·	·	·	·	·	·	·	·	·	·	·	·	·	·	·	·	·	·	·	·	·	·	·	·	•	•	•	3
Juvenile Facilities	·	·	·	·	·	·	·	·	·	·	·	·	·	·	·	·	·	·	·	·	·	·	·	·	·	·	·	•	·	1
Former Prisoners	·	·	·	·	·	·	·	·	·	·	·	·	·	·	·	·	·	·	·	·	·	·	·	·	·	·	·	•	·	1
Other																														
Capital Punishment	•	•	•	•	•	•	•	•	•	•	•	•	•	•	•	•	•	•	•	•	•	•	•	•	•	•	•	•	•	29
Deaths in Custody	·	·	·	·	·	·	·	·	·	·	·	·	·	·	·	·	·	·	·	•	•	•	•	•	•	•	•	•	•	10

NOTES: •, data collected. ·, data not collected.

SOURCE: Bureau of Justice Statistics.

and a seminannual (midyear) basis. In principle, participating agencies are asked to complete a prison admission questionnaire (NCRP-1A) for each new prisoner entry during a reporting year and to send those questionnaires (or corresponding information from facility databases) to the Census Bureau on a flow basis. A prison release record (NCRP-1B) and parole exit record (NCRP-1C) are supposed to be kept on file for each prisoner; if the prisoner is released upon the end of a sentence, the NCRP-1B is to be filled and completed. Otherwise, if the prisoner is placed on parole, the NCRP-1C is intended to be forwarded to the parole authority and sent to the Census Bureau when the person exits parole. If a parole exit results in a return to prison, both a parole exit and a new prisoner entry (1A) record are created (but not directly linked to each other; Bureau of Justice Statistics, 2007d:User Guide 3.4). In addition to basic summary counts of admissions and releases (disaggregated by gender and race), the NCRP compiles information on conviction offenses, sentence length, and completed jail time. NCRP-D records collect annual end-of-year stock counts of prisoners, though they have only been completed by at most 26 states.

In its studies of the prison population, BJS typically distinguishes between "in custody" and "under jurisdiction" counts. The state or government that has legal authority over the prisoner (under jurisdiction) may transfer physical custody of a prisoner to another government (such as to deal with prison overcrowding). In generating jurisdiction counts, BJS's definition of prison is broader than the classic penitentiary model and includes other facilities where an inmate may be held for long durations, such as halfway houses, boot camps, and treatment centers. However, its custody counts typically exclude prison inmates who may be held in local jails (again, as may be done to deal with prison crowding issues) or privately operated facilities (Sabol and Couture, 2008:9).

Beginning with calendar year 2003, responding correctional systems were provided with a Web reporting option; agencies can also submit questionnaire information on paper forms or computer media. The Census Bureau enters into specific arrangements with each state to provide NCRP data; as of 2003, 41 states (accounting for about 90 percent of the state prison population[5]), the federal prison system administered by the Bureau of Prisons, and the California Youth Authority contributed data to NCRP (Bureau of Justice Statistics, 2007d).

After about a decade of issuing an annual bulletin summarizing characteristics of both prison and jail inmates as of the middle of the preceding year (e.g., Harrison and Beck, 2006; Sabol et al., 2007), BJS began the process

---

[5]Calculation by panel staff based on 2005 prisoner counts for the nine noncontributing states (Arizona, Delaware, Idaho, Indiana, Kansas, Montana, New Mexico, Ohio, and Wyoming) reported by Sabol et al. (2007:Appendix Table 1).

of issuing separate reports for prisons and jails; the first such separate report, Sabol and Couture (2008), was released in June 2008 and summarized prisoner stocks as of mid-2007. BJS also issues an annual report on the year-end stock of prisoners (e.g., Sabol et al., 2007), that includes totals from a set of sources that only provide data on an annual, year-end basis. Separate from the NPS program, BJS receives year-end counts of prisoners from several sources. The U.S. Department of Defense Corrections Council supplies BJS with prisoner counts (disaggreated by demographic characteristics, branch of service, and basic sentence and offense information) for "persons held in U.S. military confinement facilities inside and outside of the continental United States." BJS receives similar data from the U.S. Immigration and Customs Enforcement (ICE) on persons detained for immigration violations; this includes ICE-operated facilities as well as prisoners that ICE arranges to hold in government- or privately operated facilities. Finally, correctional departments in U.S. territories and commonwealths only provide information on the year-end, annual basis (Sabol et al., 2007:10).

Census of State and Federal Correctional Facilities

Every 5–6 years, the Census Bureau has conducted the Census of State and Federal Correctional Facilities for BJS. The Census Bureau uses data provided by the American Correctional Association to update files from its most recent facility census to develop the frame for a new census.

The intended scope of the census is all correctional facilities directly administered by state governments or the federal government and that are primarily intended to house state or federal prisoners. The NACJD codebook for the 2000 correctional facility census describes its scope and specific exclusions as follows (Bureau of Justice Statistics, 2004a:5):

> The Census includes the following types of State, Federal, and private correctional facilities intended for adults but sometimes also holding juveniles: prisons, penitentiaries, and correctional institutions; boot camps; community corrections; prison farms; reception, diagnostic, and classification centers; road camps; forestry and conservation camps; youthful offender facilities (except in California); vocational training facilities; prison hospitals; and drug and alcohol treatment facilities for prisoners. . . . [It specifically excludes:] 1) private facilities not primarily for State or Federal inmates; 2) military facilities; 3) Immigration and Naturalization Service facilities; 4) Bureau of Indian Affairs facilities; 5) facilities operated by or for local governments, including those housing State prisoners; 6) facilities operated by the U.S. Marshals Service; 7) hospital wings and wards reserved for State prisoners; and 8) facilities that hold only juveniles.

Facility census data are collected through mailed questionnaires, which are intended to be filled through reference to administrative records (at either the individual facility or state corrections department level).

The census includes a battery of questions on the physical characteristics of the facility itself: age and type of the structure, physical security, capacity, operating costs, and new construction plans. The census asks for stock information on the composition of the inmate population as of a particular reference date, including breakdowns of inmates being held under contract with another state or authority and the number of juvenile (under age 18) inmates held in the facility. The census includes queries on the number and occupational category of facility staff, the number of misconduct reports filed against prison staff, and incidence of escape attempts or other disturbances. Inmate interviewing programs (described below) are based in part on facility responses to census questions on the education and health services that are provided for inmates, though the practices reported in the facility-level census have been explored in specific BJS reports such as the analysis of prevalence and treatment programs for hepatitis B and C by Beck and Maruschak (2004). Similarly, Beck and Maruschak (2001) summarize the data on mental health services (and estimated prisoner/patient counts) from the prison census while James and Glaze (2006) study the inmate-reported data from BJS's prison and jail inmate surveys described below.

BJS's report on the most recent prison census, held in 2005, is by Stephan (2008). In the 2005 administration of the survey, BJS and the Census Bureau used telephone follow-up with state corrections departments to update the list of prisons covered in 2000. The Illinois Department of Corrections did not participate in the 2005 census; the Census Bureau estimated Illinois results using both data that had been supplied in the 2000 prison census as well as data accessible on the Illinois department's website (Stephan, 2008:7).

Surveys of Inmates in State and Federal Correctional Facilities

BJS has commissioned the Census Bureau to conduct personal interviews with inmates of state prisons on an irregular basis, every 5–7 years, through the Survey of Inmates in State Correctional Facilities. More recently—first in 1991 with additional sponsorship from BOP and then again in 1997 and 2004—the Census Bureau expanded its interviewing to include prisoners in a sample of federal prisons. BJS and the Census Bureau refer to the broader collection effort as the Surveys of Inmates in State and Federal Correctional Facilities (SISFCF).

The state prisoner and federal prisoner components of the SISFCF are drawn in similar ways, though the stratification schemes differ between the two groups (Bureau of Justice Statistics, 2007h:3–8):

- The most recent BJS Census of State and Federal Correctional Facilities, updated to include known recently constructed facilities, is used as the sampling frame for the state prison group. Male- and female-only institutions are sampled separately, dividing mixed-sex facilities between the two groups. Facilities are drawn with probability proportional to institution size from strata defined by geographic region (treating the large states of California, Florida, New York, and Texas as separate strata). (Some large facilities, particularly those reporting that they provide selected health care services, are treated as self-representing and automatically included in the first stage of sampling.) A systematic sampling scheme and a random start time are then used to sequence interviews at chosen facilities. Within chosen facilities, interviewers are generally able to randomly select prisoners from a facility-provided list of inmates using a bed the previous night.

- A BOP-generated facility list is the frame for the federal prisoner portion, with facilities chosen with probability proportional to size within strata defined by facility security level (five levels for male prisons, two for female prisons). Two male-only and one female-only facilities were selected with certainty for the sample. Because of added restrictions, BOP staff served as intermediaries in arranging interviews; they selected the sample of inmates (systematically, but with oversampling of nondrug offenders) and provided it to facilities 5–7 days before interviewing.

In 2004, Census Bureau staff completed 14,499 interviews of state prisoners and 3,686 of federal prisoners. Each interview (using CAPI) is about an hour in length, including information on individual characteristics of prison inmates, current offenses and sentences, characteristics of victims, criminal histories, family background, gun possession and use, prior drug and alcohol use and treatment, and prison services.

The personal history information included in SISFCF data has been used by BJS staff in several of its "Special Report" series (see Box 1-1 for a description of this type of report). Several of these have focused on prisoner health issues (Maruschak, 2001; James and Glaze, 2006; Maruschak, 2008b) whereas others have focused on other characteristics of the convicted offender population, such as veteran status (Mumola, 2000b; Noonan and Mumola, 2007), educational attainment (Harlow, 2003), firearm use and acquisition (Harlow, 2001), and children of incarcerated parents (Mumola, 2000a; Glaze and Maruschak, 2007).

3–B.2 Jails

In terms of understanding transitions and flows, local jails are a vitally important part of the criminal justice system. The definition of "local jail" that BJS uses in describing its Annual Survey of Jails (Bureau of Justice Statistics, 2007b:4) is telling, for the sheer range of listed functions:

Local jails:

- receive individuals pending arraignment and hold them awaiting trial, conviction, or sentencing,
- readmit probation, parole, and bail-bond violators and absconders,
- temporarily detain juveniles pending transfer to juvenile authorities,
- hold mentally ill persons pending their movement to appropriate health facilities,
- hold individuals for the military, for protective custody, for contempt, and for the courts as witnesses,
- release convicted inmates to the community upon completion of sentence,
- transfer inmates for Federal, State, or other authorities,
- house inmates for Federal, State, or other authorities because of crowding of their facilities,
- relinquish custody of temporary detainees to juvenile and medical authorities,
- sometimes operate community-based programs as alternatives to incarceration,
- and hold inmates sentenced to short terms (generally less than one year).

Frase (1998:100) is more succinct, and blunt, in characterizing local jails as "the custodial dumping ground of last resort, when no other appropriate holding facility is available." As the points in the preceding definition illustrate, the incarcerated population housed in and cycling through local jails includes a mixture of short-term stays (e.g., pretrial or prearraignment holding) and long-term stays (e.g., contractual arrangements to house convicted prisoners because of crowding in state prisons). The breadth of custodial arrangements accommodated by local jails and the dynamics of the jailed population make jails a critical feature of the justice system—albeit one that defies neat definition and measurement.

Cahalan (1986:7) observes that, "apart from Census Bureau reports done at 10-year intervals, no national [statistical] reports had been done on jails" until LEAA began a program of jail surveys in 1970. As of that point, "the last Census Bureau report on jails to contain special criminal justice related information such as offenses or sentence data had been in 1933." LEAA fielded initial jail surveys in 1970, 1972, and 1978; these early efforts modeled Census Bureau practice by trying to characterize the inmate population present on the day of the survey, rather than quantifying the flow into and out of jails over the course of the year. These initial surveys gave rise to BJS's current program of jail studies, which is generally similar in structure to its core collections on prisons.

Annual Survey of Jails and Census of Jails

BJS's principal data collection on local jails, with the institution as the unit of analysis, is the Annual Survey of Jails (ASJ), which is intended to collect data on facilities that are administered (either directly or under contract to a private firm) by county and municipal governments and that hold inmates for some period after their initial arraignment (i.e., those that typically hold inmates for more than 48 hours). On an irregular basis—roughly every 5 years—the coverage of the ASJ is expanded to include a complete canvass of all known jails, and the results from this Census of Jails becomes the sampling frame for the annual ASJ.

Since the ASJ series and periodic jail census began in 1982, the Census Bureau has been engaged as the data collection agent. The work developed from experimental efforts in the 1970s, when congressional interest in correctional facility overcrowding led to a first jail census in 1970.

The core content of the ASJ and the Census of Jails is the same, including facility characteristics (structure age, capacity and average daily capacity, and staffing) and inmate demographics (age/sex/race, legal status, and length of stay). In recent years, the ASJ, like the data collections in prisons, have developed to include information on facility services and health care (including the prevalence of HIV/AIDS and tuberculosis in the inmate population).

The ASJ and the Census of Jails are still conducted principally by mailout/mailback methodology, though the Census Bureau permitted Internet and electronic reporting beginning in 1999. In ASJ (noncensus) years, a sample of jails is drawn using stratified random sampling, using strata defined by the reported average daily inmate population in the last census, with some exceptions (for instance, some jails that are regional in scope rather than serving a single jurisdiction are automatically included in the sample, as are facilities that reported housing at least one juvenile offender in the most recent census).

Stephan (2001) described the results of the 1999 Census of Jails, emphasizing comparisons with the 1993 census. Similar questionnaire items in the jail census and the NPS instrument have been used to study HIV prevalence and testing regimes among the incarcerated population (Maruschak, 2001); however, a more up-to-date, electronic-only publication on HIV/AIDS (Maruschak, 2008a) uses only the prison data.

Survey of Inmates in Local Jails

BJS's periodic Survey of Inmates in Local Jails (SILJ) collects data on the personal and family characteristics of jail inmates, past drug and alcohol use, history of physical and sexual abuse, and history of contact with the criminal justice system. The survey also probes inmates to provide information on services offered by the jail system (e.g., health care) and other jail activities and programs. The survey relies on personal interviews with a nationally representative sample of almost 6,000 inmates; as of the 2002 version of the survey, the roughly hour-length interviews are now completed using CAPI.

The survey is intended to be nationally representative of "persons held prior to trial and on those convicted offenders serving sentences in local jails or awaiting transfer to prison" (Bureau of Justice Statistics, 2006d:5). The most recent Census of Jails serves as the frame for the SILJ; in 2002, the sample of 7,750 inmates was drawn using a two-stage design, selecting jails from within strata defined on the basis of a jail's proportions of adult male, adult female, and juvenile inmates and then sampling within selected jails. Of the 7,760 names chosen for inclusion in the sample, 6,982 interviews were completed, with 263 inmates refusing to participate, 407 having exited the jail system between selection and interviewing, and 98 who could not be interviewed for medical or security reasons (Bureau of Justice Statistics, 2006d:5–6).

The SILJ is conducted periodically, if not regularly. Documentation for the 2002 implementation of the survey describes its frequency as every 5 to 6 years (Bureau of Justice Statistics, 2006d:5), though 7-year gaps have occurred. As of 2002, BJS contracts with the Census Bureau to conduct SILJ interviews.

BJS has issued both general summaries of characteristics of the jail population based on the SILJ (Harlow, 1998; James, 2004) and detailed probes of particular SILJ topic areas. For example, Maruschak (2006) details current medical problems reported by jail inmates, including assessments of whether the problems are related to fight- or accident-related injuries; Maruschak (2008b) performs a similar analysis based on prisoner survey data; and SILJ data were used in the analysis by James and Glaze (2006) of mental health problems and disorders among the incarcerated.

3–B.3　Custodial Conditions

Deaths in Custody

The Death in Custody Reporting Act of 2000 (P.L. 106-297) required that states provide the U.S. Justice Department with quarterly "information regarding the death of any person who is in the process of arrest, is en route to be incarcerated, or is incarcerated" at any correctional facility as a condition for receiving federal grant assistance.[6] The law requires that, "at

---

[6]Technically, the act was attached to the authorization for the broader Violent Offender Incarceration and Truth-in-Sentencing grant program, which dispersed $5.2 billion beginning in 1998 to the states to expand prison capacity. Correctional systems that accepted those funds had to agree to the Deaths in Custody reporting as a condition of the grant (Mumola, 2005:2).

a minimum," this information include personal characteristics (name, age, gender, race, ethnicity) and details of the death (date, time, location, and brief description of circumstances).

Coverage in Deaths in Custody reporting was added in stages, beginning with data on deaths in local jails in 2000. In 2001, state prisons were added; state juvenile correctional facilities followed in 2002, and in 2003 the program began attempting to measure deaths in the process of arrest. As described in Box 3-3, the act was up for reauthorization in 2007; H.R. 3971 reimplements the act with the added requirement of a Justice Department study of means for reducing deaths in custody. The updated legislation passed in the House of Representatives in January 2008; the Senate Judiciary Committee reported a modified version of the bill in late September 2008. Though the reauthorization is still pending, BJS has worked to expand coverage to include deaths in ICE facilities (Sedgwick, 2008:2).

Deaths in Custody data collection functions are shared by the Census Bureau and BJS (through its state Statistical Analysis Centers; see Section 4–A). On a quarterly basis, the Census Bureau collects inmate death records from each of the nation's state correctional systems (adult and juvenile) and from local jails. Data coded from these records include the deceased's personal characteristics (age, gender, and race/ethnicity), their criminal background (legal status, offense types, length of stay in custody), as well as details of the death itself (the date, time, location, and cause of each death, as well as information on autopsies and medical treatment provided for illnesses/diseases). BJS collects quarterly reports from state and local law enforcement agencies (known from the Census of State and Local Law Enforcement Agencies, as described below) on deaths incurred during the process of arrest. Though collected on a quarterly basis, reports and tabulations from the Deaths in Custody program are only reported in annual formats.

To date, BJS has used the Deaths in Custody data to produce three analytical reports on differing aspects of the data: an analysis of deaths concluded to be suicides or homicides (Mumola, 2005) was followed by a more general inventory of the medically determined causes of deaths recorded in the data (Mumola, 2007b), and finally a study making use of the newer data on deaths occurring in the process of arrest (Mumola, 2007a). BJS has since established a website page dedicated to statistical tables from the Deaths in Custody data (http://www.ojp.usdoj.gov/bjs/dcrp/dictabs.htm) which is updated with new annual counts.

Survey on Sexual Violence and National Inmate Surveys

The data collections on the incidence of rape and sexual violence in correctional facilities, as mandated by the Prison Rape Elimination Act of 2003, are described and discussed in Section 5–A.1.

3–B.4   Capital Punishment

A separate component of the NPS program, known as NPS-8, was designated in 1972 to collect annual data on prisoners under a death sentence, as well as transitions out of "death row" (e.g., through commutation or vacation of a capital sentence). By counting the number of executions performed, BJS's capital punishment program represents a continuation of a data series dating to 1930; the fuller detail on death sentencing and inmate characteristics began in 1972. The Census Bureau, as data collector for the NCRP, also collects the capital punishment data.

Part of the Capital Punishment data collection is an annual update of death penalty statutes in the various states; the Census Bureau sends a separate questionnaire to state justice departments, including questions on any actions by state supreme courts, the minimum age at which persons can be sentenced to death, and the methods of execution authorized by state law.

The codebook for a compilation of BJS's capital punishment data for the 1972–2006 time series notes several reasons why BJS's counts may be discrepant from those recorded by other authorities (Bureau of Justice Statistics, 2007c:4):

> (1) NPS-8 adds inmates to the number under sentence of death not at sentencing but at the time they are admitted to a State or Federal correctional facility. (2) If in one year inmates entered prison under a death sentence or were reported as being relieved of a death sentence but the court had acted in the previous year, the counts are adjusted to reflect the dates of court decisions. (3) NPS counts are always for the last day of the calendar year and will differ from counts for more recent periods.

Prior to data collected in 2006, BJS summarized findings from the capital punishment data in an annual bulletin (Snell, 2006); it has since switched to electronic-only dissemination of spreadsheet tables (Snell, 2007, 2008). It has also registered a combined 1973–2005 data set in the NACJD (Bureau of Justice Statistics, 2007c), covering all persons on death row since 1972, capable of analysis by state, basic demographic characteristics, capital offense type(s), and status (e.g., still awaiting execution or removed from death sentence).

3–B.5   Inventory of State and Federal Corrections Information Systems

In 1998, BJS, the National Institute of Justice (NIJ), and the Corrections Program Office jointly sponsored a study by the Urban Institute on the general state of offender-based corrections information systems at the state and federal levels. The Urban Institute was specifically tasked to describe the capacity of these systems for record linkage and electronic exchange of records. In carrying out this study, the institute obtained the assistance of the State-Federal Committee of the Association of State Corrections Administrators. The final report of this inventory of systems was issued as Bureau of Justice Statistics (1998b); to date, the study has not been repeated.

3–B.6 Probation and Parole

Some basic information on exit from parole status is collected in the NCRP (described in Section 3–B.1 above) which absorbed the former Uniform Parole Reports program in 1983. The Uniform Parole Reports series began on an experimental basis in 1966, coordinated by the National Council on Crime and Delinquency with financial support from the National Institute of Mental Health (Bureau of Justice Statistics, 2007d:4). It began by collecting data from selected state parole boards for which records were available, but developed nationwide coverage over several years, due in part to a feasibility study of yearly reporting funded by the LEAA in 1975 (Cahalan, 1986:7).

BJS sponsors an Annual Probation Survey and an Annual Parole Survey; collectively, they are described as the Probation and Parole Data Surveys. Between 1993 and 2005, BJS collected these data in-house, but it contracted with the Census Bureau as data collection from 1980 to 1992 and from 2006 to the present. The probation and parole surveys are establishment surveys, intended to be filled by agency authorities through reference to administrative records. According to the methodology note in Glaze and Bonczar (2007:9), this means that the 2006 version of the survey questionnaires was sent to 463 probation agencies[7] and 54 parole agencies. About 13 probation agencies failed to supply data and a few others provided only partial returns, necessitating imputation procedures; the state of Illinois was the only parole authority not to report in 2006. As agency-level collections, the Probation and Parole Data Surveys focus principally on aggregate counts of entries and discharges, though some data are also collected on demographic characteristics and offense types of the agencies' service population; questions are also asked about the use of procedures such as electronic monitoring. BJS staff worked to expand the coverage of the Annual Probation Survey in the late 1990s, adding 175 agencies to the survey's frame between 1995 and 2006 (Glaze and Bonczar, 2007:9).

On a one-shot basis in the early 1990s, BJS conducted a fuller study of the probation population through a contract with the Census Bureau. In 1991, the agencies mounted a Census of State and Local Probation and Parole Agencies to generate a complete inventory of agencies operated by federal, state, and local governments. This census produced facility-level

---

[7]In some states, local courts have the direct authority to supervise probationers. Hence, almost 70 percent of the 463 eligible reporting probation agencies in 2006 were concentrated in two states: 185 and 128 in Ohio and Michigan, respectively (Glaze and Bonczar, 2007:9).

data on staffing, expenditure, and basic procedures (e.g., frequency of drug or HIV testing required of probationers). However, the primary function of the census was to serve as the sampling frame for a one-shot Survey of Adults on Probation in 1995, the first nationally representative sample of probationers that had been drawn and analyzed to date. The personal interview with sampled probationers included detailed questions on drug and alcohol use, criminal history, and the extent of their contact with their supervising probation authorities. Save for a legislative mandate under the Prison Rape Elimination Act of 2003 to query a sample of the probation and parolee population about the incidence of sexual violence during imprisonment, the 1995 survey is BJS's only personal-interview measurement of community corrections to date.

3–B.7 Recidivism

In 1983 and 1994, BJS tracked large samples of released prisoners for 3 years. BJS described the need for these studies as being motivated by "widespread demand for information on the topic of recidivism" (Bureau of Justice Statistics, 2002:1):

> Among the many information requests that come to the U.S. Department of Justice each day—from departments of corrections, elected officials, policy makers, the media, members of the general public—one of the most frequent is for facts regarding recidivism. Legislators drafting community notification laws, for example, wish to know how often released sex offenders commit a new sex offense. Departments of corrections need to learn how the recidivism rate in their State compares to the national rate. Of special interest to the FBI is the extent to which released sex offenders become involved in criminal activity in States other than the State in which they had served time. This is important information relevant to the development of a national DNA registry.

The databases compiled in the 1983 and 1994 studies drew from criminal history information recorded on "rap sheets" (Records of Arrests and Prosecutions) and derived multiple measures of recidivism or resumption of criminal activity. The 1983 study tracked about 16,000 prisoners released from 11 state corrections systems; the 1994 study reached 15 states and 38,624 prisoners. In both cases, the sample of states was purposive—that is, based on a state's willingness and ability to cooperate—while the samples of prisoners within states was generally drawn based on stratification by most severe conviction offense.

At the time these recidivism studies were conducted, BJS researchers had no access to the national criminal history record databases compiled by the FBI (see Section 4–B for subsequent developments). Participating correction departments turned over lists of all prisoners released in the reference year; BJS drew its samples and returned a list of identifiers to the state departments, asking for computerized rap sheets for those prisoners. Separately, BJS also provided the set of identifiers for sampled prisoners to the FBI, asking it to query its databases—particularly useful to get information on offenses committed outside the state that released the prisoner. The state and FBI records were combined to form a master database—in 1994, one that included 6,427 variables on the 38,624 prisoners, 6,336 of which provided criminal histories for up to 99 "arrest cycles" per prisoner (Bureau of Justice Statistics, 2002:6–7).[8]

The four recidivism measures derived for each prisoner were rearrest, reconviction, resentence to prison, and return to prison with or without a new prison sentence. The BJS report on the 1994 study (Langan and Levin, 2002:2) follows the description of these four measures with a disclaimer that the measures are likely to be underestimates:

> To an unknown extent, recidivism rates based on State and FBI criminal history repositories understate actual levels of recidivism. The police agency making the arrest or the court disposing of the case may fail to send the notifying document to the State or FBI repository. Even if the document is sent, the repository may be unable to match the person in the document to the correct person in the repository or may neglect to enter the new information. For these reasons, studies such as this one that rely on these repositories for complete criminal history information will understate recidivism rates.

BJS conducted the recidivism studies in-house, with assistance in data collection and processing from the Regional Justice Information Service. Funding for the 1994 study was received, in part, from the FBI and the Corrections Program Office within the Office of Justice Programs (Langan and Levin, 2002:16).

On a one-time basis, in 1986, BJS drew from criminal history records to track a sample of convicted felons for 3 years upon their entry into probation. This collection generated estimates of the percentage of probationers who were rearrested, reconvicted, or reimprisoned for new crimes during the study period.

3–C LAW ENFORCEMENT

In its final report, the President's Commission on Law Enforcement and Administration of Justice (1967:10) was struck by the difficulty in studying "law enforcement" as a unified entity, where policy changes made on high directly affect the public experience at the street level:

---

[8]Ten prisoners in the 1994 sample had more than 99 arrest cycles, and one had 176 different arrests on record. In these cases, the earliest arrests were dropped to fit the 99-maximum limit of the database (Bureau of Justice Statistics, 2002:7).

> At the very beginning of the process—or, more properly, before the process begins at all—something happens that is scarcely discussed in lawbooks and is seldom recognized by the public: law enforcement policy is made by the policeman. For policemen cannot and do not arrest all the offenders they encounter. It is doubtful that they arrest most of them. A criminal code, in practice, is not a set of specific instructions to policemen but a more or less rough map of the territory in which policemen work. How an individual policeman moves around that territory depends largely on his personal discretion. . . . Every policeman [is] an interpreter of the law.

In a sense, it is difficult to draw a clear conceptual line between "crime statistics" and "law enforcement statistics" because the two concepts are so intertwined; much of the task of law enforcement is identifying and responding to crime. Indeed, law enforcement agencies are a major provider of statistical data on crime. Still, for purposes of this report, we can define "law enforcement statistics" as those that describe the activity and social organization of law enforcement agents and agencies, where social organization includes organizational structure, resources, personnel, policies, and tactics. Under this rubric, the mobilization of the police by citizens and the response of police to crime events would be considered part of law enforcement statistics. Law enforcement and the more general concept of "crime statistics" intersect when the police serve as the source of data to identify and characterize crimes.

As we review in this section, and discuss elsewhere in this report, BJS's data collections to date in the area of law enforcement (see Table 3-6 for a collection timeline) have heavily emphasized a top-level, management and administration focus. Though BJS has framed collections related to special-purpose agencies such as campus law enforcement departments or medical examiners offices, the data content is administrative in focus, describing workforce levels, available resources, and general workload.

Like the NCVS, BJS's law enforcement data collections share some substantive overlap with components of the FBI's UCR program; the law enforcement aspects of the UCR are summarized in Box 4-4.

3–C.1 Law Enforcement Management and Administrative Statistics

The core BJS data collection in the area of law enforcement is the Law Enforcement Management and Administrative Statistics (LEMAS) survey of agency administrators that has been conducted roughly every 3 years since 1987. Most recently, BJS has used the Police Executive Research Forum (PERF) as the data collection agent for LEMAS.

Langworthy (2002:23) observes that "the LEMAS survey has its roots in salary surveys conducted both by the Fraternal Order of Police (FOP) and the Kansas City Police Department (KCPD)," which were conducted

Table 3-6 Bureau of Justice Statistics Data Collection History and Schedule, Law Enforcement, 1981–2009

Series	81	82	83	84	85	86	87	88	89	90	91	92	93	94	95	96	97	98	99	00	01	02	03	04	05	06	07	08	09	Total
Law Enforcement Management and Administrative Statistics (LEMAS) Survey	·	·	·	·	·	·	•	·	·	•	·	·	•	·	·	·	•	·	•	•	·	·	•	·	·	·	•	·	·	8
Community Oriented Policing Supplement	·	·	·	·	·	·	·	·	·	·	·	·	·	·	·	·	•	·	•	·	·	·	•	·	·	·	·	·	·	3
Census of State and Local Law Enforcement Agencies	·	·	·	·	·	•	·	·	·	·	·	•	·	·	·	•	·	·	·	•	·	·	·	•	·	·	·	•	·	6
Special Agency and Service Agency Censuses																														
Federal Law Enforcement Agency Census	·	·	·	·	·	·	·	·	·	·	·	·	•	·	·	•	·	•	·	•	·	•	·	•	·	•	·	•	·	8
Campus Law Enforcement Survey	·	·	·	·	·	·	·	·	·	·	·	·	·	·	•	·	·	·	·	·	·	·	·	•	·	·	·	·	·	2
Census of Law Enforcement Training Academies	·	·	·	·	·	·	·	·	·	·	·	·	·	·	·	·	·	·	·	·	·	•	·	·	·	•	·	·	·	2
Census of Medical Examiner and Coroner Offices	·	·	·	·	·	·	·	·	·	·	·	·	·	·	·	·	·	·	·	·	·	·	·	•	·	·	·	·	·	1
Census of Publicly Funded Forensic Crime Labs	·	·	·	·	·	·	·	·	·	·	·	·	·	·	·	·	·	·	·	·	·	•	·	·	•	·	·	·	·	2
National Survey of DNA Laboratories	·	·	·	·	·	·	·	·	·	·	·	·	·	·	·	·	·	•	·	·	•	·	·	·	·	·	·	·	·	2
Survey of Law Enforcement Gang Units	·	·	·	·	·	·	·	·	·	·	·	·	·	·	·	·	·	·	·	·	·	·	·	·	·	•	·	·	·	1
State Police Traffic Stop Survey	·	·	·	·	·	·	·	·	·	·	·	·	·	·	·	·	·	·	•	·	•	·	·	•	·	·	·	·	·	3
Traffic Stop Statistics	·	·	·	·	·	·	·	·	·	·	·	·	·	·	·	·	·	·	·	·	·	•	·	·	·	•	·	·	·	2

NOTES: •, data collected. ·, data not collected.

SOURCE: Bureau of Justice Statistics.

on an annual basis beginning in the early 1950s. Over the years, the two collections took different approaches, with the FOP survey emphasizing more complete coverage of police departments and the KCPD effort targeting large-jurisdiction departments (approximately 40) but expanding the range of questions. After the KCPD was forced to discontinue its collection, it collaborated with the Police Foundation and PERF on two general surveys of police operational and administrative practices in 1977 and 1981. BJS commissioned a study in 1983–1984 on the utility of a recurring survey of police agencies; that study "established that there was considerable demand for comparative police organizational data captured on a recurring basis from both the police practice and research communities" (Langworthy, 2002:23–24, summarizing Uchida et al., 1984). Additional background on recent and historical survey series of law enforcement agencies is given by Maguire (2002).

As the name suggests, the focus of the survey is on organizational and administration matters. In addition to acquiring counts of sworn and civilian employees and information on budgetary resources, the LEMAS survey instrument queries agencies about whether they follow certain policies or have specific technical resources. For example, the instrument asks about use of academies and special curriculum for training new recruits and equipment provided to officers (e.g., distribution of weapons or armor to officers and placement of computers in patrol cars). To show the basic nature of the questionnaire, 2 of the 11 pages of the 2003 LEMAS questionnaire are excerpted in Figures 3-3 and 3-4. As the figures suggest, the LEMAS instrument is an establishment survey that is intended to be filled out with relatively little need to refer to available records; questions are generally multiple choice.

Under the current LEMAS framework, agencies with 100 or more sworn officers as of the most recent Census of State and Local Law Enforcement Agencies (see next section) are always included in the sample as self-representing units. In 2003, this included 574 local police departments, 332 sheriffs' offices, and the 49 state police agencies. A stratified random sample (by type of agency, size of service population, and number of sworn officers) of agencies with fewer than 100 sworn personnel makes up the rest of the LEMAS sample, as non-self-representing units. In 2003, 2,199 agencies were selected for inclusion. An additional 25 agencies had been designated for inclusion but had either "closed, outsourced their operations, or were operating on a part-time basis," ruling them out of scope. Of the 3,154 agencies contacted by mail in 2003, 2,869 responded to the survey, for a 91 percent response rate (Bureau of Justice Statistics, 2006c:4–5).

As of the 2003 administration of the LEMAS survey by PERF, agencies were allowed to respond to the mailed questionnaire by any of several modes: mail, fax, or Internet. In the case of Internet responses, entries could be typed into a fillable PDF form; however, Internet respondents were re-

CJ-44L   **2003 SAMPLE SURVEY OF**   ID NUMBER
                          LAW ENFORCEMENT AGENCIES

SECTION V - COMMUNITY POLICING

28. During the 12-month period ending June 30, 2003, what proportion of agency personnel received at least eight hours of community policing training (problem solving, SARA, community partnerships, etc.)? Mark (■) one choice per line. If your agency did not conduct training for a particular type of employee, please check 'None.' If your agency did not have a particular type of employee for the specified time period, please mark 'NA.'

	All	Half or more	Less than half	None	NA
New officer recruits	☐	☐	☐	☐	☐
In-service sworn personnel	☐	☐	☐	☐	☐
Civilian personnel	☐	☐	☐	☐	☐

29. During the 12-month period ending June 30, 2003, which of the following did your agency do? Mark (■) all that apply.

☐ Actively encouraged patrol officers to engage in SARA-type problem-solving projects on their beats (If yes, please specify the approximate percentage of patrol officers engaged in these projects during the 12-month period ending June 30, 2003.)

(Specify percentage: [| |]%)

☐ Conducted a citizen police academy

☐ Maintained or created a formal, written community policing plan

☐ Gave patrol officers responsibility for specific geographic areas/beats

(Specify percentage of officers: [| |]%)

☐ Included collaborative problem-solving projects in the evaluation criteria of patrol officers

☐ Trained citizens in community policing (e.g., community mobilization, problem solving)

☐ Upgraded technology to support the analysis of community problems

☐ Partnered with citizen groups and included their feedback in the development of neighborhood or community policing strategies

☐ None of the above

30. Does your agency's mission statement include a community policing component?

☐ Yes   ☐ No   ☐ NA - Agency does not have a mission statement

31. During the 12-month period ending June 30, 2003, did your agency have a problem-solving partnership or written agreement with any of the following? Mark (■) all that apply.

☐ Advocacy groups

☐ Business groups

☐ Faith-based organizations

☐ Local government agencies (non law enforcement agencies)

☐ Other local law enforcement agencies

☐ Neighborhood associations

☐ Senior citizen groups

☐ School groups

☐ Youth service organizations

☐ None of the above

32a. During the 12-month period ending June 30, 2003, did your agency conduct or sponsor a survey of citizens on any of the following topics? Mark (■) all that apply.

☐ Public satisfaction with police services

☐ Public perception of crime/disorder problems

☐ Personal crime experiences of citizens

☐ Reporting of crimes to law enforcement by citizens

☐ Other (please specify) [   ]

☐ Did not survey the general public - SKIP to Section VI

b. For which purposes does your agency use the information described in Q32a? Mark (■) all that apply.

☐ Allocating resources targeted to neighborhoods

☐ Evaluating agency performance

☐ Evaluating officer performance

☐ Evaluating program effectiveness

☐ Prioritizing crime/disorder problems

☐ Providing information to patrol officers

☐ Redistricting beat/reporting areas

☐ Training development

☐ Other (please specify) [   ]

☐ None of the above

1748328597   Page 7

Figure 3-3 Law Enforcement Management and Administrative Statistics questionnaire, 2003, p. 7

CJ-44L

2003 SAMPLE SURVEY OF LAW ENFORCEMENT AGENCIES

ID NUMBER

SECTION VI - EMERGENCY PREPAREDNESS

33a. Does your agency have a written plan that specifies actions to be taken in the event of terrorist attacks? (Include emergency operation plans that would be applicable to such an attack.)

☐ Yes   ☐ No - SKIP to Question 34

b. Does your agency's plan include mutual aid or cooperative agreements between city, county, transit, public works, and/or other agencies?

☐ Yes   ☐ No

34. Do the public safety agencies operating in or nearby your jurisdiction (including your agency) use a shared radio network infrastructure that achieves interoperability?

☐ Yes   ☐ No

35. As of June 30, 2003, did your agency have any of the following types of emergency response equipment? Mark (■) all that apply.

☐ Personal Protective Equipment (PPE)

☐ Chemical detection equipment

☐ Radiological detection equipment

☐ Biological detection equipment

☐ Chemical/biological decontamination equipment

☐ Explosives detection equipment

☐ None of the above

36. In which of the following terrorism preparedness activities did your agency engage during the period ending June 30, 2003? Mark (■) all that apply.

☐ Partnership with culturally diverse communities

☐ Public anti-fear campaigns

☐ Dissemination of information to increase citizen preparedness

☐ Community meetings on homeland security/preparedness

☐ Increased sworn officer presence at critical areas

☐ None of the above

37. As of June 30, 2003, how many personnel did your agency have assigned to a multi-agency terrorism task force?

	Assigned Full-time	Assigned Part-time
a. Sworn personnel with general arrest powers		
b. All other employees		

1336328590   Page 8

38. Of the total number of actual FULL-TIME personnel, how many are intelligence personnel with primary duties related to terrorist activities? If none, enter '0.'

	Sworn	Non-sworn
Intelligence personnel with primary duties related to terrorist activities		

SECTION VII - EQUIPMENT

39. Does your agency supply or give a cash allowance to its regular field/patrol officers for the following? Mark (■) all that apply.

	Supplied	Cash allowance	Neither
Primary sidearm	☐	☐	☐
Backup sidearm	☐	☐	☐
Body armor	☐	☐	☐
Uniform	☐	☐	☐

40. Which types of sidearms are authorized for use by your agency's field/patrol officers? Mark (■) all that apply.

	On-duty weapon		Off-duty sidearm
Semiautomatic:	Primary sidearm	Backup sidearm	Off-duty sidearm
10mm	☐	☐	☐
9mm	☐	☐	☐
.45	☐	☐	☐
.40	☐	☐	☐
.357	☐	☐	☐
.380	☐	☐	☐
Other caliber (please specify)	☐	☐	☐
Any semiautomatic, as long as they qualify	☐	☐	☐
Revolver	☐	☐	☐

41. Are your agency's uniformed field/patrol officers required to wear protective body armor while in the field? Mark (■) only one response.

☐ Yes, all the time

☐ Yes, in some circumstances (e.g., serving warrants)

☐ No

42. Enter the number of animals regularly maintained by your agency for use in activities related to law enforcement. If none, enter '0.'

Dogs   **Horses**

Figure 3-4 Law Enforcement Management and Administrative Statistics questionnaire, 2003, p. 8

quired to type the ID number from the printed questionnaire received by mail into the electronic form.

BJS has reported results from the LEMAS survey in large publications covering major segments of the sample—agencies with 100 or more officers in the 2000 administration of the survey (Reaves and Hickman, 2004) and separate *Sheriffs' Offices* (Hickman and Reaves, 2006b) and *Local Police Departments* (Hickman and Reaves, 2006a) reports from the 2003 data. LEMAS data have also spawned specific BJS reports on the adoption of community-oriented policing practices (Hickman and Reaves, 2001), the frequency of citizen complaints of police use of force (Hickman, 2006), and long-term trends specific to large-city police departments (Reaves and Hickman, 2002b).

3–C.2 Census of State and Local Law Enforcement Agencies

Conducted on a 4-year cycle, the primary purpose of the Census of State and Local Law Enforcement Agencies (CSLLEA) is to produce the sampling frame for the main LEMAS survey. It also provides the frame for some of the special-agency censuses described in the next section. The CSLLEA is sometimes known, and is archived in the NACJD, as the Directory of Law Enforcement Agencies (or the Directory Survey) for its comprehensive focus, providing data on all state and local law enforcement agencies. Its intent is to gather information on "all police and sheriffs' departments that were publicly funded and employed at least one full-time or part-time sworn officer with general arrest power" (Bureau of Justice Statistics, 2003a).

The 2000 version of the questionnaire, administered by the Census Bureau as data collector, is reproduced in full in Figures 3-5 and 3-6. As a frame- or directory-building operation, the CSLLEA is a short, two-page questionnaire. The questionnaire includes standard items on the number of sworn and civilian personnel and budget level (with a specific question on drug asset forfeiture); question 5 on functions "perform[ed] on a routine basis" is one that can be used to determine the presence of some of the policies or practices that may be queried in greater detail in the special-agency censuses.

Between the 2000 and 2004 administrations of the CSLLEA, BJS changed data collection agents for the census, switching from the Census Bureau (2000) to the National Opinion Research Center (2004; NORC). In both years, responses were permitted by mail, fax, or Internet. In 2004, NORC and BJS developed the contact list for the CSLLEA by updating the 2000 directory with lists of agencies requesting an Originating Agency Identifier (ORI) from the FBI since 2000, as well as lists provided by Peace Officer Standards and Training offices.

Although its principal purpose is internal—to provide the basis for subsequent surveys—BJS has issued bulletins following the completion of the CSSLEA, providing general statistics on the comparative size and workforce of agencies (Reaves and Hickman, 2002a; Reaves, 2007).

3–C.3 Special-Agency and Service-Agency Censuses

BJS has conducted several data collections on special-focus law enforcement agencies (e.g., police forces maintained by colleges and universities) as well as agencies that support law enforcement in various ways (e.g., medical examiners' and coroners' offices). These special agency censuses tend to follow the basic mold and cover information similar to the LEMAS survey, though they are not "branded" as a part of or supplement to LEMAS. They typically tend to be censuses that are intended to capture data on the agencies of a particular type and, as such, function largely as a directory or catalog of agencies. Though this directory-building function could be the basis for follow-up sample surveys (asking, perhaps, more extensive questions on agency policies, practices, and experiences), this typically has not been done; in reference to these data collections, the "survey" label is used when a "census" label might be more appropriate.

Survey of Campus Law Enforcement Agencies

The original 1995 Survey of Campus Law Enforcement Agencies was motivated by concern over the coverage of college police departments in the CSLLEA and LEMAS. By their design and their focus on law enforcement agencies affiliated with governmental units, campus police forces commonly fall out of the LEMAS scope. "Because LEMAS includes only a small number of agencies serving public colleges and universities in its sample and does not include any of those at private institutions," BJS concluded that a special survey of college campuses was warranted (Bureau of Justice Statistics, 1997c). BJS worked with the International Association of Campus Law Enforcement Administrators (IACLEA) in developing the 1995 survey; contact information for an IACLEA representative was included on the 1995 questionnaire, though BJS handled the data collection in-house.

The collection focuses on 4-year institutions with 2,500 or more students, excluding the U.S. military academies, professional schools, and for-profit schools. For the 2004 administration, the scope was increased to include 2-year public colleges with enrollments of 10,000 or more (Reaves, 2008). Similar to LEMAS content, data are collected on agency personnel, expenditures and pay, arrest powers (e.g., whether limited to on-campus incidents), operations, equipment, computers and information systems, policies, and special programs; college-specific questions ask for enrollments of full- and part-time students as both undergraduates and graduates.

CJ-38S

OMB No. 1121-0240: Approval Expires 05/31/2003

| RETURN TO | U.S. Census Bureau
1201 East 10th Street
Jeffersonville, IN 47132-0001 | FORM **CJ-38S**
(6-29-2000)
2000 CENSUS OF STATE AND LOCAL
LAW ENFORCEMENT AGENCIES
Law Enforcement Management and
Administrative Statistics | U.S. DEPARTMENT OF JUSTICE
BUREAU OF JUSTICE STATISTICS
AND ACTING AS COLLECTION AGENT
U.S. DEPARTMENT OF COMMERCE
ECONOMICS AND STATISTICS ADMINISTRATION
U.S. CENSUS BUREAU |

(Please correct any error in name, mailing address, and ZIP Code above)

Agency Internet Home Page address:
(If none, mark (X) here ☐)

Agency central e-mail address for citizen use:
(If none, mark (X) here ☐)

INFORMATION SUPPLIED BY

Name			Title		
POSTAL ADDRESS ▶	Number and street or P.O. box/Route number		City	State	ZIP Code
PHYSICAL ADDRESS ▶	*If different from postal address – Number and street*		City	State	ZIP Code
E-MAIL ADDRESS ▶					
TELEPHONE ▶	Area code	Number	Extension	**FAX NUMBER** ▶ Area code	Number

Enter the year the agency began operation with sworn personnel ______________

IMPORTANT — *Please read the instructions below prior to completing the questionnaire.*

- If any of the following conditions apply, you do not need to complete this questionnaire. Mark (X) the appropriate box and return survey using the enclosed postage paid envelope.
  - ☐ Agency is no longer in existence
  - ☐ Agency contracts or "outsources" to the agency listed below for performance of all services –

    Full name of the agency that performs these services

    ____

  - ☐ Agency employs only part-time officers AND the total combined hours worked for these officers averages less than 35 hours per week
  - ☐ All of the officers within the agency volunteer their time (i.e., are unpaid)
  - ☐ Agency is private (i.e., not operated with funds from a state, local, special district, or tribal government)

GENERAL INFORMATION

- Please submit your data by using the web-reporting option at **harvester.census.gov/csllea,** mail your completed questionnaire to the U.S. Census Bureau in the enclosed postage-paid envelope, or FAX, (each page) toll-free to **1–812–218–3304 before July 28, 2000.**
- Please retain a copy of the completed survey for your records.
- If you have any questions, call **Theresa Reitz** toll-free at **1–800–352–7229,** or email to **csllea@census.gov**

INSTRUCTIONS

- If the answer to a question is "not available" or "unknown," write "DK" in the space provided.
- If the answer to a question is "not applicable," write "NA" in the space provided.
- If the answer to a question is "none" or "zero," write "0" in the space provided.
- When exact numeric answers are not available, provide estimates and place an asterisk (*) next to the figure.

Figure 3-5   Census of State and Local Law Enforcement Agencies questionnaire, p. 1

1. What type of government operates this agency?
Mark (X) only one.

☐ State ☐ Township ☐ Tribal
☐ County or Parish ☐ Regional ☐ Special
☐ Municipal ☐ School district district or authority

2. Which of the following law enforcement services did your agency provide on a regular basis during the 12-month period ending June 30, 2000? *Mark (X) all that apply.*

Criminal investigation for:
☐ Homicide
☐ Arson
☐ Other crimes
☐ Crime prevention
☐ Drug law enforcement
☐ First response to criminal incidents
☐ Patrol services
☐ Responding to citizen calls/requests for service
☐ Traffic law enforcement
☐ None of the above

3. Are the law enforcement services provided by your agency normally limited to a special jurisdictional area (e.g., airports, parks, schools, etc.)?

☐ Yes – *Specify area* ↗

____

☐ No

4. Does your agency PRIMARILY perform enforcement or investigation activities related to a specific category of laws (e.g., agricultural, alcoholic beverage, gaming, natural resources, etc.)?

☐ Yes – *Specify category of laws* ↗

____

☐ No

5. Which of the following functions did your agency perform on a routine basis during the 12-month period ending June 30, 2000? *Mark (X) all that apply.*

☐ Providing court security
☐ Serving civil process
☐ Operating one or more jails
☐ Executing arrest warrants
☐ Participating in a multi-agency drug task force
☐ Operating a training academy
☐ Dispatching calls for service
☐ Search and rescue operations
☐ Tactical operations (SWAT)
☐ None of the above

6. Enter the number of facilities or sites, SEPARATE FROM HEADQUARTERS, operated by your agency as of June 30, 2000. *If none, enter 0.*

	Number
a. District/Precinct stations	
b. Fixed neighborhood/community substations	
c. Mobile neighborhood/community substations	

7. Enter the number of AUTHORIZED FULL-TIME SWORN paid agency positions on June 30, 2000.

[ ]

8. Enter the number of ACTUAL full-time and part-time paid agency employees during the pay period including June 30, 2000. *Full-time employees are those regularly scheduled for 35 or more hours per week. If none, enter 0.*

	Full-time	Part-time
a. Sworn personnel with general arrest powers		
b. Officers without general arrest powers		
c. Nonsworn employees		
d. TOTAL *(Sum of lines a+b+c)* . .		

9. Of the total number of FULL-TIME sworn personnel with general arrest powers, entered in 8a, enter the number of uniformed officers whose REGULARLY ASSIGNED DUTIES included responding to citizen calls/requests for service. *If none, enter 0.*

[ ]

10. Of the total number of FULL-TIME sworn personnel with general arrest powers, entered in 8a, how many served as: *If none, enter 0.*

a. Community Policing Officers, Community Resource Officers, Community Relations Officers, or other sworn personnel specifically designated to regularly engage in community policing activities [ ]

b. School Resource Officers, School Liaison Officers, or other sworn personnel whose primary duties are related to school safety [ ]

11. Of the total number of FULL-TIME sworn personnel with general arrest powers, entered in 8a, how many performed the following duties as their PRIMARY job responsibility? *Count each officer only once. If none, enter 0.*

	Number
a. Patrol duties	
b. Investigative duties (e.g., detectives)	
c. Jail-related duties.	
d. Court security duties.	
e. Process serving duties	

12a. Enter your agency's total operating budget for the 12-month period that includes June 30, 2000. *If data are not available, provide an estimate and mark with an asterisk(*). Include jails administered by your agency. Exclude building construction costs and major equipment purchases.*

$ [ ]

b. Which 12-month period best reflects the budget amount entered in 12a? *Mark (X) only one.*

☐ Calendar year ☐ Fiscal year

13. Enter the total estimated value of money, goods, and property received by your agency from a drug asset forfeiture program during calendar year 1999. *If no money, goods or property were received, enter 0.*

$ [ ]

FORM CJ-38S (6-29-2000) Page 2

Figure 3-6 Census of State and Local Law Enforcement Agencies questionnaire, p. 2

To date, the Survey of Campus Law Enforcement Agencies has only been performed in 1995 and 2004; descriptions at the time of the initial 1995 collection imply that it was hoped that the survey could be repeated as early as 1997, but this was apparently not done until 2004.

Federal Law Enforcement Agency Census

BJS's Federal Law Enforcement Agency Census, conducted every 2–3 years since 1993, concentrates on federal agencies with general law enforcement and criminal investigation authority. In the 2004 version of the survey, this broad definition covered 65 federal agencies, including 27 offices of inspector general at cabinet departments or independent agencies; however, it was not meant to cover the U.S. armed forces, and the Central Intelligence Agency and the Transportation Security Administration's Federal Air Marshal program were excluded "because of classified information restrictions." Agencies were asked to report on the number of officers assigned to various duties, such as police patrol or security and protection. Officer counts are meant to include "personnel with Federal arrest authority who were also authorized (but not necessarily required) to carry firearms while on duty" (http://www.ojp.usdoj.gov/bjs/pub/ascii/fleo04.txt).

Forensic and DNA Crime Laboratories

In 1998–1999, NIJ provided funding for BJS to conduct the National Study of DNA Laboratories, as part of NIJ's larger DNA Laboratory Improvement program. Conducted in-house by BJS staff, the survey of DNA laboratories was repeated in 2001; in this second administration, BJS supplemented its frame from the 1998 wave by adding laboratories that had applied for grants from NIJ and checking against lists of participants in the FBI's Combined DNA Index System repository (Steadman, 2002:2, 7). The survey focused on publicly operated forensic crime laboratories that perform DNA analyses. According to its NACJD codebook (Bureau of Justice Statistics, 2003c)

> The survey included questions about each lab's budget, personnel, procedures, equipment, and workloads in terms of known subject cases, unknown subject cases, and convicted offender DNA samples. The survey was sent to 135 forensic laboratories, and 124 responses were received from individual public laboratories and headquarters for statewide forensic crime laboratory systems. The responses included 110 publicly funded forensic laboratories that performed DNA testing in 47 states.

In 2002 and 2005, BJS set out to conduct a fuller Census of Publicly Funded Forensic Crime Laboratories, including those that may not perform DNA testing. In addition to information on the range of services provided by the laboratories, a major focus of the censuses was on workload and backlog of pending cases. Though labeled as the 2002 and 2005 censuses, the data collection for these studies was actually conducted in 2003–2004 and 2006–2007, respectively, for facilities known to exist in the nominal (2002 or 2005) year. In 2002, the Survey Research Laboratory of the University of Illinois at Chicago was awarded a grant to conduct the census; in 2005, the data collection grant was won by Sam Houston State University. In both cases, the universities consulted with the American Society of Crime Laboratory Directors on questionnaire content and development of frames and mailing lists (Bureau of Justice Statistics, 2005b, 2008b). The 2002 and 2005 studies were summarized in BJS reports by Steadman (2002) and Peterson and Hickman (2005), respectively.

Census of Law Enforcement Training Academies

To date, BJS has twice conducted a Census of Law Enforcement Training Academies, obtaining information on the number and types of staff employed at these training facilities, their budget and funding sources, the number of trainees, and their general policies and practices. In addition to these basic organizational data, the survey collected information on training curriculum issues critical to current law enforcement policy development; for instance, questions asked whether the nature of terrorism and tactics to respond to terrorist attacks is a part of the training program. The initial 2002 administration of the collection was partially funded by the Justice Department's Office of Community Oriented Policing Services (COPS); the COPS office also provided input on the questionnaire. The collection was later repeated in 2006.

The 2002 census found "a total of 626 law enforcement academies operating in the United States [that] offered basic law enforcement training to individuals recruited or seeking to become law enforcement officers. This includes 274 county, regional, or State academies, 249 college, university, or technical school academies, and 103 city or municipal academies" (Hickman, 2005b:1). The report summarizing the collection notes only that the list of agencies "was compiled from a variety of sources, including professional associations, State law enforcement training organizations, and existing law enforcement data collections" (Hickman, 2005b:21)

BJS contracted with PERF to conduct the data collection and consulted with the International Association of Directors of Law Enforcement Standards and Training (IADLEST); IADLEST was also enlisted to provide a supporting letter to bolster participation and help with nonresponse follow-up efforts.

Census of Medical Examiner and Coroner Offices

Conducted once to date, in 2004, the Census of Medical Examiner and Coroner Offices was conducted by RTI International under contract to BJS; the National Association of Medical Examiners and the International Association of Coroners and Medical Examiners were enlisted to help with the development of the questionnaire and to encourage individual offices to respond to the query. The Centers for Disease Control and Prevention (CDC) generated an initial list of offices (because CDC regularly compiles morbidity and mortality data from state and local offices) to contact in the survey, and the list and contact information was updated by RTI.

RTI developed mixed-mode response options for the census; in addition to mailed paper questionnaires, individual coroner offices were permitted to respond through an online website or by fax. In all, 1,998 offices responded to the census (85.9 percent response rate).

Much like the other special-agency censuses, the medical examiner census focused on administrative characteristics such as staffing levels, expenditures, and workload; general findings from the collection are reported by Hickman et al. (2007). One particular line of inquiry in the data collection concerned the number of unidentified human remains in the custody of each office. Curiously, when BJS developed a special "Fact Sheet" on unidentified human remains (Hughes, 2007), it did so after BJS obtained access to the FBI's Unidentified Person File (a voluntary reporting system) and the National Center for Health Statistics' National Death Index. Through the National Death Index, BJS derived a time series for the span 1980–2004, but the fact sheet made no attempt to compare the latest of these estimates (based on individual death records reported by state vital statistics offices) with the agency-level totals from BJS's own census of coroner offices.

Special-Agency Censuses Under Development

In 2007 and 2008, BJS filed Information Collection Review packages to the U.S. Office of Management and Budget (OMB) for several additional special-agency censuses. The Census of Law Enforcement Aviation Units follows up on findings from the 2003 LEMAS survey that about 250 service units provide helicopter or fixed-wing aircraft service for state and local law enforcement agencies. BJS also filed a request to obtain clearance from OMB to conduct the Survey of Law Enforcement Gang Units. The supporting statement for that collection indicates that the data collection is being initiated, in part, because of a department-level antigang initiative ("one of the current top priorities for the Attorney General and the Department of Justice is the development of more effective programs to prevent gang violence and enforce anti-gang laws when such violence does occur").

3–C.4　Police Traffic Stop Data

NCVS Supplement: Police-Public Contact Survey

As described in Section 3–A.2, the PPCS supplement to the NCVS grew directly out of a legislative mandate to study "excessive force by law enforcement officers" (P.L. 103-322).[9] Faced with this mandate, BJS adopted a strategy that it would later use—albeit on a much larger scale—when organizing data collections to respond to the Prison Rape Elimination Act of 2003 (see Section 5–A.1). BJS convened a workshop on police use of force in May 1995 to highlight challenges in systematic measurement of the phenomenon (McEwen, 1996). BJS then began with an administrative, facility-level study; BJS and NIJ jointly funded a study by the International Association of Chiefs of Police that contacted 110 agencies in 1995 and about 30 agencies in both 1996 and 1997. From those contacts, it was concluded that "the police use of force rate was 4.19 per 10,000 responded-to-calls for service, or 0.0419 percent," in 1996 (Henriquez, 1999:21; see also Fyfe, 2002). At the same time, it also set into motion a plan to directly gather data through direct survey interviewing. The PPCS was constructed and fielded on a pilot basis in 1996, evolving into a triennial collection. In its building of the PPCS, BJS approached the problem of public interactions with police more broadly than the "excessive force" text of the act envisioned. The pilot PPCS (and its successors) was fielded "with the goal of better understanding the types and frequency of contacts between the police and the public, and the conditions under which force may be threatened or used" (U.S. Government Accountability Office, 2007:7).

As we discuss further in Section 5–A.2, disputes over the release of data from the 2002 PPCS, and in particular the evidence it presented about differential treatment by race during traffic stops, led to the termination of a BJS director.

State Police Traffic Stop Data Collection Procedures

BJS staff have also periodically compiled what might be considered metadata related to traffic stops. In 1999, 2001, and 2004, staff have contacted the nation's 49 primary state police agencies[10] with a questionnaire asking about *policies* for recording data on race and ethnicity of persons stopped on traffic violations. This effort does not collect, and does not intend to collect, actual traffic stop records or even counts of traffic stops, but merely whether demographic data are routinely recorded and whether those data are electronically accessible.

---

[9]The same law also requires the attorney general to "publish an annual summary" of these data (42 USC § 14142).

[10]"Hawaii does not have a state police agency" (Bureau of Justice Statistics, 2006c:3).

3–D ADJUDICATION, PROSECUTION, AND DEFENSE

Detailed as they are, the legal duties of BJS do not explicitly mention the judicial processing of criminal trials. However, BJS's general charge to statistically document the "operations of the criminal justice system at the Federal, State, and local levels" does clearly give BJS the mandate to generate statistics on the operations of the courts, because the courts are an integral part of the system. This is a task that is as difficult as it is unusual, from an operational standpoint: a small federal agency tasked with being a data collector on a separate branch of government with highly decentralized operations that vary greatly by locality.

Article III of the U.S. Constitution established the federal court system, specifically creating the U.S. Supreme Court and reserving to Congress the authority to create lower federal courts. The modern federal judiciary includes 94 U.S. District Courts and 13 U.S. Courts of Appeals, as well as U.S. Bankruptcy Courts and Courts of Claims and International Trade.[11] The federal courts have primary jurisdiction in cases involving federal laws and treaties, as well as disputes between multiple states. Data from the federal court system are collected and disseminated by the Administrative Office of the U.S. Courts.

The powers and areas of jurisdiction that are not explicitly assigned to the federal courts are the province of the state court system. Individual state constitutions and laws create the network of courts of original jurisdiction for civil and criminal cases, as well as appellate structures. States vary greatly in their court organizational structures, in the number of original courts, and in the number of appellate layers. Though many of the states have a single court of last resort (e.g., the state supreme court) and a single intermediate court of appeals, this is not a universal rule: as of 2007, 11 states have no intermediate court of appeals, so that appeals from district and other trial courts are appealed directly to the state supreme court.[12] Likewise, some states have two courts of last resort, one for civil and one for criminal cases.

In addition to appellate structure, individual state court systems also vary in the degree to which specific legal matters are distributed to special-jurisdiction courts, such as family courts, juvenile courts, and probate courts.

---

[11]The U.S. Court of Veterans' Appeals, the U.S. Court of Military Appeals, and the U.S. Tax Court are considered "Article I" or legislative courts because they have been created by Congress but are not vested with full judicial power to decide questions of federal and constitutional law.

[12]Those 11 states are Delaware, Maine, Montana, Nevada, New Hampshire, North Dakota, Rhode Island, South Dakota, Vermont, West Virginia, and Wyoming. In recent years, the Nevada Supreme Court has argued for the creation of an intermediate appellate court, given its steadily increasing workload and legal mandate to hear and consider all cases filed (Supreme Court of Nevada, 2007). Since 1999, the seven-member Supreme Court has dealt with its workload by dividing into three-judge panels, rotating between Carson City and Las Vegas, to hear many cases.

The state court system is sufficiently complex that even the most basic summary of the scope of the system—the number of "courts" it encompasses—is difficult to characterize. The National Center for State Courts (NCSC), with which BJS works on various collections, estimates that the system includes approximately 16,000 "courts" (LaFountain et al., 2007:9):

> However, this number may be somewhat misleading and is not derived from any universally agreed upon definition. For example, Texas, the second-largest state, considers each judgeship in the state to be a court and thereby reports over 3,300 trial courts statewide. Conversely, California, the largest state, has 58 superior courts in its trials system.

State courts are also inherently difficult to conceptualize and measure because their basic structures and jurisdictions are subject to change. The example of California is useful again; the state's current 58 county-level superior courts were reformed between 1998 and 2001, following passage of a constitutional amendment via 1998's Proposition 220. The amendment permitted counties to consolidate the operations of then-existing mixed-jurisdiction superior and municipal trial counts in a single, unified superior court with jurisdiction over all civil and criminal cases.

The measurement of activity in the judicial branch, and particularly the state court system, has been a long-standing challenge, and previous attempts by the Census Bureau and the FBI to do so have been relatively short-lived. In the 1930s and early 1940s, the Census Bureau attempted such measurement, and the introduction to the Census Bureau's 1933 "Judicial Criminal Statistics" report explained the goal and expressed great hope for the collection (quoted in Cahalan, 1986:4–5):

> It is the purpose of the Census Bureau, through cooperation with the several States, to develop a national system of collecting judicial criminal statistics which will be mutually advantageous to the States and the Federal government. . . . It is hoped that eventually each State will adopt the Census forms and classifications. If this is done, one report for the court will suffice for the State and for the Federal government, the statistics of different States will be compiled on the same basis, and needless duplication of work and expense will be avoided.

The Census Bureau's intent was to collect, "by offense, the number of persons prosecuted, the disposition made of prosecutions, and the sentences imposed on convicted persons" (Beattie, 1959:584). However, the Census Bureau cited the major difficulty in obtaining comparable data from the states and incomplete responses (from at most 32 states) in discontinuing the data series in the early 1940s. Likewise, the FBI UCR program asked police departments to submit information about judicial disposition of arrests beginning in 1955, but the practice was abandoned after 1977 (Cahalan, 1986:5).

Although court information systems have improved over time, the measurement of even basic parts of the adjudication and prosecution systems presents a formidable challenge. An unusual government document—five agencies, including BJS, coauthoring a two-page summary—illustrates the point: It describes the difficulties involved in comparing case processing statistics even at the federal level.[13] The summary attributes the incomparability of processing statistics across data sources to fundamental definitional differences, from the classification of offenses and definition of "defendants" to differences in labeling types of case dispositions (U.S. Department of Justice, 1998).

Table 3-7 illustrates the collection dates for BJS's data series in the area of adjudication, and the balance of this section describes the principal series. Two collections—the Civil Justice Survey of State Courts and BJS's work with the Federal Justice Statistics Resource Center—have already been discussed at more natural points in the narrative, in Section 2–C.2 and Box 3-4, respectively. Several of BJS's projects related to adjudication have been conducted with or by NCSC; in particular, NCSC's Court Statistics Project is the source of monitoring data on caseloads and completed cases within the state court systems. We describe NCSC and the project more completely in Box 3-2.

3–D.1   National Judicial Reporting Program

BJS contracts with the Census Bureau to compile court record data from felony trial courts in a sample of counties through the National Judicial Reporting Program (NJRP), a biennial collection. The NJRP provides national estimates of the demographic characteristics, conviction offense type(s), and sentence imposed. When selected to participate in the sample, jurisdictions can provide records data in a variety of formats (electronic and paper, with electronic submissions accounting for 97 percent of data received in the 2004 NJRP), which the Census Bureau then keys, codes, and formats. "State courts were the source of NJRP data for about 44% of the 300 counties sampled [in 2004]. For other counties, sources included prosecutors' offices, sentencing commissions, and statistical agencies" (Bureau of Justice Statistics, 2007e:5).

For the 2002 and 2004 NJRP (which used the same sample), the sample of counties was drawn by assigning each state a "cost factor"—with values 1 (low), 3 (moderate), or 5 (high)—based on the estimated cost of collecting data in those counties in 2000. These cost factors were then used in combination with county populations from the 2000 census to define 20 strata,

---

[13]The coauthoring agencies were the Administrative Office of the U.S. Courts, the Executive Office for the U.S. Attorneys, BOP, the U.S. Sentencing Commission, and BJS. Because the document was issued with "U.S. Department of Justice" as the header, that label is used as the author in the citation.

Table 3-7 Bureau of Justice Statistics Data Collection History and Schedule, Adjudication, 1981–2009

Series	81	82	83	84	85	86	87	88	89	90	91	92	93	94	95	96	97	98	99	00	01	02	03	04	05	06	07	08	09	Total
Federal Justice Statistics	●	●	●	●	●	●	●	●	●	●	●	●	●	●	●	●	●	●	●	●	●	●	●	●	●	●	●	●	●	29
National Judicial Reporting Program	·	·	●	·	●	●	·	●	·	●	·	●	·	●	·	●	·	●	·	●	·	●	·	●	·	●	·	●	·	14
State Court Annual Caseload Statistics	·	·	·	·	·	·	·	·	·	·	·	·	·	●	●	●	●	●	●	●	●	●	●	●	●	●	●	●	●	16
State Court Processing Statistics	·	·	·	·	·	·	·	●	·	●	·	●	·	●	·	●	·	●	·	●	·	●	·	●	·	●	·	●	·	11
Census of State Court Prosecutors	·	·	·	·	·	·	·	·	·	·	·	·	·	·	·	·	·	·	·	·	·	·	·	·	·	·	●	·	·	1
Civil Justice Survey	·	·	·	·	·	·	·	·	·	·	·	●	·	·	·	●	·	·	·	·	●	·	·	·	●	·	·	·	·	4
National Prosecutors Survey	·	·	·	·	·	·	·	·	·	●	·	●	·	●	·	●	·	·	·	·	●	·	·	·	●	·	●	·	·	7
National Survey of Indigent Defense Services	·	●	·	·	·	●	·	·	·	·	·	·	·	·	·	·	·	·	●	·	·	·	·	·	·	·	●	·	·	4
Prosecution of Felony Arrests	·	●	·	·	·	●	●	●	·	·	·	·	·	·	·	·	·	·	·	·	·	·	·	·	·	·	·	·	·	4
State Court Organization	·	·	·	·	·	·	●	·	·	·	·	·	●	·	·	·	·	●	·	·	·	·	·	●	·	·	·	·	●	5

NOTES: ●, data collected. ·, data not collected.

SOURCE: Bureau of Justice Statistics.

Box 3-2 The Court Statistics Project and the National Center for State Courts

Much of the Bureau of Justice Statistics' (BJS's) work with the state courts in the Court Statistics Project has been conducted by the nonprofit National Center for State Courts (NCSC). NCSC was founded in 1971 on the recommendation of Chief Justice Warren Burger at a national conference of the judiciary. Since 1978, it has been headquartered in Williamsburg, Virginia, and is governed by a Board of Directors elected by state court administrators and chief justices. The NCSC's mission is to improve state court administration by serving as a clearinghouse of information, including training and development of performance standards.

NCSC's Court Statistics Project (called the State Court Statistics Project by BJS) began in 1978; it receives financial support from BJS for its work on collecting data on state court caseload. Results from the Court Statistics Project are posted and maintained on the NCSC website; the URL http://www.courtstatistics.org redirects browsers to the specific site. The core reports from the Court Statistics Project are the annual *Examining the Work of State Courts* (LaFountain et al., 2007) and *State Court Caseload Statistics* (Court Statistics Project, 2006), both of which are now maintained and updated in electronic format on the website. Both of these report series are branded and identified as NCSC or Court Statistics Project reports and not BJS reports, though a BJS logo is included and extensive links to related BJS reports are included in the electronic documents.

The Court Statistics Project compiles data on cases filed and disposed in state appellate and trial courts from the 50 states, the District of Columbia, and Puerto Rico. Coverage of trial court caseloads is not limited to criminal cases; for instance, entries for civil cases, traffic and other violations, domestic relations, and juvenile cases are also recorded. However, data in the project are limited to aggregate counts of cases and not specific characteristics of cases.

BJS also periodically sponsors a Survey of State Court Organization that is conducted by NCSC, with assistance from the Conference of State Court Administrators; it is conducted every 5–7 years, most recently in 2004. Though the survey includes requests for administrative counts (e.g., number of judges and personnel), the emphasis of the study is on changes to the structure of the state courts. The 2004 survey, in particular, attemped to query court systems about their processing of domestic violence cases and their adoption of specialized courts to handle certain case types. Results from the survey are summarized in regular BJS reports on *State Court Organization* (Rottman and Strickland, 2006; Langton and Cohen, 2007), which draw extensively on NCSC data and the flowcharts that the center maintains to illustrate the court structures in individual states.

Among other activities, the NCSC worked with BJS on a study of habeas corpus petitions filed by state prisoners in federal court challenging their sentences on the basis of violations of constitutional rights (Hanson and Daley, 1995). NCSC also initiated and organized a 10-year effort that led to the publication of Trial Court Performance Standards (TCPS). The TCPS effort was conducted with funds from the Bureau of Justice Assistance, and the formal result of the work is a four-volume report issued by an advisory commission organized by NCSC. The standards are intended to give individual courts a metric to compare their own activities within such performance areas as ensuring access to justice and public trust and confidence. The TCPS initiative is described more fully in an NCSC-hosted website (http://www.ncsconline.org/D_Research/tcps/index.html), and Keilitz (2000) describes remaining challenges in converting the standards into practices, including the fuller development of statistical measures and indices.

which were constructed to give the 75 largest counties—which "account for a disproportionately large amount of serious crime in the Nation"—"a greater chance of being selected than the remaining counties." The final sample consisted of 300 counties (out of 3,141 county-level equivalents in the United States), 58 from among the largest 75 counties; some selected counties that declined to participate were replaced by other counties. On the basis of this sample, Census Bureau staff examined case-level data for 471,646 convicted felons sentenced in 2004, of which about 70 percent originated in the most populous counties (Bureau of Justice Statistics, 2007e:4).

BJS summarizes findings from the NJRP on sentence length in its bulletin series *Felony Sentences in State Courts* (e.g., Durose and Langan, 2007) and, earlier, *Felony Sentences in the United States* (e.g., Brown and Langan, 1999). In recent incarnations, spreadsheet tables of key results are presented on the BJS website at http://www.ojp.usdoj.gov/bjs/abstract/scscfst.htm.

3–D.2   State Court Processing Statistics

Originally developed in 1982 and known, through 1994, as the National Pretrial Reporting program, the State Court Processing Statistics (SCPS) program provides data on the criminal justice processing of felony defendants in a sample of large counties. The program prospectively tracks felony defendants from charging by the prosecutor until disposition of their cases or for a maximum of 12 months. Data are obtained on demographic characteristics, arrest offense, criminal justice status at time of arrest, prior arrests and convictions, bail and pretrial release, court appearance record, rearrests while on pretrial release, type and outcome of adjudication, and type and length of sentence. In at least one instance, the standard SCPS program has been augmented to cover special case types: in 1998, records were drawn from 40 large urban counties on juveniles facing felony charges in adult criminal courts (Strom et al., 1998; Rainville and Smith, 2003).

The documentation for a 1990–2004 compilation of SCPS data on the NACJD (Bureau of Justice Statistics, 2007g:3) summarizes the data collection's content:

> This data collection effort was undertaken to determine whether accurate and comprehensive pretrial data can be collected at the local level and subsequently aggregated at the national level. The data contained in this collection provide a picture of felony defendants' movements through the criminal courts. Offenses were recoded into 16 broad categories that conform to the Bureau of Justice Statistics' crime definitions. Other variables include sex, race, age, prior record, relationship to criminal justice system at the time of the offense, pretrial release, detention decisions, court appearances, pretrial rearrest, adjudication, and sentencing. The unit of analysis is the defendant.

The sampling scheme for the SCPS is unusual; in what follows, we describe the procedure and counts used in the 2004 SCPS, but the general design of the collection has been similar in previous years. The intended universe that the program is meant to reflect is "felony court filings during the month of May in even numbered years from 1990–2004 in the 75 most populous counties in the United States" (Bureau of Justice Statistics, 2007g:4). Why the 75 most populous counties is not entirely clear (though the 75 largest counties are said to "account for more than a third of the United States population and approximately half of all reported crimes," and availability of records is also likely a factor), nor why May is the targeted month. The sampling scheme designed by the Census Bureau calls for 40 counties to be chosen from the 75 in the first stage, 10 with certainty ("because of their large number of court filings") and the others drawn from three strata based on their levels of filings. The court system in each of the chosen counties was asked to provide "data for every felony case filed on selected days during" May 2004; the high-filing 10-counties chosen with certainty were only asked to provide 5 days' worth of filings, while other counties were asked for 10 or 20 days' worth of filings. In the end, data were collected for 15,761 felony cases, out of an estimated 57,497 May 2004 cases in all 75 large counties (Bureau of Justice Statistics, 2007g:4–5).

Since the inception of SCPS, BJS has contracted with the Pretrial Justice Institute (PJI) as the data collector for the program. Founded in 1977, PJI was known as the Pretrial Services Resource Center until 2007. BJS opened the SCPS data collection contract to competition in 2006 and PJI was again selected as the collector. According to PJI's website, PJI "completely redesigned the internal project management processes, moved to online data collection and submission, and for the first time, accepted large data sets extracted from jurisdictions' information management systems" between 2006 and 2008.[14]

In early 2008, BJS issued a "redesign solicitation" request for proposals, asking for bids to "re-conceptualize SCPS to take into account the increasing levels of automation in state courts and other enhanced collection mechanisms that have occurred since the late 1980s" (CFDA No. 16.734). The solicitation also candidly describes "several important limitations" to the current SCPS:

> First, the SCPS project covers case processing in the Nation's 75 most
> populous counties. It does not have the capacity to make national or
> county level inferences about felony case processing or pretrial release.
> Secondly, the current SCPS sampling strategy of selecting only 40 of
> the Nation's 75 most populous counties and requesting participating
> SCPS jurisdictions to provide less than a whole month of felony filing

---

[14]See   http://www.pretrial.org/AnalysisAndResearch/StateCourtProcessingStatistics/Pages/
default.aspx.

> data (e.g., 5 or 10 business days) introduces certain levels of sampling
> error into the data collection process. Lastly, SCPS does not collect
> several key data elements that potentially play a crucial role in pretrial
> decision-making. The decision to restrict the SCPS sample to 40 of
> the Nation's 75 most populous counties, confine felony filing data to
> less than a whole month, and limit the types of data collected were
> primarily driven by time and cost restraints and the difficulties inherent
> in obtaining court case processing data.

Specific goals called for in the redesign are to "develop and test alternative sampling strategies that allow for periodic modular enhancements of SCPS" and to take "advantage of automated systems of case management, state criminal history depository programs, and administrative jail systems." PJI's website acknowledges that its bid, in partnership with Justice Management Institute and the National Association of Pretrial Services Agencies, was selected in September 2008.[15] SCPS data collection will be suspended in 2009 during the redesign, the first break in the series since 1984.

BJS reports based on SCPS data include Cohen and Reaves (2006), Reaves (2006), and Cohen and Reaves (2007).

3–D.3   National Prosecutors Survey

The National Prosecutors Survey (NPS) asks chief prosecutors in state court systems to report management information, resources, and policies; in its content and focus, it is very analogous to the LEMAS survey. The codebook for the NACJD filing of the 2005 NPS summarizes the content as follows:

> The [NPS's] purpose was to obtain detailed descriptive information on
> prosecutors' offices, as well as information on their policies and prac-
> tices. Variables cover staffing, funding, special categories of felony pros-
> ecutions, caseload, juvenile matters, work-related threats or assaults, the
> use of DNA evidence, and community-related activities, such as involve-
> ment in neighborhood associations.

Most recently, BJS has used the NORC to collect NPS data.

In 2005, the NPS sample was drawn from a frame of 2,400 prosecutorial districts handling felony cases that was assembled by the Census Bureau. Districts were grouped into five strata on the basis of 2004 population estimates; "within each stratum, districts were systematically selected for the sample," yielding a final sample of 310 offices, 307 of which responded to the mail survey (Bureau of Justice Statistics, 2007f:4).

The periodicity—and the scope—of the prosecutor survey have been irregular. First conducted in 1990, the survey was performed every 2 years

---

[15]PJI also indicated that it plans to partner with the Urban Institute specifically to improve the sampling strategy underlying SCPS; see http://www.pretrial.org/TechnicalAssistance/Pages/OurProjects.aspx.

but, more recently, it has been performed every 5 years or so. Moreover, the NPS has most frequently been conducted as a sample of prosecutor offices but also, occasionally, as a full census. The 2001 collection was the first intended to be a complete enumeration of all prosecutors' offices (on the order of 2,400), whereas the 2005 version was a sample of 310 offices. BJS's supporting statement in requesting clearance for the 2007/2008 version of the NPS (see Box 5-4; ICR 200704-1121-004) reflects the confused nature of the collection; though generally maintaining the "National Survey of Prosecutors" nomenclature, it also refers to conducting this version as a "National Census of Prosecutors" or "National Census of State Court Prosecutors." The need—or even potential use—for this collection to support more detailed, targeted surveys in subsequent years is not mentioned in BJS's argument. Though it approved the collection, OMB chided BJS for the uncertain periodicity and requested feedback on "the magnitude and nature of change identified from 2001 and 2005 to the present," as well as fuller articulation of the utility of the resulting data, prior to future collections.

Because of the "census" nature of the 2007/2008 survey, BJS scaled back the level of information requested in the most recent administration of the survey, with the objective of capping the burden on responding prosecutor offices at 30 minutes. Previous versions of the survey went into somewhat fuller detail, including questions on handling of cases involving juveniles and civil actions filed against prosecutors.

3–D.4   National Survey of Indigent Defense Systems

In 1996, BJS published a "Selected Findings" report (Smith and DeFrances, 1996) that pieced together the limited glimpses of what its existing data systems revealed about the use of indigent defense systems: that is, the use of court-appointed legal representation and legal defender services by defendants unable to afford them. In particular, the report drew from questions on the NPS (on the availability of public defender or assigned counsel arrangements in their jurisdictions) and BJS's prison and jail inmate surveys (asking about the inmates' own representation).

On the basis of this first effort, BJS sponsored NORC to conduct a National Survey of Indigent Defense Systems; it was BJS's first structured survey of such defense agencies since two studies in the early 1980s. Though "National" in label, the survey was restricted to agencies within the 100 most populous counties in the United States, as of 1997 population estimates. The survey was conducted in two stages in 1999–2000; an initial "county survey" sent to county governments asked for their assistance in identifying indigent criminal defense programs in their area, and the more detailed 141-question "program survey" was then sent to identified programs. In instances where the first survey suggested that such programs were solely administered by the state government, the program survey was sent to the appropriate state department. DeFrances and Litras (2000) summarize the survey as a whole, while DeFrances (2001) focuses on the 21 states that are the sole funders of indigent defense services.

The 2007 BJS Census of Public Defender Offices covered publicly funded indigent defense offices, not the full range of service providers.

3–E   OTHER DATA COLLECTIONS

For the sake of completeness, Table 3-8 summarizes the data collection years of miscellaneous activities that do not fall neatly into the major categories covered in this chapter. Some of these are related to BJS's criminal history improvement grant programs, which we describe more fully in Chapter 4.

3–F   ASSESSMENT OF THE PORTFOLIO

It is not difficult to find references to and uses of the major BJS data series in the academic literature. It is also clear that BJS data series and results frequently garner the attention of one critical audience and user—the U.S. Congress—as we describe in Box 3-3. BJS's data collection programs range widely in their scope and universe size—from the sprawling and nationally representative NCVS to the subset of law enforcement agencies that operate dedicated gang units—and so vary in their level of expense. They vary in methodology from hand-coding of paper court dockets to online questionnaire completion by facility administrators.

One thing that we think clear from a review of BJS's entire data collection portfolio is that it is a solid body of work, generally well justified by public information needs or required by law. It represents a better-than-good-faith effort by the agency to marshal data relevant to an astoundingly large substantive mandate, given that fiscal resources typically have been less than commensurate. Within its resources and the topics it has chosen to address, BJS has done well in the sense that nothing in its portfolio is obviously frivolous, wasteful, or inconsistent with its legal mandates. Certainly, however, not all of BJS's individual data series are equally influential, and there are some important topics (such as those described in Section 2–C) on which BJS currently collects little or no data.

Our review of the existing data collections of the Bureau of Justice Statistics yields some basic observations about the major topic segments of the portfolio:

- BJS's key data series in the area of *victimization*, the NCVS, is its most expensive, most flexible, and most scrutinized collection. It is also, ar-

Table 3-8 Bureau of Justice Statistics Data Collection History and Schedule, Criminal History Improvement and Miscellaneous Studies, 1981–2009

Series	81	82	83	84	85	86	87	88	89	90	91	92	93	94	95	96	97	98	99	00	01	02	03	04	05	06	07	08	09	Total
Criminal Justice Agency Survey	●	●	●	●	●	●	●	●	●	●	●	●	●	●	●	●	●	●	●	●										20
Domestic Violence Processing																						●								1
Domestic Violence Recidivism																							●							1
Expenditure and Employment Statistics	●				●	●	●	●	●	●	●	●	●	●	●	●	●	●	●	●	●	●	●	●	●	●	●	●	●	26
Firearm Inquiry Statistics																			●	●	●	●	●	●	●	●	●	●	●	11
Inventory of Correctional Information Systems																				●										1
Justice Assistance Data Survey										●									●											2
National Study on Campus Sexual Assault																		●												1
Offender-Based Transaction Statistics			●	●	●	●	●	●	●	●																				8
Survey of Cybercrime on Businesses																						●				●				2
Survey of State Criminal History Information Systems							●		●		●		●		●		●		●		●		●		●					10
Survey of State Procedures on Firearm Sales															●	●	●	●	●	●	●	●	●	●	●					11

NOTES: ●, data collected. ·, data not collected.

SOURCE: Bureau of Justice Statistics.

Box 3-3 Congressional Uses of Bureau of Justice Statistics Data

BJS reports and data are frequently cited in congressional debates and statements. In many instances, the citation is used to establish some basic fact, as grounding for new legislative initiatives; accordingly, they are frequently referenced in "whereas" or "findings" clauses. For instance, in the 110th Congress, H.R. 4611—the End Racial Profiling Act of 2007—includes among its prefatory findings a basic recounting of findings from the 2002 Police-Public Contact Survey supplement to the NCVS:

> (12) A 2005 report of the Bureau of Justice Statistics of the Department of Justice on citizen-police contacts that occurred in 2002 [(Durose et al., 2005)], found that, although Whites, Blacks, and Hispanics were stopped by the police at the same rate–
>
> (A) Blacks and Hispanics were much more likely to be arrested than Whites;
>
> (B) Hispanics were much more likely to be ticketed than Blacks or Whites;
>
> (C) Blacks and Hispanics were much more likely to report the use or threatened use of force by a police officer;
>
> (D) Blacks and Hispanics were much more likely to be handcuffed than Whites; and
>
> (E) Blacks and Hispanics were much more likely to have their vehicles searched than Whites.

(Section 5–A.2 describes the referenced report more fully, and the controversy that surrounded it; Judiciary Committee chairman John Conyers cited specific figures from Durose et al. (2005) in his remarks on introducing the bill [*Congressional Record*, December 13, 2007, p. 2576].) Likewise, H.R. 5654—the Families Beyond Bars Act of 2008—begins noting that:

> Congress finds as follows:
>
> (1) The Bureau of Justice Statistics estimates that 1,500,000 children in the United States have at least one incarcerated parent, and an estimated 10,000,000 more individuals have at least one parent who was incarcerated at some point during the individual's childhood.
>
> (2) In 2006, the Bureau of Justice Statistics estimated that 75 percent of incarcerated women were mothers, two-thirds of whom were mothers of children under the age of 18, and an estimated 32 percent of incarcerated men were fathers of children under the age of 18. . . .
>
> (4) The Bureau of Justice Statistics estimates that children with imprisoned parents may be almost 6 times more likely than their peers to be incarcerated.

BJS findings are used similarly in floor debates. One such example arose during debate on H.R. 3992, the Mentally Ill Offender Treatment and Crime Reduction Reauthorization and Improvement Act of 2008, on January 23, 2008. The remarks of Rep. Bobby Scott (D-Va.) included the following paragraph (*Congressional Record*, January 23, 2008, p. 426):

> A 2006 report by the United States Department of Justice Bureau of Justice Statistics entitled "Mental Health Problems of Prison and Jail Inmates" suggests that the criminal justice system has become, by default, the primary caregiver of the most seriously mentally ill individuals. [The report referenced is James and Glaze (2006) and uses data from the Survey of Inmates in State and Federal Correctional Facilities, 2004, and the Survey of Inmates in Local Jails, 2002.] The bureau reports that over one-half of the prison and jail population of this country is mentally ill. More specifically, 56 percent of State prisoners, 45 percent of Federal prisoners, and 64 percent of jail inmates have some degree of mental illness.

Occasionally, the findings from one BJS report lead to new suggestions for data collection and analysis. The BJS recidivism work mentioned in the Second Chance Act of 2007

(continued)

Box 3-3   (continued)

(see Box 3-5) is one such example. Another is the debate on reauthorizing the Death in Custody Reporting Act, first enacted in 2000, in early 2008. In debate on H.R. 3971, both the chairman and ranking minority member of the House Judiciary Subcommittee on Crime argued for improved data collection. Chairman Bobby Scott (D-Va.) outlined the state of knowledge before the original act and basic findings from BJS's original data collection under the act (*Congressional Record*, January 23, 2008, pp. 428–429):

> Before the enactment of the Death in Custody Act of 2000, States and localities had no uniform requirements for reporting the circumstances surrounding the deaths of persons in their custody, and some had no system for requiring such reports. The lack of uniform reporting requirements made it impossible to ascertain how many people were dying in custody and from what causes, although estimates by those concerned suggested that there were more than 1,000 deaths in custody each year, some under very suspicious circumstances. . . .
>
> Since the enactment [of the Death in Custody Reporting Act] in 2000, the Bureau of Justice Statistics has compiled a number of statistics detailing the circumstances of prisoner deaths, the rate of deaths in prison and jails, and the rate of deaths based on the size of various facilities and so forth. But the most astounding statistic reported since the enactment of the [act] is the latest Bureau of Justice statistics report dated August 2005, which shows a 64 percent decline in suicides and a 93 percent decline in homicides in custody since 1980 [(Mumola, 2005)]. Those statistics showing a significant decline in the death rate in our Nation's prisons and jails since stricter oversight has been in place suggest that the oversight measures, such as the Death in Custody Reporting Act, play an important role in ensuring the safety and security of prisoners who are in the custody of State facilities.

Other congressional requests in enacted laws for new BJS data collection efforts include "a study of the criminal misuse of toy, look-alike and imitation firearms, including studying police reports of such incidences [and reporting] on such incidences relative to marked and unmarked firearms" (1988; P.L. 100-615) and the addition of questions on disabilities (1998; P.L. 105-301) and crimes against seniors (2000; P.L. 106-534) to the NCVS.

Of course, many such suggestions in legislative bills are never enacted. Prior to the establishment of the U.S. Department of Homeland Security, the proposed Barbara Jordan Immigration Reform and Accountability Act in the 107th Congress (H.R. 3231) would have markedly increased BJS's scope. In abolishing the existing Immigration and Naturalization Service and distributing its functions elsewhere in the Justice Department, it would have created a separate Office of Immigration Statistics within BJS. The bill passed in the House but did not advance beyond referral to the Senate's Judiciary Committee.

Examples of proposed new inquiries in bills introduced in the 110th Congress include H.R. 259, which would create "a task force within the Bureau of Justice Statistics to gather information about, study, and report to the Congress regarding, incidents of abandonment of infant children." This would involve "collecting information from State and local law enforcement agencies and child welfare agencies regarding incidents of abandonment of an infant child by a parent of that child," including "the demographics of such children and such parents" and "the factors that influence the decision of such parents to abandon such children." Similarly, H.R. 3187 would require BJS to "conduct a study to determine the extent to which methamphetamine use affects the demand for (and provision of) oral health care in correctional facilities," including statistical information on the financial impact of "meth mouth" treatment on corrections budgets.

guably, the agency's most underutilized collection, with its capacity to achieve its potential undercut by scarce resources, diminishing sample size, and—to a degree—a lack of innovation in analyzing and promoting the data. Though much methodological research was conducted in the early years of the survey and again during the late-1980s redesign, a major threat to the NCVS is stagnation in content and methodology.

- BJS's data series in *corrections* are a good and successful example of a well-designed and integrated system of collections, carefully delineating information to be obtained by different methodologies (personal interviewing and facility or administrative records) at certain time intervals for a range of facilities (prisons, jails, support agencies). The set of collections is designed such that "censuses" build the frame for subsequent, more detailed "surveys," in a way that is blurred in other areas of the BJS portfolio. Going forward, the challenge for BJS's efforts in the general area of corrections is one of expanding its coverage to include prisoner reentry issues, generally, and improving on its previous solid (but infrequent) studies of recidivism.

- BJS's work in *law enforcement* is hindered by a sharp and overly restrictive focus on management and administrative issues; its analysis of law enforcement generally lacks direct connection to data on crime, much less providing the basis for assessing the quality and effectiveness of police programs. It is also in the area of law enforcement, with the proliferation of numerous special-agency censuses and little semblance of a fixed schedule or interconnectedness of series, where the need for refining the conceptual framework for multiple data collections is most evident.

- Critique of BJS's work in *adjudication* is more a reflection on the general difficulty of measurement in the justice system than a criticism of BJS. Information systems in state court systems—and, indeed, the structure and jurisdiction of those courts—vary strongly in their accessibility and sophistication. The dominant impression that comes from looking at BJS's statistical series in the courts is that of the agency (with its data collection partners) doing the best it can with what it has. That said, there are numerous areas where improvement is needed: bolstering the adjudication series' basis in statistical sampling and patching important gaps in statistical coverage of the justice system funnel (particularly declinations to prosecute and out-of-court settlements) would dramatically upgrade the relevance and utility of BJS's data series.

Generally, our major concerns with the shape of BJS's data portfolio and our suggestions for improvement can be grouped under two broader themes that we discuss more completely in the balance of this section:

- BJS's data collections are mainly cross-sectional in nature and focus on relatively narrow, individual parts of the justice system. The coverage that is attained through these cross-sectional series is extensive, but *knowledge of longitudinal flows and progressions through (and out of) the justice system is comparatively scant.*

- Reflecting its cross-sectional nature, the major fault of BJS's data collection portfolio is not that any individual component is deficient but that *the portfolio lacks a sense of integration and cohesiveness.* New data series that have been added are generally important, but how they fit within broader conceptual frameworks and what they uniquely contribute to knowledge of crime and justice is not always well articulated.

3–F.1   Lack of Longitudinal Series

Finding 3.1: BJS currently gathers data about the criminal justice system but it does so on an institution-by-institution basis (police, courts, corrections) using varying units of analysis (crimes, individuals, cases) and sometimes varying time periods and samples. This approach provides good cross-sectional assessments of parts of the system, but makes it difficult or impossible to answer questions about the flow of individuals from arrest through eventual exit from the system. Yet people exit the system at many different stages in ways that are ill-understood but consequential for the effectiveness and fairness of criminal justice system processes. The cross-sectional approach misses the interfaces between the institutions, such as the large but unknown number of individuals who are arrested but not prosecuted.

The elegance of the funnel model of the criminal justice system is its longitudinal, progress-over-time structure and the way that it focuses attention on the system as a whole. However, as the coverage bars in Figure 2-2 make clear, *there exist no longitudinal data that actually follow the flow of individuals (or cases) through all steps of the system.* BJS develops and maintains a large set of data series that describe the basic features of the sequence of events in the criminal justice system. These data series cover various dimensions of law enforcement, prosecution and pretrial services, adjudication, sentencing and sanctions, corrections, and recidivism. A complex set of institutions operating at the local, state, and federal levels is covered by these data series, generating a wealth of information about the staffing and caseloads of those institutions. However, these data are generally cross-sectional "slices" of information at various points in the justice system that do not permit an assessment of experiences in the system as a whole, from initial contact (arrest) through placement in correctional supervision to, perhaps, reentry into the community.

BJS has developed one major continuing data resource that permits examination of some of the processes that influence felony case dispositions from case filing through sentencing. The BJS-sponsored Federal Justice Statistics Resource Center (see Box 3-4) uses the "defendant-case" as its unit of analysis, and the data files linked in this collection indicate the flow of these "defendant-cases" from step to step. However, these data have the significant limitation of including only federal matters, not those in the state courts which handle the vast majority of criminal and civil justice cases.

On a one-time basis in 1988, BJS mounted an effort that linked different stages of the justice system. BJS traced background and demographic information for about half of the cases in a sample of 33 large urban counties that involved a murder charge (brought in 1988 or earlier) and that were disposed during 1988. The resulting sample of 9,576 murder defendants contained information on the circumstances of the crime, the relationship between victim and defendant, and the disposition of the case; the sample was meant to be representative of the circumstances surrounding murder cases in the 75 most populous counties of the nation. The data were summarized by Dawson and Boland (1993), and the data set (Bureau of Justice Statistics, 1996) was used in subsequent analyses: on murder involving family members and dependents (Dawson and Langan, 1994) and the particular circumstances of spousal murder cases (Langan and Dawson, 1995).

The ability to follow persons from initial contact with the arrest through exit from the system is important for understanding the fairness and effectiveness of the criminal justice system at all stages of its operations. Hence, developing the longitudinal structure of BJS data should be a high priority. Although the creation of new data collections, explicitly designed to collect information on longitudinal flows, is one approach to improvement in this area, it is also important to note that some important progress along these lines can be made without new series, by working within the framework of BJS's existing data series and data archives.

Emphasize Flows in Current BJS Series

Within its existing data series, BJS could make strides to provide some empirical insight on gross flows in the system through improvements we suggest elsewhere in this report. BJS could more effectively use its surveys, particularly the NCVS, to examine points of contact throughout the justice system; the PPCS is a useful model in this regard by examining public interaction with law enforcement, but targeted modules could also query about experiences with adjudication or correctional systems. BJS's one-time effort to study victim, incident, and offender characteristics of murder cases that were adjudicated and disposed in 1988 in large urban counties (Bureau of Justice Statistics, 1996) is a possible model that could be considered in revis-

Box 3-4 The Federal Justice Statistics Resource Center of the Urban
Institute

The Bureau of Justice Statistics (BJS) contracts with the Urban Institute to maintain the Federal Justice Statistics Resource Center (FJSRC) website, an attempt to consolidate available data from federal agencies and courts. By combining series, the intent is to describe all steps of the processing funnel (see Figure 2-1) for suspects and defendants faced with federal charges. Significantly, the project does not cover processing in state courts, but it does attempt to make definitions consistent with those used in BJS's collections on state court processing.

The FJSRC takes a "defendant-case"—the combination of a defendant (either a person or a corporation) and a particular case—as a unit of analysis. The Urban Institute's FJSRC staff receive regular extracts from the case management systems of participating federal agencies, corresponding to different stages of the criminal justice process:

- *Arrest*—The U.S. Marshals Service Prisoner Tracking System includes arrests made by all federal law enforcement agencies (e.g., Customs and Border Protection, Bureau of Alcohol, Tobacco, and Firearms, and the Marshals Service itself) and bookings by the Marshals Service. Separate data are obtained from the Drug Enforcement Administration's Defendant Statistical System.

- *Prosecution*—FJSRC works with data from the Executive Office for U.S. Attorneys Central System and Central Charge files to create six analysis files: matters filed, matters concluded, cases filed, cases terminated, charges filed, and charges disposed.

- *Pretrial Release*—Data from the U.S. Probation and Pretrial Service System documents any pretrial hearings, detentions, and releases of federal defendants between the time of an initial interview to disposition in district court. FJSRC uses extracts from these data to form three analysis files: defendants interviewed, investigated, or otherwise entering pretrial services; defendants terminating periods of pretrial supervision; and defendants under active pretrial supervision.

- *Adjudication*—Separate analysis files on cases filed, cases terminated, and cases pending for each year are derived from the Criminal Master Files of the Administrative Office of the U.S. Courts.

- *Sentencing*—The U.S. Sentencing Commission's Monitoring Data Base is used to extract information on sentences reviewed under the terms of the Sentencing Reform Act of 1994; sentencing data may not be fully complete, because they are limited to cases obtained by the commission.

- *Appeals*—The Administrative Office of the U.S. Courts obtains docket information on appeals filed and appeals terminated from the U.S. Courts of Appeal.

- *Corrections*—Federal Bureau of Prisons data are processed to form annual analysis files for three cohorts: offenders entering prison, offenders imprisoned, and offenders released from prison. The Post-Conviction Federal Probation Supervision Information System of the U.S. Probation and Pretrial Service System is mined to produce files for three similar cohorts: persons entering active probation supervision, persons under supervision, and persons terminating supervision (whether successfully or unsuccessfully).

Though branded "a project of the Bureau of Justice Statistics," the FJSRC online presence is hosted on Urban Institute servers at http://fjsrc.urban.org. The site includes capability to construct simple tables based on individual analysis files for each year.

(continued)

Box 3-4 (continued)

Subpages on "Publications" lead to links to reports on the main BJS website, particularly the *Compendium of Federal Justice Statistics* (Bureau of Justice Statistics, 2006a), *Federal Criminal Justice Trends* (Motivans, 2006), and—most recently—*Federal Justice Statistics* (Motivans, 2008). Some reports using the Federal Justice Statistics Program, such as Sabol et al. (2000) on offenders returning to federal prison after a first release, are authored by Urban Institute staff (and credited as such) but are released as BJS bulletins or special reports, whereas others are prepared by BJS staff (e.g. Scalia, 1996, 1999, 2000, 2001).

---

ing the SCPS series, making at least occasional special efforts to follow cases forward (through disposition) or backward (to recover victim and incident data). Similarly, improvement of knowledge about the justice "system" as a whole would benefit from focused attention on "leaks" and diversions in the justice system model such as declinations to prosecute cases and out-of-court settlement arrangements; we discuss such issues in more detail below in Section 3–F.3. In Section 4–B.3, we discuss one other major effort that is within BJS's grasp for understanding longitudinal flows—making use of the criminal history record databases that it supports through the National Criminal History Improvement Program for research purposes.

BJS's correctional data collections have generally emphasized *stocks* of incarcerated populations: inmate counts and demographic breakdowns at annual or midyear levels and more detailed cross-sectional inmate-level information from the inmate surveys. Over the past 30 years, this stock information has been a critical policy interest as the growth in the correctional population has been propelled by increases in prison admission rates and increases in time served (Blumstein and Beck, 1999, 2005). These trends have involved a large rise in admission rates for drug offenses, and large increase in time served for violent offenses. An ideal set of correctional data series should yield high-quality stock information—specifically, yearly counts of the jail, parole, and prison population for each state, for detailed demographic groups.

> *Recommendation 3.1:* **BJS's goal in providing statistics from basic administrative data on corrections should be the development of a yearly count of correctional populations capable of disaggregation and cross-tabulation by state, offense categories, and demographic groups (age, race, gender, education).**

However, ideal corrections data would include at least as much attention as *flows* and transitions in the correctional population as it does stocks and levels. The current National Corrections Reporting System has been used to estimate admission and release rates (entry and exit) but not transition rates at each stage of criminal processing: arrest to conviction to commitment to prison to parole release (unsupervised status), and so on. It would be useful for BJS to explore ways to use its existing data (with, perhaps, slight modification) to regularly produce estimates of transition rates as a counterpart to its regular stock data.

> *Recommendation 3.2:* **BJS should produce yearly transition rates between steps in the corrections process capable of disaggregation and cross-tabulation by state, offense categories, and demographic groups.**

Although the corrections data are a prominent example of an area where greater emphasis on flows would be beneficial, the same guidance also applies to other changes in status in the justice system. These include filing of charges (transition from law enforcement operations to adjudication) and conviction (transition from court processing to correctional handling).

Facilitate Linkage in Existing Data Sets

In terms of improvements that can be made in the data-processing and archival process, it should be noted that BJS and its public data warehouse, the NACJD, have taken a number of steps to increase the utility and accessibility of the data. Through the creation of multiyear compilations for major series, they have made it easier to link BJS data sets over time and with each other to add value to the information. NACJD staff also developed a "crosswalk" file (Bureau of Justice Statistics, 2004d) that approximates the linkage between the FBI's ORI "geography" (law enforcement agency jurisdictions, used in UCR and National Incident-Based Reporting System data) and standard geographic boundaries; this file facilitates linkage of LEMAS, UCR, Census Bureau, and other data. Such steps to make it easier to work with BJS data files and facilitate rich analyses can and should be taken. Specifically, a standard-format NCVS could be assembled across the entire time series to facilitate long-term trend analyses of these data; the same could be done for other data series including the jail and prison inmate surveys or LEMAS. Moreover, linkage of individuals in the NCRP across years would be very useful in approximating a recidivism study; the public cannot do this because they do not have access to inmate identifiers but it could be done for BJS by the NACJD. The linking of individual-level data collections in the corrections area such as the NCRP or the inmate surveys to many of the facility-based data collections, such as the Census of Adult Correctional Facilities, would leverage the data in both series.

The information needed to create these linkages—linking units across time and collections—is generally available within the series' structures, but this and other information is not now available because of fear of violating confidentiality. To be sure, methods of releasing link-capable data sets that protect the confidentiality of respondents is a major challenge, along with the basic logistics of linkage. To be most useful, agreements must be reached between BJS and its data providers—particularly the Census Bureau—as to how to make linked data available to the public without making the process so irksome as to make it unworkable. Greater use of the Census Bureau's dedicated research centers may be useful in this regard, but the cumbersomeness of the Census Bureau's access process is discouraging to potential users; new approaches to these issues should be pursued.

> *Recommendation 3.3:* **BJS should explore the possibilities of increasing the utility of their correctional data collections by facilitating the linkage of records across the data series. For example, the ability to link records from the Recidivism Studies or from NCRP to the Census of Adult Correctional Facilities (CACF) would increase the ability to understand how correctional facilities contribute to recidivism.**

Develop Additional Panel Surveys

The ideal tool for studying longitudinal experiences in the justice system would be a data series directly designed for that purpose. However, the major impediment to creating such a series is the same reason why extensive longitudinal information does not exist in current data. A basic, logical need in order to approach an ideal measurement of experiences in the system is a systematic "tracking number" attached to a single person (or, more generally, actor) at all stages in the system. To be effective in studying all types of judicial resolutions, such a tracking number would have to be attached to a defendant at a very early stage and maintained. To get a handle on dynamics of the courts, a systematic tracking number assigned near the time of filing might be sufficient, yet a number would practically have to be assigned at booking in order to study the full pretrial mechanisms.

Such a tracking number does not currently exist; indeed, common identifiers generally do not exist between broad steps of the process (such as linking court records with later corrections records). This problem is exacerbated by state-to-state variation in the quality and completeness of electronic case management systems. The lack of a tracking number that facilitates use of a person as the unit of analysis—logging all charges attached to an individual and following what happens to those charges throughout the system—obviously hinders longitudinal studies. But it is also part of the basic flaw we have already noted in BJS's adjudication portfolio: the lack of a tracking number complicates the problem of selecting nationally representative samples of judicial proceedings. At this time, given the current level of automation at the various levels of courts, there is basically no alternative to examination of individual "jackets" or file records in the courts.

Though the lack of a tracking number is a formidable challenge, we suggest that assessment of the fairness and quality of the justice system would be substantially improved by longitudinal studies:

> *Recommendation 3.4:* **BJS should develop an approach to measure the experiences of individuals through the criminal justice system on a prospective, longitudinal basis, beginning as early as practicable in the process (arrest) and ending with their eventual exit (ranging from early dismissal of charge through completion of sentence).**

Practically, the most feasible approaches for developing panel studies that track the same individuals over time lie at either "end" of the justice system funnel model. On the input end, the flexible survey vehicle that is the NCVS could serve as the input to a follow-on study of crime victims' experiences. Very little research has made use of the current panel structure of the NCVS and its repeated interviews at the same household address; in part, this is due to the cumbersome structure of the data files as well as the use of the address as the unit of analysis rather than the individual. An add-on to the NCVS could serve as the starting point for a concerted effort to follow a sample of individuals (even if they move from an NCVS-sample household) over time for a pure analysis of their experiences with other parts of the justice system and any subsequent victimizations; current data systems based on the justice system funnel model tend to lose focus on victims of crime after the decision to report or not report incidents to police.

> *Recommendation 3.5:* **BJS should develop an approach to measure the victimization experiences of individuals on a prospective, longitudinal basis, beginning from a focal victimization and following the victim forward in time measuring subsequent victimizations and possible consequences of victimization. The NCVS may be used to recruit respondents to a panel survey of crime victims.**

BJS's survey of prison inmates could be the springboard to a parallel panel survey:

> *Recommendation 3.6:* **BJS should develop a panel survey of people under correctional supervision to understand how individuals move between institutional and community settings, and to understand the social contexts of correctional supervision.**

The respondents might enter the panel survey through the 5-yearly SIS-FCF. An initial survey could mirror and expand content in the existing inmate survey about the prisoner's history of offending and experience in the courts. Those sampled prisoners within, say, a year of their release date could be reinterviewed annually over the next 3 years. The panel component of the survey might begin with an interview immediately prior to prison release. After release, the survey might measure aspects of community supervision, correctional programming (including educational and vocational training received while under supervision in addition to drug programming), offending and other risky behaviors, victimization, housing, family relationships, and employment. With this population, survey attrition is a formidable challenge. The survey interviews might thus be linked to administrative records to provide additional information about further contacts with police and corrections. The panel would be refreshed with each new inmate survey.

3–F.2   Lack of Conceptual Frameworks

Our review of BJS data programs in this chapter demonstrates the wide range and breadth of BJS's data collections; for a small statistical agency, BJS's level of quality output is certainly impressive. That said, the second broad critique we raise concerning BJS's portfolio is that it lacks a sense of cohesiveness in some respects. The volume of BJS's data holdings is such that it can be overwhelming; the differences between different data series, and the unique value of a particular series to describe specific phenomena, are not always immediately clear. Parts of BJS's data portfolio have an unmistakable—and unfortunate—"scattershot" feel to them, whether because the interrlationships between series are not clear or because they seem to lack a well-expressed and common technical basis. Generically, we characterize these problems as lack of conceptual frameworks across the full suite of data series, within broad topic areas, and even within highly related series.

Develop a Blueprint of Existing Data Collections

BJS's mapping of its data series to the justice system funnel (Figure 2-2) that it presented to the panel is a particularly interesting document because it is a first step toward something that is absent from BJS's strategic plan (Bureau of Justice Statistics, 2005a) and other planning documents: an articulation and assessment of the extent to which steps and processes in the justice system are covered and explained by BJS data.

BJS's current strategic plan cites the number of data series that the agency produces—whether on an ongoing basis or as a special request as two of the benchmark measures by which BJS evaluates its own effectiveness (Bureau of Justice Statistics, 2005a:17):

> **Core and recurring series conducted**   The number of data collection series scheduled to be conducted during a particular calendar year and the number actually conducted.
>
> **Special analyses conducted**   BJS periodically conducts special collections or analyses for specific purposes, such as a collaborative effort

with other Federal agencies or fulfilling a congressional mandate. The number of special analyses conducted is maintained as an indicator of the utility of specific datasets for unanticipated requirements.

That BJS can cite its "maintain[ance of] over three dozen major statistical series" (Bureau of Justice Statistics, 2005a:1) as a measure of the agency's vibrancy and activity is certainly true, to a point. However, that specific claim in the strategic plan has a second clause—that the three dozen series are "designed to cover every stage of the American criminal and civil justice system"—that is not fully expressed, save for reference to the broad topic areas of the series. Specifically, the plan does not explain:

- The unique design features of specific data collections, their methodology (even in capsule form, such as personal interview, facility interview, or reference to facility or administrative records), or their capacity to describe actions at multiple stages of justice system processing;

- Goals for key activities and programs;

- Priorities across the programs, including the identification of coverage gaps or the development of specific data resources to fill them;

- Milestones for key programs, such as the implementation of census-updated samples in the NCVS or developments in securing corrections data from frequently nonresponding jurisdictions; or

- Evaluative criteria for "success" of individual data series and topic-area groups of data series (separate from evaluative criteria for the agency as a whole).

Accordingly, we recommend:

Recommendation 3.7: **To be useful, a BJS strategic plan must articulate a blueprint of interrelated data collection and product activities, including both current and potentially new data products. This blueprint would be used to evaluate new opportunities.**

For data collections that are in development, such a blueprint would detail the steps necessary to carry out the work and the timing of the steps; this would include any pilot or small-scale collection used to assess the feasibility of the full collection. Of course, we recognize that this concept for a "strategic plan" may not necessarily square with the templates for strategic plans that may be imposed on individual agencies by their parent departments or by government-wide standards. Specific nomenclature aside, what we recommend is that BJS expand on its mapping of data series to the justice system sequence of events as a first step in such a "blueprint" planning document.

Core-Supplement Designs for Major Surveys and the Scope of Law Enforcement Data

One of the class of designs we suggest as a possibility for the NCVS in our interim report (National Research Council, 2008b:90) is a core-supplement framework:

> Some surveys have a set of questions that are consistently asked of all respondents, sometimes labeled the "core." The full survey questionnaire contains core questions and a rotating set of supplement questions. Scheduled supplements allow topical reports from the survey, enriching the breadth of reports. These supplements might change over time, to reflect the changing nature of crime.

As we described in the interim report, the United Kingdom's Home Office has adopted a core-supplement strategy for the British Crime Survey (BCS), its analogue to the NCVS. Though we did not embrace a full adoption of the BCS methodology—in particular, we are reluctant to abandon the NCVS's repeated panel design, with multiple contacts of the same household, for the BCS's cross-sectional sample—we think that there is much to be gained from a core-supplement strategy. Accordingly, we reiterate and reaffirm a recommendation from our interim report (National Research Council, 2008b:Rec. 4.3):

> *Recommendation 3.8:* **BJS should make supplements a regular feature of the NCVS. Procedures should be developed for soliciting ideas for supplements from outside BJS and for evaluating these supplements for inclusion in the survey.**

(It follows that similar outreach beyond BJS for possible topic supplements—and funding for said supplements—would also be beneficial for BJS's other major data series.)

The SCS is a good example of the kind of topic supplements BJS should seek for the NCVS. The supplement provides information on a class of crime and violence that is a clear issue of continuing public concern, and so new SCS data are regularly awaited and analyzed in concert with related non-BJS data sources. Functionally, the SCS is a useful example because it wins the attention of (and a source of funding from) an executive department other than the Justice Department, and both BJS and the National Center for Education Statistics benefit from methodological and technical interchange in planning and designing the supplement.

Moreover, we think that a concerted effort to refine a relatively small core set of questions and build support for regularly scheduled supplements would benefit other BJS surveys besides the NCVS. Based on BJS's presentations to the panel, it is clear that BJS is giving the collection of law enforcement–related data a higher priority in its portfolio. We suggest that applying a core-supplement framework to LEMAS and related surveys would have the benefit of correcting the particularly scattershot appearance of the numerous special-agency censuses and surveys that BJS conducts, from campus law enforcement agencies to police aviation units. A core-supplement approach to the NCVS and the LEMAS survey, among others, would also be instrumental in permitting BJS to expand beyond the management and administration focus that prevails in its current collections.

On the first point—imposing an organizational framework on BJS's law enforcement surveys—our suggestions are intended to put a structure on BJS's law enforcement collections such as exists in its corrections data. The corrections data mix various approaches, asking administrative-type census queries of all agencies while collecting a much wider range of items in the inmate surveys. A major difference with the corrections data is that the basic unit of analysis changes between the collections—institutions or correctional agencies in the census-type studies, individual inmates in the surveys—whereas BJS's law enforcement surveys are all focused on institutions or agencies as respondents.

The problem with the numerous special-agency censuses is not that they are too costly; they are relatively low in cost because of their targeted nature and finite universes. Nor, to be clear, is it that their focus on smaller numbers of specialized agencies necessarily makes them less important. They provide value in filling in some major gaps left in the broader CSLLEA, notably the array of federal law enforcement agencies; collections such as the campus law enforcement surveys are important steps in more complete understanding of the prevalence and powers of security services maintained by nongovernmental institutions. Instead, the primary weakness of the special-agency collections is that the appearance of myriad, not-obviously-connected data series contributes to a perception that BJS is distracted and trying to do too much at the same time. Moreover, the special-agency inventories appear to serve two basic objectives—a genuine collection of information on policies, procedures, and resources of highly specialized agencies and a "frame-building" function to update and maintain an inventory of law enforcement–related offices—neither of which is fully articulated. Unlike the corrections arena, these law enforcement "censuses" develop survey frames as the basis for administering more detailed information on a representative sample yet do not obviously result in actual surveys; content is geared toward high-level characteristics and multiple-choice categories are geared more toward quick questionnaire completion times rather than furthering knowledge on law enforcement. And, from a frame-building perspective, the collections suffer from the appearance of being ad hoc measures, without (generally) a clear idea of if or when the information will be used in a later administration of the census or will feed into larger efforts such as the general LEMAS survey or the CSLLEA.

Finding 3.2: The multitude of scattershot "census" studies of specific law enforcement agency types (e.g., campus law enforcement, medical examiners, training academies) detracts from the appearance of a coherent measurement program in the area of law enforcement. Instead, the impression left is that these "censuses" are sporadic inventories or catalogs of particular agency types with no obvious internal consistency.

We suggest that the main LEMAS survey be recast as a core-supplement design—identifying a core set of questions to provide critical information on a timely basis while offering the flexibility to add supplemental questions to query local departments about emerging issues. However, in this case, "supplement" should be interpreted to include both expansion of *sample*—for instance, to expand the collection of information from medical examiner offices on a systematic basis—as well as expansion of content through topic modules. The effect of this effort would be the more aggressive development of a LEMAS "brand" (or, better, a general law enforcement statistics "brand") within BJS's portfolio, creating and reinforcing the position that the collections are part of a cohesive whole.

The expansion of sample to include specialized agencies of a particular type need not be annual—and, indeed, would likely not be annual; what we suggest is development of a calendar for these special collections (and, as appropriate, negotiation of ongoing sponsorship arrangements with other Justice Department and government agencies). By corollary, we suggest that if slots cannot be developed and found for one of these special-agency collections within a 5-year time horizon (and that there is little prospect for repeating the collection within 10 years), then its value is likely to be so limited that it should be discontinued.

Recommendation 3.9: **To maximize both utility and timeliness of information, the LEMAS survey should be conducted as core-supplement design in the context of a continuous data collection.**

Recommendation 3.10: **To improve the utility of censuses of law enforcement agencies, BJS should develop an integrated conceptual plan for their periodicity, publish a 5-year schedule of their publication, and integrate their measurement into the LEMAS as supplements.**

The adoption of a core-supplement strategy for LEMAS is consonant with a recommendation by a predecessor National Research Council panel, the Committee to Review Research on Police Policy and Practices, which recommended that BJS implement "an enhanced, yearly version" of the current LEMAS survey (National Research Council, 2004a:107). In particular, that committee noted that "the research utility of the survey would be enhanced by ensuring that a panel of consistently surveyed agencies be maintained within the framework of the survey sample." The committee further recommended attention to the quality of the census or directory survey that serves as the LEMAS sampling frame, and that BJS conduct follow-up studies of the validity of agency responses to LEMAS queries—all of which remain useful and sound suggestions that would helpfully develop a research and evaluation base for BJS programs (as we discuss further in Chapter 5).

The second point we make in suggesting a reorganization of the law enforcement surveys is that it is an important starting point in expanding the scope of BJS's collections in the general area of law enforcement. A look at BJS's portfolio (and Figure 2-2) leaves the unfortunate impression that the state of knowledge about "law enforcement" generally can be equated with the head- and resource-count totals in the LEMAS survey and agency censuses. Law enforcement statistics within BJS have been largely defined by the specific LEMAS data collection vehicle, and not a substantive definition of the activities and actors that constitute law enforcement. It would benefit BJS and the consumers of their data if law enforcement were defined substantively and all available data collections were used to illuminate this area of the criminal justice system.

Data of the sort produced by the current LEMAS and BJS's other law enforcement collections are valuable to states and localities for comparative purposes, for planning and for justifying requests for grants and assistance, and for maintaining accreditation as standing agencies. Information on other jurisdictions that have taken a particular approach—the use of tasers or other nonlethal-force technologies, for instance—make it easier to justify adoption of those technologies in similar jurisdictions. (Of course, to be most helpful in this regard, current data are more compelling than those that may be 3 to 4 years old.) Knowledge of other law enforcement agencies that have adopted particular new approaches also gives late adopters the chance to learn from the experiences of their earlier-adopting peers. Police agencies are also interesting from the organizational standpoint because they are not purely static; there are "births" and "deaths" among agencies that are important to understand, such as the merger of separate city and local departments in Charlotte and Mecklenburg County, North Carolina, and Indianapolis and Marion County, Indiana. They are also useful in providing context on current events or policy matters; for instance, current studies of the resources for and demands on campus law enforcement are important contexts in developing policy responses to college campus shootings. Likewise, documentation on the current status and operational backlogs faced by forensic crime laboratories is important for state and local policy makers as crime scene evidence weighs larger in the public imagination and in court proceedings.

That said, use of the existing LEMAS-type data for advancing knowledge about law enforcement and the administration of justice has generally been limited. One important exception is Wilson (2005), who used LEMAS data from 1997 and 1999, in conjunction with other survey data of police organizations, to analyze the adoption and implementation of community policing (COP) techniques. His models suggest that implementation was most strongly related to whether the police department is located in the western United States and the age of the police department as an organization; receipt of federal funding to support COP implementation was found to be only a weak predictor of effective adoption.

Useful though LEMAS data are for benchmarking and cross-agency comparison, BJS's challenge in furthering the law enforcement part of its portfolio is generating more information on law enforcement that is more relevant to its practitioners than a narrow focus on management permits. A characteristic of BJS's reports on its LEMAS-type data that is particularly telling is that they consistently stop short of drawing linkages to *crime* data; agencies are also asked relatively few questions about counts or characteristics of incidents, with questions oriented more to the presence or absence of policies. To be sure, this sidesteps the major problems of assuming an evaluative (or even regulatory) role of individual departments, but it also causes BJS reports to be silent on the most basic notions of *effectiveness* of police policies or personnel decisions. The sharp management focus—absent a connection to crime data—thus generates numerous facts but not always insights. For example, the report of the most recent campus law enforcement census (Reaves, 2008) reveals the universities with the largest number of sworn officers and the degree of implementation of blue-light emergency phones and in-field computers, but it makes no attempt to assess the relationship between agency staffing levels and either the number of reported crimes or service calls. Though it does usefully devote a page to the mechanisms by which campus crimes are reported to the U.S. Department of Education under the 1990 Clery Act,[16] it raises but does not quantify some interesting features such as the nature of relationships and interactions between campus

---

[16]Provisions requiring that postsecondary education institutions compile and disclose regular statistics on campus crime and security were first written into law in 1990's Crime Awareness and Campus Security Act (P.L. 101-542). Institutions are formally required to produce annual counts of crimes reported to campus security authorities or to local police agencies, whether the incidents occurred on campus, on noncampus buildings or property, or on relevant public property. These disclosures, and summary statements of security programs, are required as a condition for participation in federal financial aid programs. In 1998, the crime reporting provisions were renamed the Jeanne Clery Disclosure of Campus Security Policy and Campus Crime Statistics Act in memory of a Lehigh University freshman who was murdered in her campus residence hall room in 1986. The Clery Act is codified at 20 USC § 1092(f) and the U.S. Department of Education's rules for compliance with the act are promulgated at 34 CFR § 668.46. See U.S. Department of Education, Office of Postsecondary Education (2005) for additional detail.

law enforcement agencies and the state or local police forces within their area.

One example of a law enforcement data collection that could address a wider range of issues about the effectiveness of police work and the nature of police interactions with the public is that suggested by our predecessor Committee to Review Research on Police Policy and Practices (National Research Council, 2004a:163, 164). That panel recommended that BJS be given support to "develop and pilot test in a variety of police departments a system to document information applications of police authority." Such a system would provide data on police activities that stop short of invoking the criminal process (and the later stages of the funnel model), such as "simply making their presence or interest known to potential troublemakers, stopping and questioning them, persuading, advising, commanding, or threatening them, or referring problems to other agencies." The committee described the difficulties involved in creating such a system as "truly daunting," but nonetheless noted its potential value in developing a complete picture of police activities. A related direction for expanding coverage of policing activity more generally—and a possibility for a LEMAS supplement as an initial step—is to collect data on the use and extent of private security agencies and processes.

Suggesting broader and more detailed data collection from law enforcement agencies is easy, but implementing such a suggestion is far from easy. A Justice Research and Statistics Association (2003) summary of a series of focus groups organized by the Illinois and Pennsylvania state Statistical Analysis Centers identified four principal and perennial obstacles to "buy-in" by law enforcement agencies to wider data collection efforts:

- Inadequate resources for departments to assemble responses and comply with multiple data collections;

- Increased demands on time;

- Fear of negative publicity, particularly if new data are not strictly comparable to old data; and

- Continual changes in direction in collection and use of data, and proliferation of data collection requests to address the "next high visibility problem."

Cognizant of these constraints, we suggest the creation of a major, new law enforcement–related data set—but one that draws from existing resources—in Section 4–C.4.

Short of dramatically expanded collection directly from law enforcement and increasing the number of survey questionnaires that departments are expected to complete, there is much that BJS can do to expand its data on policing issues. Part of BJS's work in the law enforcement area should be making more effective analytical use of existing data systems—the measures collected by the FBI's UCR program and BJS's own NCVS—to inform law enforcement. For example, the Law Enforcement Officers Killed and Assaulted data collected by the FBI could be used in concert with LEMAS data to say more about deaths of and assaults on law enforcement officers. (The same might be said about the Deaths in Custody program described in Section 3–B, but that data collection is very recently begun and it may need further development.) BJS should also seek ways to exploit relevant non–Justice Department data in its analyses, including the aforementioned campus crime data compiled by the U.S. Department of Education.

However, returning to a main point of this section, a critical area for BJS to improve its information on law enforcement is making better use of NCVS to study related issues through structured and recurring supplements. Using the NCVS to study law enforcement is decidedly not a novel concept; indeed, the original statutory authority to start the National Crime Survey in the first place was a clause approving the collection of data on "the condition and progress of law enforcement" in the United States. Generally, Lynch (2002:62–63) argues that the NCVS "can tell us a great deal about the performance of the police industry and citizens' perceptions of it." In particular:

> The [NCVS incident form] includes questions on whether the police found out about the victimization incident, and, if so, whether they responded when called, their response time, and the various activities they engaged in at the scene. These activities include taking a report, gathering evidence, interviewing witnesses, and notifying the victim about further processing of the case. . . . [With this detail, NCVS] data can be used to identify the subpopulations and situations involved when police mobilization and service (or the perceptions of police service) differ. Among the most important distinctions to be made is the type of victimization that prompted the call for service.

The NCVS has the added advantage of allowing analysts to "define their own classes of crime"—such as "domestic violence, crime at school or at work, crime in the neighborhood, crime in public places, interracial crime, and intraracial crime"—that may be "more meaningful" than the common UCR classifications (Lynch, 2002:63). Hence,

> Some of the specific issues that [NCVS-based analyses of police issues] could address include:
> - Changes in the percent of criminal victimization reported to the police by racial and ethnic group and type of crime, with important attributes of the crime (e.g., degree of injury or amount of loss) and the victim (e.g., age) held constant.
> - Changes in the percent of reported criminal victimization events to which the police responded by sending an officer who made contact with the victim.

- Changes in the activity of the police at the scene and afterward by type of crime and demographic group, holding constant other relevant attributes of the event.
- Changes in the recovery of stolen property by demographic group, type of theft, and other attributes of the victim and the offense.
- Changes in the outcomes of victimization events, e.g., injury or loss, by whether the police were mobilized and by their actions after mobilization.
- Generic area estimates of mobilization and service that would identify areas according to their size and position in the metropolitan area, e.g., towns of 10,000 to 25,000 people outside of a large city, or central cities of between 50,000 and 75,000 people.

BJS's fielding of the PPCS is arguably its richest and most probing collection related to law enforcement behaviors and actions (we discuss the supplement further in Section 5–A.2). The use of the NCVS as a "citizen's survey" and an indirect measure of the effects of justice system actors on the public at large should be further developed.

> *Recommendation 3.11:* **The NCVS (and its supplements) should be more effectively used as a tool for studying law enforcement, both in terms of the types of crime that are reported (and not reported) to police and the action that results from the reporting of a crime (e.g., the Police-Public Contact Survey).**

Enhance Technical Framework Within Series—Sampling and Adjudication

A basic trait, and problem, with BJS's data collections in adjudication can be phrased very simply:

> **Finding 3.3:** BJS's current approach to data collection in adjudication lacks an effective basis in sampling.

Blunt though this statement is, we do not intend it to be interpreted as being unduly harsh. As we noted above, the reason for this lack of an effective basis in sampling is fairly clear: The collections developed from having to work with select jurisdictions for which records could be made available for analysis. BJS and its data collection providers must continue to work within the confines of available records and the confines of court information processing systems that may vary greatly within and between states.

To their credit, BJS and its providers are candid in their methodological notes and reports about the design of the court record collections. In particular, NCSC attached a useful graphical device—miniature state-level maps with colored shading indicating those states providing the relevant records—to every trend line in its extensive reports from the ongoing Court Statistics Project (LaFountain et al., 2007). Convenient though this is (and certainly superior to extended footnotes), flipping through the multiyear trends in caseloads for particular types necessarily means coming across dispute types with near-complete (aggregate counts of incoming cases by year or counts of judges, covering over 40 states) and minimal coverage (local ordinance violations, reported by about 5 states).

Where BJS has been able to design collections in adjudication, the combination of available resources have yielded designs whose representativeness is questionable at best. The "sampling" scheme of the SCPS series is such that it is intended to be representative of "felony court filings during the month of May in even numbered years from 1990–2004 in the 75 most populous counties in the United States" (Bureau of Justice Statistics, 2007g:4). That is to say, SCPS data are not nationally representative and the program's design—undoubtedly driven by the ability and willingness of jurisdictions to participate—makes it difficult to make generalizations to all felony filings. Though the most populous counties account for the bulk of felony filings, it is unclear how representative they are of the complete national experience. Moreover, the arbitrary selection of May as the target month raises the possibility of seasonality or other temporal effects in case filing that might make May unrepresentative of the rest of the year.

Variety in the nature and development of state court record systems is a long-standing concern and an obstacle that cannot be wished away overnight. That said, processing systems continue to develop, and BJS and its partners need to be aware of the state of development in those systems so that samples of courts (and their records) can be chosen more rigorously, avoiding potential biases that may be induced by overemphasizing states where access to records is convenient as well as those that may be due to temporal effects in filing.

> *Recommendation 3.12:* **As court records become more accessible through computerized case management systems, BJS should implement more rigorous methods of probability sampling in its adjudication series.**

In our assessment, BJS currently has good access to state court systems through its collaboration with NCSC and other data providers, as well as through contacts through the BJS-funded state Statistical Analysis Centers. Improvements in the adjudication series depend on cultivating and extending those partnerships.

> *Recommendation 3.13:* **To inform future revisions to its adjudication portfolio and to more efficiently acquire and work with court data in the future (including longitudinal analysis), BJS should develop a research program to build representative samples of courts and to assess strategies for collection of case records.**

The steps that BJS has taken to redesign the SCPS program during 2009—with the explicit goal of improving the sampling structure from the current month-of-May-for-large-counties plan, and ideally supporting sub-national estimation—are very heartening in this regard. In recent years, it was unclear whether BJS or its data collection contractor, PJI, had made a concerted effort to assess the ability of local jurisdictions to participate (and sustain participation) in SCPS data collection; such an assessment should surely accompany a reconceptualization of SCPS. In turn, this work could be instrumental to the development of a wider jurisdiction-based data collection system linking cases or persons across decision points in time.

Developing a more rigorous sampling for primary data collection in adjudications is a top priority. In line with the principles and practices expected of statistical agencies (see, in particular, Section 5–B.7), a secondary but essential step for BJS is to evaluate the quality of the information it obtains through SCPS and related collections. Particularly to the extent that case characteristics are coded from documentary files by court administrative staff, completing forms that describe the transactions regarding an individual defendant, it would be useful to implement periodic evaluation and verification procedures. SCPS documentation does not indicate the use of audits or spot checks for completeness and accuracy of provided records, and such information is important to building the credibility of data series.

3–F.3 Improving Statistical Coverage of the Justice System

We have already noted in Section 2–C certain missing topics in BJS's statistical coverage of events in the justice system, and our comments on building a framework for BJS's law enforcement collections in Section 3–F.2 also point out areas in which the topic coverage of the agency's existing surveys can be improved by adding supplements. We close our assessment of BJS's overall portfolio by noting areas within the scope of the agency's existing collections where expanded coverage and greater depth would be beneficial.

However, the same point that we made in introducing Section 2–C applies here: A discussion such as this risks descending into a "wish list" for a statistical agency facing mounting costs and a flat budget, for there are always areas of interest under the general heading of crime and justice where more data are highly desirable. The intention here is to encourage strategic thinking on some particularly high-priority areas while emphasizing the ways in which existing data collections can be used and adapted. Such efforts, we believe, would expand the constituencies for BJS data, inform policy, and draw BJS into valuable relationships with other public agencies.

The two areas we discuss in more detail in this section are acute needs where objective information from BJS would have high value. One stems from the basic conceptual flaw in the funnel model of justice processing, which tacitly treats correctional supervision as an ending state; however, the challenges of prisoners exiting corrections and reentering the general population will be critically important to policy makers in the coming years. The other is one of the most substantial "leaks" in the funnel: cases that drop out of the system because prosecutors decline to pursue legal proceedings or when settlements are reached out of court.

Reentry and Recidivism

BJS correctional data have made vital contributions to research on trends and disparities in criminal punishment in the United States and on the consequences of increasing incarceration rates. One research literature has used the NCRP/NPS series (see Section 3–B.1) to study trends in imprisonment and corresponding effects on the economy. National time series of imprisonment rates have been associated with trends in crime and the economy (this research is discussed by Chiricos and Delone, 1992, and Harcourt, 2006); an alternative design has examined panels of states (e.g., Bridges and Crutchfield, 1988; Jacobs and Carmichael, 2001). Similar research has been directed at the social dimensions and consequences of incarceration rates, studying prison admission rates calculated from the NCRP for panels of detailed demographic groups (Western et al., 2006). National and state imprisonment series have also been used extensively to study the effects of incarceration on crime rates; recent contributions include Levitt (1996), Useem et al. (2001), and Johnson and Raphael (2006a).

Another thread of research has shifted from studying the scale and effect of aggregate levels of imprisonment to examining variation in imprisonment in the population. BJS has a long-standing interest in racial disparities in incarceration, publishing long historical series on state-level prison admission rates for blacks and whites (Langan, 1988), and regularly publishing imprisonment rates for blacks, whites, and Hispanics. The Surveys of Inmates of State and Federal Correctional Facilities have been used to construct detailed incarceration rates by age, race, sex, and levels of schooling (Western, 2006). A widely cited BJS study has extended the usual focus on incarceration rates, using the inmate surveys to estimate lifetime risks of incarceration (Bonczar and Beck, 1997; Bonczar, 2003). Life-table estimates of these lifetime risks showed that African American men, at current levels of incarceration, face a 28 percent of chance of going to state or federal prison. The BJS report on lifetime risks spurred other research using the Surveys of Inmates that estimated more detailed figures for specific birth cohorts and at different levels of schooling (Pettit and Western, 2004). The analysis has been extended further to study children's risk of parental incarceration (Wildeman, 2009).

Against this backdrop, the two recidivism studies conducted by BJS in 1983 and 1994 have been influential in structuring research on the relationship between crime and incarceration—and hinting at a major looming challenge for policy makers. The studies, which followed two cohorts of prison releasees in selected states for 3 years and recorded their subsequent patterns of arrest and reincarceration, demonstrated that around 60 percent of those coming out of state prison were rearrested within 3 years of prison release. These special studies of recidivism yielded several widely cited BJS reports. The recidivism microdata were also made publicly available, and have been widely studied by researchers and policy analysts (e.g. Solomon et al., 2005; Travis and Visher, 2005).

About 700,000 people are now released annually from state and federal prison (Sabol et al., 2007:1). Another 5 million are currently under some kind community supervision either on parole (800,000) or on probation (4.2 million) (Glaze and Bonczar, 2007:2). The most recent estimates of recidivism, from 15 states in 1994, suggest that about two-thirds of prison releasees will be rearrested within 3 years, and a quarter will be reincarcerated with a new sentence (Langan and Levin, 2002). The significant growth of imprisonment rates over the past several decades has highlighted the policy and social science challenges presented by historically large cohorts of released prisoners and has increased the urgency of developing techniques for community reintegration in order to deter recidivism.

Between the 1994 recidivism study and BJS's recent reorganization to elevate "recidivism, reentry, and special projects" as a program priority, BJS's direct role in reentry issues has been outpaced by research efforts mounted by other units in the Justice Department and external researchers. For example, NIJ funded a major evaluation of the Serious and Violent Offender Reentry Initiative (SVORI), a collaborative grant program funded by five cabinet departments that instituted reentry programs in 69 sites around the country.[17] SVORI programs included in-prison training programs prior to release as well as postrelease programs; programs included substance abuse and mental health treatment, housing assistance, and faith-based programs. The multisite evaluation is intended to identify effective approaches; findings from the evaluation in progress are described by Lattimore et al. (2005) and Lattimore et al. (2004). Other researchers have begun to intensively study the consequences of incarceration for the employment, family, and health outcomes of men and women released from prison; see, for example, Pager (2003), Lopoo and Western (2005), Kling (2006), and Johnson and Raphael (2006b).

---

[17]Specifically, the Departments of Education, Health and Human Services, Housing and Urban Development, Justice, and Labor contributed funding to SVORI programs. The NIJ-funded Multi-site Evaluation of SVORI Programs was administered by RTI International and the Urban Institute.

Though its direct work in the field has been limited to date, BJS can and should be a major source of quantitative information on prisoner reentry, recidivism, and community-based supervision issues. Nearly 80 percent of prisoners are released to community supervision, and so data collections on prisoner reentry would significantly extend the coverage of parolees. More than this, regular data collections on reentry and recidivism would advance the core charge of compiling statistics on crime and analyzing its correlates. In the current period of historically unprecedented incarceration rates, reentry and recidivism data would also offer valuable information about the social impacts of incarceration, an area now regarded as of pressing policy significance. Data collection efforts should evolve in response to social trends and their policy context; developing a program on reentering prisoners reflects the new reality of large cohorts of releases, and the significance of these cohorts for crime and social cohesion in the general population.

Recommendation 3.14: BJS should mount a feasibility study of the flow of individuals between correctional supervision and community settings. Repeated interviews of samples of about-to-be-released prisoners that track their successes and failures in reintegrating with the community would enhance understanding of this critical policy issue.

In early 2008, the Second Chance Act became law, instituting and authorizing a wide variety of reentry programs. Included in its provisions is a section defining a role for BJS data collections; see Box 3-5. The act permits BJS to routinize its earlier recidivism studies, calling for BJS to conduct them on a triennial basis. In the legislative context, of course, authorization is different from appropriation, and the act's language that BJS "may conduct research" rather than "shall" stops short of a direct mandate. Still, the act is an important signal that correctional programming will likely gain renewed support over the coming decade, and that a BJS role is both expected and required. The act also authorizes NIJ to carry out research studies along these lines; exactly how large efforts such as a major recidivism study would be divided between and administered by a statistical agency (BJS) and a research agency (NIJ) would need to be carefully determined.

Although BJS's coverage of custodial correctional populations is strong and it has a fairly complete picture of annual flows of persons moving in and out of prison, its current coverage of the population released from incarceration is seriously incomplete at present. Hence, higher priority to these issues and congressional authorization are both welcome developments. As BJS approaches the problems of recidivism and prisoner reentry—significant frontier issues that should weigh heavily in the agency's strategic planning—we suggest some possible topics that form part of this planning and shape expanded data collections:

Box 3-5 The Second Chance Act of 2007

Signed into law on April 9, 2008, and codified as 42 USC § 17551, the Second Chance Act of 2007 directs that the Bureau of Justice Statistics (BJS):

> may conduct research on offender reentry, including—
>
> (1) an analysis of special populations (including prisoners with mental illness or substance abuse disorders, female offenders, juvenile offenders, offenders with limited English proficiency, and the elderly) that present unique reentry challenges;
>
> (2) studies to determine which offenders are returning to prison, jail, or a juvenile facility and which of those returning offenders represent the greatest risk to victims and community safety;
>
> (3) annual reports on the demographic characteristics of the population reentering society from prisons, jails, and juvenile facilities;
>
> (4) a national recidivism study every 3 years;
>
> (5) a study of parole, probation, or post-incarceration supervision populations and revocations; and
>
> (6) a study concerning the most appropriate measure to be used when reporting recidivism rates (whether rearrest, reincarceration, or any other valid, evidence-based measure).

In debating the act on April 12, 2007, bill sponsor Edward Kennedy (D-Mass.) cited existing BJS corrections data series in arguing for the bill's merits (*Congressional Record*, pp. 4430–4431):

- "Large prison populations and high recidivism rates place heavy burdens on prisons, communities, and taxpayers. Of the 2.2 million persons housed in prisons today—an average annual increase of 3 percent in the past decade—97 percent will be released into the community. Overcrowding continues to plague the system. State prisons are operating at full capacity and sometimes as much as 14 percent above capacity, and Federal prisons are 34 percent above capacity. In 2005, prison populations in 14 States rose at least 5 percent. Recidivism and inadequate reentry programs add to the problem. Over 600,000 prisoners are released each year, but two-thirds of them are arrested again within 3 years."

- "According to a recent Bureau of Justice Statistics report, of the approximately 50 percent of prisoners who met the criteria for drug dependence or abuse, less than half participated in drug treatment programs since their admission to prison."

- "The Bureau of Justice Statistics reports that only 46 percent of incarcerated individuals have a high school diploma or its equivalent. The limited availability of education and vocational training programs exacerbates the problem. Only 5 percent of jail jurisdictions offer vocational training, and 33 percent of jurisdictions offer no educational or vocational training at all."

- *Reinstitute sample survey of probationers and parolees:* As described in Section 3–B.6, BJS's current coverage of persons under community supervision is limited to administrative data collected through the Annual Probation and Parole Surveys administered to supervising agencies. These surveys provide counts of probationers and parolees disaggregated by race, sex, and offense category. However, these administrative data are extremely limited for studying reentry and recidivism; they provide little information about the conditions of supervision or the circumstances of success or failure in individual cases, and they yield no information about past criminal history. Actual person interviewing with a sample of probationers has only been conducted once by BJS, in 1995, yet it is this kind of rich information on the experiences of those who have already reentered the community that would be most valuable in shaping emerging reentry strategies. Occasional surveys of the parole and probation populations would improve understanding of the process of reentry, recidivism, and successful reintegration into the community.

A reinstituted representative sample of probationers and parolees should elicit data on risk factors (schooling, social background, health status, and criminal history, for example), the conditions of community supervision and its intensity, and participation in assistance programs. Information on the spatial distribution of parole and probation populations (e.g., distance from previous "home" communities prior to incarceration and limitations on geographic mobility) would also likely advance understanding of recidivism and reentry. The current administrative data also do not speak to the circumstances of arrest, revocation, conviction, or incarceration of parolees and probationers. Regular statistics are collected on arrest and recommitment to prison, but there is little detail on technical violations or the administrative procedure of parole and probation revocation; a survey program, complemented by revision of the content of the administrative survey questionnaire, could help fill these gaps.

Survey costs for this population are likely to be substantial, and meeting these costs will likely require long-range planning and partnering with other agencies. Still, because the population of released prisoners is now so large, agencies within the Departments of Health and Human Services, Education, Housing and Urban Development, Labor, or Veterans Affairs (as well as other statistical agencies such as the Census Bureau and the Bureau of Labor Statistics) may have shared interests in the probation and parole populations and be a source of input and funding.

- *Routinize the national recidivism studies:* As acknowledged in the debate on the Second Chance Act and exemplified by a direct request in the act's language, the BJS recidivism studies of 1983 and 1994 have made major contributions to understanding of postprison experiences of state prisoners and should be conducted on a more regular basis. Recidivism studies are also important by significantly expanding the empirical scope of the BJS data collections by including those who have completed sentences and are no longer under any kind of supervision at the time of their return to incarceration. The previous recidivism studies were based on a large and complex record linkage effort that joined correctional records to arrest and court data. Although exemplary, the paradigm could be pushed further with either survey interviews or linked records for several terms of prison incarceration. (This was typically infeasible in the recidivism study because of the relatively short 3-year follow-up period.) Survey data from released prisoners would be particularly informative about the social context of recidivism, describing in greater detail the economic and social situation of those coming out of prison.

- *Measuring jail flows:* A point inherent in our suggestion to improve measurement of transitions in BJS's existing corrections data (Section 3–F.1) is worth reemphasizing here. Not as much is known about those persons passing through the nation's jail system as BJS's data reveal about the federal and state prisons. While about 700,000 people annually enter and exit prison, some 10 million people are estimated to pass through local jails in a given year. Although jail incarceration is likely common for released prisoners and releasees might cycle in and out of jail before returning to prison, there are no national statistics to document the pattern. Likewise, relatively little is known about the frequency with which the same individuals go in and out of the jail system—for instance, whether frequent contacts with law enforcement and numerous short spells spent in jails or police lock-ups constitute a de facto form of community supervision.

- *Alternative approaches to studying the unsupervised population:* Just as prison sentences expire, so too do sentences of probation or other community supervision. As rules for probation supervision change (in part due to state efforts to grapple with growing costs of corrections), the size of the released and unsupervised population is growing. Some may argue that those who have "maxed out" of prison or corrections supervision fall outside the statistical jurisdiction of BJS, but the point remains that the unsupervised population may be at high risk of rearrest, and their criminal histories place them at risk of an array of diminished life chances. If BJS data collections are going to be significantly informative about recidivism and reentry, the released unsupervised population should be contemplated as targets for new data collection. Unlike parolees, the unsupervised present acute difficulties for data collections. Here, linking criminal justice to noncriminal justice (say, social welfare) administrative records, may provide one promising path for data collection.

- *Studying the demographic significance of the penal system:* While the topics for BJS strategic planning might speak to the challenges of understanding recidivism and reentry, they also speak to the broader demographic significance of the penal system. As an institutionalized influence on the life course and spatial mobility, the penal system now commands a large demographic influence that is mostly hidden from the nation's statistical system. Though prisoners and other institutionalized populations were counted in the 2000 decennial census and are included in the group quarters component of the Census Bureau's American Community Survey, nothing is known of the geographic areas from which they originate or much about their lives immediately before institutionalization. By trying to capture the flow of people into and out of prison, and their return, the reentry and recidivism perspective highlights the significant influence of the prison on basic population processes.

The ideas discussed here suggest a variety of new BJS activities focused on formerly-incarcerated men and women. In the current climate of tight budget constraints, and short of the authorizations in the Second Chance Act actually yielding significant appropriations of new funds, the birth of extensive new data collections seems unlikely. Still, BJS should think opportunistically about (a) partnering with other agencies, (b) linking records across databases, and (c) augmenting existing administrative surveys of probation and parole agencies.

First, BJS should study the possibility of partnering with other statistical agencies. Such partnerships would provide two kinds of benefits. Items about involvement in the criminal justice system could be added to household and other surveys, expanding understanding of the reach of the justice system in the noninstitutionalized population. For example, questions about prior arrests or incarceration could be asked of the large samples in the American Community Survey, the Current Population Survey, or the NCVS. Because of the missions (and already large scope) of these surveys, such additional queries would not be highly detailed; however, they would provide general indicators at the national level and present the opportunity for some disaggregation by geography and demographic subgroups. In return, the BJS is also uniquely placed to provide detailed information about the institutionalized population. Questions about education, health status, aging, or demography, for example, could all be of interest to other agencies that have largely focused on the noninstitutionalized population. We return to this point in Section 5–B.11.

Second, record linkage holds great promise for expanding understanding of how people move in and out of institutional settings, pass through the formal labor market, and use social services. Those released from custodial supervision may be easier to track through their contacts with criminal justice and other public agencies than through surveys. Thus linking administrative records may yield special benefits in the study of recidivism and reentry. Such a system could connect, for example, records from different files of the NCRP, providing longitudinal records of movements from prison to parole, and return to prison. In contrast to the special studies of recidivism among release cohorts in 1983 and 1994, linked records of NCRP would provide an automatic and ongoing measure of reimprisonment and successful parole completion. In its most ambitious implementation, a system of linked records could join criminal justice to social service data such as unemployment insurance records or welfare enrollment. A broad linked system of administrative records would provide help to place released prisoners in a much broader social context, providing measures of their legitimate earnings, poverty status, and use of social services.

There are two main obstacles to exploiting the potential of record linkage for gathering data on recidivism and reentry. Severe practical difficulties are associated with matching records from different databases for the same individual. Because identifiers differ across databases, record linkage is expensive and prone to error. A unified system of identifiers for a range of databases—say, all BJS correctional microdata collections—would unlock the potential of record linkage. The practical challenges to a unified system are substantial but we urge BJS to explore concrete steps in this direction. The other obstacle to large-scale record linkage, particularly linking criminal justice to social service records, relates to privacy protections. Because of the sensitivity of the linked data, BJS and cooperating agencies would need to take special steps to protect the confidentiality of records. Some kind of institutional review, monitoring data security and research data centers, may provide a process for ensuring the privacy of individuals recorded in a linked system of administrative records.

Finally, administrative surveys of correctional, probation, and parole agencies could be expanded to explicitly incorporate policy interest in recidivism and reentry. Administrative surveys have been relatively inexpensive and accurate sources of counts of different correctional populations. The surveys could speak more directly to interests in recidivism and reentry by obtaining more detailed information about program participation and conditions of supervision. Enrollment counts of correctional populations in specific programs, and program spending and staffing information would help measure the resources applied to reintegration and criminal desistance. Some of this information is already reflected in surveys of correctional institutions, though spending on different categories of correctional programs is largely unmeasured. To advance understanding of the criminal justice system as an institutionalized influence on population processes (births, deaths, and migration), the administrative surveys could usefully collect more detailed demographic data. In addition to information about race and sex, which is currently reported for the prison, probation, and parole populations, a more detailed survey could obtain population counts by race and sex for given age and education groups. Spatial data describing, say, counties of origin and destination for entering and released prisoners would also help map the spatial distribution of the criminal justice system's reach into the community; the accuracy of those data would, of course, have to be evaluated.

Understanding Prosecution and Declination Decisions

The adjudication phases of the crime funnel model are areas of major "leaks" that are not well understood or measured. The percentage of cases that actually go to trial (enter the formal court system, where they might be more readily followed and tracked) can be small, and that percentage can vary strongly by jurisdiction and by type of court. Mechanisms for resolutions and alterations through "bargaining" at various stages are not well understood or studied: "charge bargaining" between attorneys prior to filing, "plea bargaining" as alternative to trial, "sentence bargaining" during or after trial.

The prosecution function or component in the funnel includes the charging decision (including plea bargaining), filing decision, the pretrial custody decision, tests of evidentiary strength, and all pretrial motions (e.g. discovery). In addition to information on these decisions, a useful statistical system describing the prosecutorial function would contain data on the social organization of the prosecutorial and the defense process. This would include resources, such as the number of staff, but also the way in which those staff are assigned and organized. Is the chief prosecutor elected or appointed? Is the staff specialized by stage of litigation or crime type? Is the indigent defense bar staffed by public defenders or court-appointed attorneys? Since the nation has a state and federal justice system, statistical systems describing the prosecutorial function would address both levels.

Currently BJS describes the state-level prosecution function with two different but related data collections—SCPS and NPS—which we summarized in Section 3–D; federal prosecution is described by the Federal Justice Statistics Program. There was a one-time survey of indigent defense in 1999. Some observations on the ways in which these collections cover (and do not cover) important parts of the prosecutorial function follow:

- The SCPS series does not cover a number of the decisions included in the prosecutorial function. The most important omission is the declination decision wherein the prosecutor decides not to file charges on arrests brought by the police or from some other source. This decision is not reviewable by anyone, and it is the single greatest source of prosecutorial discretion. Commenting on 1996 data from the NJRP that preceded SCPS, Forst (2000:26) lamented that:

  > [The program] gives no information about cases rejected or dropped by prosecutors; more than a few people might like to

know why over 80% of all arrests for motor vehicle theft fail to end in conviction, and why about 60% of all arrests for robbery and burglary fail as well.

On one hand, since the data collection in SCPS begins at filing, all of the decisions made prior to filing are lost. On the other hand, many important decisions made after filing are captured in this data collection. For those cases filed, it is possible to assess the amount of charge mobility that occurs from arrest to filing, through adjudication and sentencing. There is also extensive information on pretrial custody decisions and status, time that it takes to complete stages of processing, as well as some criminal history data.

- The key limitation of the NPS in understanding trends in prosecution is that it is to prosecution what the LEMAS survey is to law enforcement: a strictly partial look at basic administrative and management information. As an establishment survey, it has the same potential respondent selection problems (effects) as other surveys; there is also a certain inherent amount of noncomparability across prosecutorial units, because one can serve a single county and another an entire state. For what information it does provide on the dynamics of personnel and workload in prosecutor's offices, the NPS has suffered as a measurement device because of its unstable periodicity (shifting from a 2-year to a 5-year cycle). It has also been unstable in the degree to which it has been conducted and treated as a sample or a census, as we described in Section 3–D; in the "census" years, content is particularly pared back, excluding all but the basic administrative questions.

- The Federal Justice Statistics Program operated by the Urban Institute with BJS sponsorship links administrative records across decision point in the federal justice system. It does provide the "flow" data that the "funnel" promises, and it does cover more of the decisions made by prosecutors than the SCPS series (including the declination decision). However, it is strictly limited to the federal justice system.

There are senses in which prosecution and prosecutorial decisions should be amenable to data collection efforts, among them the fact that basic concepts and definitions are relatively invariant. In broad strokes, Forst (2000:22) observes that changes in prosecution "have mostly followed rather than led developments outside the prosecution domain. The basic nature and goals of prosecution, the role of the victim as witness in a matter between the state and defendant, the essential steps in processing cases through the courts and systems of public accountability have all remained fundamentally unchanged over the past 30 years." However, a major reason for this resistance to procedural change—the relative insularity of prosecutors, as opposed to the police and elected officials whose work has a larger profile—also serves to create a culture that works against openness in providing data. By and large, "the prosecutor's work is invisible to the public at large" (Forst, 2000:23), and the prevailing inclination is to keep things that way. Another reason for prosecutorial insulation is adherence to a basic maxim of their adversarial culture: "Do not divulge the particulars of your case to anyone who is not in a position to help you win it." Accordingly, prosecutors "typically see little to gain and considerable risk in divulging any information that is not required by law" (Forst, 2000:25).

As a means to "improve the systems by which prosecutors are held accountable" and to make the operations of their offices more transparent, Forst (2000:42) argues for:

> the annual publication of uniform office performance statistics and a formal periodic survey of all who depend on prosecutors: victims, witnesses, judges, police, defense bar, and the general public. Private sector organizations have long used surveys to obtain systematic feedback about the effectiveness of service delivery, including measures of consumer satisfaction about specific elements of service. Police departments and other public agencies are turning increasingly to such assessment systems, and so can prosecutors. An effort along these lines should be coordinated by the Bureau of Justice Statistics to minimize political adulteration of the system, perhaps in collaboration with national associations of district attorneys and state attorneys general. Such an idea seems no more farfetched than that of a uniform crime reporting system with the cooperation of virtually all 20,000 independent police departments in the United States, a program that has been operating for most of the twentieth century.

That vision remains far off. However, as noted above in Section 3–D, BJS's investment in redesigning the SCPS program raises interesting possibilities. In working with local jurisdictions to participate in SCPS-type collections, it will also be important to assess whether prosecutor's offices may be able to provide similar types of information. This is particularly the case if methods to work with electronic submissions from court and prosecutor databases continue to develop.

As a short-term measure—and consonant with our advice to expand the concept of "law enforcement" data beyond the strict management focus of LEMAS (Section 3–F.2)—BJS should consider low-cost means to gather at least some procedural information in its existing NPS. A question or set of questions asking for basic counts of resolutions reached by the prosecutor's office within some time window—ideally, broken down to include cases handled through alternative dispute resolution techniques such as mediation—would provide a partial picture of prosecutorial activity, but a fuller one than currently exists.

– 4 –

———————

State and Local Partnerships

Tʜᴇ Uɴɪᴛᴇᴅ Sᴛᴀᴛᴇs has a significant national justice system, composed of the federal court and penal systems and numerous federal law enforcement agencies. However, the vast majority of the activity related to crime and justice occurs at the subnational level. Most crime is pinpointed geographically, and much of the response to crime is handled by police, courts, and correctional facilities at the state, county, and municipal levels. The Bureau of Justice Statistics (BJS) thus shares with many of its fellow federal statistical agencies the challenge that it is a national government agency tasked to measure phenomena that are inherently local in nature.

To meet this challenge, it has been common for federal statistical agencies to forge partnerships with state and local governments. These partnerships vary in their level of formality, in their goals and objectives, and in the fiscal resources dedicated to them on both the federal and state sides. In some cases, they are as basic as establishing regional offices to make interactions with local authorities more convenient and to serve as a venue for dissemination of information; in others, the federal agency and individual state governments are essentially equally committed to the partnership, jointly funding and staffing data collection operations. Federal-state partnerships also include models where the federal agency role is principally one of coordination and compilation, directly accumulating data provided by local authorities and piecing together national files.

Under its current legal authority, BJS has at least two distinguishing characteristics in terms of its work with state and local governments, relative to its peers in the federal statistical system. One is the boldness with which state and local issues are written into BJS's legal mandate: in no uncertain terms,

165

BJS's authorizing language mandates that "[BJS] shall give primary emphasis to the problems of State and local justice systems" (42 USC § 3731), and its list of legal duties is replete with reference to performing studies "at the Federal, State, and local levels" (see Box 1-2). The second is BJS's explicit charter—inherited from the functions of the former Law Enforcement Assistance Administration and consistent with the function of BJS's parent Office of Justice Programs (OJP)—to provide direct financial and technical assistance to local governments and agencies, rather than solely conduct data collection functions.

It follows that an assessment of BJS's programs and functions must pay particular attention to the agency's interactions with state and local governments, evaluating the effectiveness of these partnerships and contemplating the role of BJS's grant programs for local authorities. In this chapter, we discuss the centerpiece of BJS's State Justice Statistics (SJS) program—BJS's network of state Statistical Analysis Centers (SACs; Section 4–A)—and directly compare BJS's work with the states to the models of federal-state cooperation in other parts of the federal statistical system. We then turn to BJS's principal grant program, the National Criminal History Improvement Program (NCHIP; Section 4–B).

This chapter on state and local partnerships is also the most logical place to explore in depth federal and state roles in the compilation of one of the longest-standing statistical series in the criminal justice system—albeit not one administered by BJS. For decades, state and local police departments have supplied crime count data to the U.S. Federal Bureau of Investigation (FBI) as part of the Uniform Crime Reporting (UCR) program. As part of the panel's charge to examine BJS's relationship to other data-gathering entities in the U.S. Department of Justice, we discuss the current and future state of the UCR and BJS's role relative to that of the FBI in this series; this discussion is in Section 4–C.

4–A   STATE JUSTICE STATISTICS: THE STATISTICAL ANALYSIS CENTER NETWORK

BJS's network of state-based SACs actually predates the creation of BJS in its current form. The same section of the 1968 Omnibus Crime Control and Safe Streets Act that authorized the new Law Enforcement Assistance Administration (LEAA) to collect statistical information also directed the LEAA to (P.L. 90-351 § 515(c)):

> cooperate with and render technical assistance to States, units of general local government, combinations of such States or units, or other public and private agencies, organizations, or institutions in matters relating to law enforcement.

The LEAA's National Criminal Justice Information and Statistics Service started the Comprehensive Data Systems (CDS) program in 1972, providing the earliest funds for establishment of SACs. As summarized by the Justice Research and Statistics Association (2008):

> The CDS guidelines established six objectives for the SACs:
>
> - provide objective analysis of criminal justice data, including data collected by operating agencies;
> - generate statistical reports on crime and the processing of criminal offenders in support of planning agencies;
> - coordinate technical assistance in support of the CDS program in the state;
> - collect, analyze, and disseminate management and administrative statistics on the criminal justice resources expended in the state;
> - promote the orderly development of criminal justice information and statistical systems in the state; and
> - provide uniform data on criminal justice processes for the preparation of national statistical reports.

Ten states established SACs or designated existing agencies in SACs in 1972; by 1976, the SAC network had grown to 34 states and a nonprofit association—the Criminal Justice Statistics Association—developed to support and coordinate SAC activities. In 1991, the association renamed itself the Justice Research and Statistics Association (JRSA). BJS assumed responsibility for the SAC program in 1979 as LEAA was phased out (to be replaced by OJP) and it has retained this role since, as reflected in the authorizing legislation.

The initial focus of SACs was to coordinate state-level data collection, act as a statistical clearinghouse, and assist the federal government with justice statistics series through contributions of state data or statistics. SAC involvement in statistical collections has historically focused and remains on state-level needs, although either direct participation in BJS programs (e.g., contributing data) or indirect participation (facilitating participation by other state agencies) has remained central to the program.

As of 2008, all states and several U.S. territories had a designated SAC, though their forms and functions vary; Box 4-1 describes the types of structures that exist in the current SAC network.

The funding program for SACs was reformulated from a clearinghouse orientation to a research-oriented program in 1996 under the leadership of then-BJS Director Jan Chaiken. The SJS program has since provided BJS an avenue for fostering data collection and analysis in areas consistent with BJS and Justice Department priorities. This approach has been particularly useful in collecting data on emerging local or regional issues, which often are of greater concern in specific parts of the country and states before becoming

---

Box 4-1   State Statistical Analysis Center Network

The Justice Research and Statistics Association (JRSA) defines state Statistical Analysis Centers (SACs) as state-level units or agencies "that use operational, management, and research information from all components of the criminal justice system to conduct objective analyses of statewide and systemwide policy issues" (http://www.jrsa.org/sac/aboutsacs.html). Individual SACs receive financial support through the Bureau of Justice Statistics' (BJS's) State Justice Statistics Program, as does JRSA; some SACs receive additional funding from their state governments. JRSA is a nonprofit organization of SAC directors that provides a central staff and coordination effort for SAC activities; it hosts an annual research conference (with BJS funding), publishes the journal *Justice Research and Policy*, and maintains databases of SAC activities.

SACs vary in their organizational structure and standing with respect to the state government, falling into a few basic categories:

- *Independent State Justice Information Center:* For instance, the Illinois SAC (created in 1977) is the research and analysis unit of the Illinois Criminal Justice Information Authority (ICJIA). The ICJIA was created as an independent state agency by executive order in 1982, inheriting functions from a former Illinois Law Enforcement Commission. Other states following similar models as of 2008 are Alabama and Arkansas.

- *Unit of State Crime Commission or Planning Commission:* For example, the Kansas SAC is a unit of the state sentencing commission and Montana's is part of the state Board on Crime Control; both Georgia and the District of Columbia have SACs affiliated with the Criminal Justice Coordinating Council in those jurisdictions. Other states following this model include Arizona, Indiana, Louisiana, Maryland, Nebraska, Oregon, Pennsylvania, Rhode Island, and Utah.

- *Unit of State Law Enforcement Agency or Justice Department:* For this most common organizational structure, examples include California (branch of the state justice department), Missouri (part of the state highway patrol), and Tennessee (unit of the Tennessee Bureau of Investigation). Echoing BJS's administrative position in the U.S. Department of Justice, both the Minnesota and South Carolina SACs are part of an Office of Justice Programs in those states' public safety departments; similarly, Idaho's SAC is housed in the Office of Planning, Grants, and Research of the Idaho State Police. Other states following this basic model include Colorado, Florida, Hawaii, Kentucky, Massachusetts, New Hampshire, New Jersey, New York, North Carolina, North Dakota, Ohio, South Dakota, Virginia, and West Virginia.

- *Unit of Other State Agency or Entity:* Examples include the SACs in Delaware (part of state Office of Management and Budget), Iowa (Department of Human Rights), Oklahoma (Legislative Service Bureau), Washington (Office of Financial Management), and Wisconsin (Office of Justice Assistance). The Texas SAC was reestablished, after some absence, by executive order in 2007, and is housed in the Office of the Governor.

(continued)

Box 4-1 (continued)

- *Affiliation with Academic Department or Institute:* In some states, the SAC is directly affiliated with a university or university-affiliated research institute, and the SAC director or staff may be faculty members. The states organized in this manner (with their host institutions) are Alaska (University of Alaska Anchorage), Connecticut (Central Connecticut State University), Maine (University of Southern Maine, in partnership with state Department of Corrections), Michigan (Michigan State University), Mississippi (University of Southern Mississippi), Nevada (University of Nevada, Las Vegas), New Mexico (University of New Mexico), Vermont (the non-profit Norwich Studies and Analysis Institute, affiliated with Norwich University), and Wyoming (University of Wyoming).

SACs also operate in the Northern Mariana Islands and Puerto Rico.

---

a national issue. States continue to assist BJS as a liaison for data collection efforts, enhancing state and local analytical efforts and analyses which demonstrate the utility of various systems (e.g., National Incident-Based Reporting System, criminal history records etc.).

Data and policy priorities of BJS are reflected in the substantive areas under which SACs may apply for SJS program support. The "themes" in the annual SJS solicitation include issues related to BJS initiatives in an array that also allows states to focus on problems of more immediate state, local or regional concern. Themes from the 2008 SJS program for SAC solicitation are enumerated in Box 4-2.

4–A.1 State Partnerships in the Federal Statistical System

In examining BJS's partnerships, it is useful to consider some exemplars from other statistical agencies; we describe some of these arrangements in the following list. In doing so, we emphasize that this is a selective list meant to describe a range of approaches, rather than a complete canvass of federal-state partnerships, and that no assessment of the quality of the data produced by the systems is implied. Unless otherwise indicated, cost information in the following list is from the fiscal year 2008 edition of the U.S. Office of Management and Budget (OMB) publication *Statistical Programs of the United States Government* (U.S. Office of Management and Budget, 2007):

- The *Behavioral Risk Factor Surveillance System* (BRFSS) is a collaborative data collection system maintained by the Behavioral Surveillance Branch of the Centers for Disease Control and Prevention (CDC). We described the BRFSS in our first report (National Research Council, 2008b:Box 4-1 and Section 4–B) as the basis for one possible design alternative for the National Crime Victimization Survey (NCVS). As a federal-state partnership, the BRFSS follows a contracting model:

Box 4-2 State Justice Statistics Program Themes, 2008

(1) **Deaths in Police Custody Reporting—Obtaining statewide data on deaths occurring in the process of arrest or in pursuit of arrest.** [The Bureau of Justice
Statistics (BJS)] continues to request assistance from State [Statistical Analysis
Centers (SACs)] to obtain specified data on these deaths and report them quarterly
to BJS. [Applicants] wishing to address this theme may utilize SJS to establish a
long term reporting process, rather than a one time study.

(2) **Prison rape and victimization in confinement facilities—improving quality of
administrative data involving criminal acts within adult and juvenile facilities.**
[BJS] continues to encourage SACs to examine the quality of their State administrative records and where feasible provide recommendations for the improvement of
the quality and accuracy of these data.

(3) **Criminal justice system crisis planning.** The SAC may wish to pursue research
or data collection to support criminal justice system planning for dealing with major
crises, disorders, or other catastrophic incidents. [Examples include] prisoner relo-
cation and/or alternative housing needs [and] backup records systems in the courts
or other entities.

(4) **Increased Web access to data.** SJS funds could be used by the SAC for Internet
infrastructure development, enhancements, and linkages, including building a World
Wide Web site, computer support, and preparing reports for dissemination via the
Internet.

(5) **Performance measurement.** SJS funds could be used by the SAC to help States
develop and improve performance measures and the tools available to agencies to
assess progress in addressing public safety and administration of justice goals.

(6) **Analyses utilizing a State's criminal history records.** BJS encourages SACs to
utilize the State's criminal history records for research purposes. In particular, the
SAC may wish to seek SJS funds to support studies of:
   a) Patterns of criminal behavior such as sex offending, stalking, or domestic
      violence;
   b) Arrests, prosecutions and convictions for firearms-related offenses;
   c) Prisoner and/or probationer recidivism, including rates of rearrest, reconvic-
      tion, and return to correctional custody;
   d) The implementation and/or impact of programs such as drug courts, prisoner
      reentry initiatives, or specialized probation programs; or
   e) The implementation and/or impact of a State's criminal history record improve-
      ment activities.

At most one topic may be proposed in this thematic area. *Funds may be requested
to establish the technical capacity to conduct criminal history records-based research.* The application must either state that the applicant is also the State's
administrator of [National Criminal History Improvement Program (NCHIP)] funds or
include a letter or memorandum of endorsement from the State agency administering NCHIP funds.

(7) **Statewide crime victimization surveys.**

(8) **Analysis of the uses of new or emerging biometric technologies to improve
the administration of criminal justice.** SJS funds may be used by the SAC to support research which describes and examines the uses of new or emerging biometric
technologies (DNA evidence collection/analysis, facial recognition, etc.) to improve
the administration of criminal justice in a State.

(9) **Research using incident-based crime data that are compatible with the National Incident-Based Reporting System (NIBRS).** [SJS] funds under this theme
may be used to examine the utility of linking NIBRS incident reports to a State's
criminal history records for research purposes.

(continued)

Box 4-2 (continued)

(10) **Data collection and/or research examining a special topical area:**

(a) **Minority overrepresentation in the criminal or juvenile justice systems.**
. . .

(b) **Civil justice.** SJS funds may be used by the SAC in developing estimates of the number and characteristics of tort, contract, and real property cases and the dispositions of those cases for both adjudicated and settled civil matters. The longer term objective might be to estimate change over time within the State in the nature of case issues, judgments, and awards and to evaluate the impact of civil justice reforms such as capping punitive awards or medical malpractice mediation boards.

(c) **Cybercrime.** SJS funds could be used by the SAC to examine the magnitude and consequences of computer crime and identity theft and fraud.

(d) **Human trafficking.**

(e) **Justice issues in Indian Country.**

(f) **Criminal activity in U.S. border areas.**

(g) **Violent crime in schools.**

(h) **The impact of substance abuse on State and/or local criminal justice and public health systems.**

(i) **Family violence and/or stalking.**

(11) **Evaluation of prisoner reentry initiatives and programs.**

(12) **Other theme or topic identified by the SAC.** SJS funds may be used by the SAC to support research examining another theme or topic provided the application is accompanied by persuasive documentation and justification that the subject is a top priority for the State's Governor or criminal justice policy officials.

SOURCE: Excerpts from Bureau of Justice Statistics (2008f); emphases in the original.

the CDC executes contracts with state health agencies, paying for them to conduct monthly telephone interviews and administer a core questionnaire. In return, the states get processed returns in terms of state-level estimates (by design, the BRFSS is an amalgam of state samples and is not meant to be a nationally-representative sample); they also have the latitude to add their own topic supplements. This form of partnership—in which the federal agency exercises strong control over content and resulting data but pays the states to provide data collection—is expensive. Of the $453.1 million estimated to be spent in fiscal year 2008 on purchasing statistical services from state and local governments, CDC's $162.2 million (not including activities of the National Center for Health Statistics [NCHS], which is a component of CDC) is the largest single share.

- The *Vital Statistics* program of NCHS compiles information from birth and death certificates that have been collected by state health (vital registration) departments. NCHS's fiscal year 2008 allocation for statistical purchases from state governments is about $18.6 million—lower spending relative to the BRFSS because of the different relationship between the federal agency and the localities. The law directs that (42 USC § 242k(h)(1)):

> There shall be an annual collection of data from the records of births, deaths, marriages, and divorces in registration areas. The data shall be obtained only from and restricted to such records of the States and municipalities which [NCHS] determines possess records affording satisfactory data in necessary detail and form. . . . Each State or registration area shall be paid by [NCHS] the Federal share of its reasonable costs (as determined by [NCHS]) for collecting and transcribing (at the request of [NCHS] and by whatever method authorized by [NCHS]) its records for such data.

Hence, the collection costs of vital statistics data are largely assumed at the state level (with some reimbursement of "the Federal share of the reasonable cost"). NCHS's costs involve compilation and processing, as well as the promulgation of standards. This model has its difficulties, because disagreements over the costs of providing the data can put a damper on participation. In recent years, NCHS has struggled to achieve cooperation by state health departments (and the county health departments and local facilities coordinated by the states) in adopting new standard formats for birth and death certificates. Participation in supplying divorce records declined sufficiently that NCHS abandoned their collection in 1996, and the most recent comprehensive study by NCHS of marriage and divorce information from vital records is based on 1989–1990 data (see http://www.cdc.gov/nchs/mardiv.htm).

- The *Bureau of Labor Statistics* allocation for purchases of statistical services from states and localities was the second largest among statistical agencies in fiscal year 2008, estimated at $96 million. BLS operates federal-state cooperative arrangements through its regional offices; limited contracts for collection and sharing of employment data between BLS and individual states had been crafted as early as 1916, but the establishment of the regional offices in 1942 formalized those arrangements (Hines and Engen, 1992). Today, the BLS Federal-State Cooperative Programs (administered through six regional offices) encompass a number of labor market information programs, including surveys and records collections on occupational safety and health.[1]

---

[1] In recent years, BLS has switched from operating eight regional offices to six: its Kansas City, Missouri, regional office was merged with the Dallas office and the operations of its Boston and New York offices were consolidated into a Boston (Northeast) office. However, New York City still retains a Regional Office for Economic Information and Analysis headed by a regional commissioner.

As an operational model, BLS's Federal-State Cooperative Program is closer to the contract-driven BRFSS model than the vital statistics example, involving contractual agreements to collect and transfer specific data series.

The Census Bureau employs numerous mechanisms to work with state and local governments, including the coordination of field activities through 12 regional offices. Two of its major partnerships are of particular interest because they share a similar structure with the BLS model. Between 1967 and 1973, the Census Bureau formalized loose arrangements with state agencies to create the Federal-State Cooperative Program for Population Estimates, under which states supply some of the raw information (vital statistics data on births and deaths, estimates of prison population, and other records) needed to update decennial census information and generate intercensal population estimates. By 1979, a parallel Federal-State Cooperative Programs for Population Projections was forged to build collaboration with the states on the production of population forecasts.

- As an agency, the *National Agricultural Statistical Service* (NASS) of the U.S. Department of Agriculture (USDA) dates back to the 1961 formation of a Statistical Reporting Service, but its roots—and partnership with states—are more extensive. Wisconsin was the first state to enter into a memorandum of understanding (MOU) for data gathering and sharing with USDA in 1917; over the years, similar MOUs were executed with state agriculture departments, land grant universities, and other agricultural entities, and all states currently have an MOU on file with NASS (Dantzler, 2008). The defining characteristic of NASS's state partnerships is the high degree to which labor and other resources are shared, by means of a third party. In 1972, NASS established an agreement with the National Association of State Departments of Agriculture (NASDA) under which NASDA bears the principal costs of data collection, including salaries and travel expenses of about 3,700 field interviewers. Field work is coordinated through NASS's 46 field offices,[2] which are staffed by a mix of federal/NASS employees (675 total) and state "cooperator" employees (151 total). Because of the intermediary role of NASDA, OMB tabulations indicate that NASS purchases no statistical services directly from state and local governments, but rather from a private-sector entity ($29 million, out of NASS's $167.7 million total estimated budget).

BJS's state partnerships do not correspond neatly with any of these organizational models. The NCHIP and related grant programs provide rela-

---

[2]All states are covered by the NASS-state partnerships; however, the New England states are coordinated through a single regional office in Concord, New Hampshire.

tively unfettered funding to the states, not the more formal data collection contracts executed by the other agencies. In its SAC network, BJS's system bears some similarities to the NASS arrangement, including the presence of a third-party coordinator (NASDA for NASS and JRSA for the SACs), but is neither as formal nor intensive as the state agricultural arrangements. In large part, the more-limited role of the SACs is dictated by basic logic and the breadth of BJS's scope: in the criminal justice arena, there is no ideal, single, state-level point of contact with which a strong data collection arrangement can be brokered, because of the differences in state justice organizations. That is, state police or public safety departments may be distinct from state corrections departments, which in turn are distinct from state court systems, which are distinct from state victims' offices and other related agencies, all of which are distinct from their local or large-city equivalents.

As noted in our first report (National Research Council, 2008b:60), BJS has provided technical support to state and local agencies as well as broker- ing partnerships to disseminate and collect data. This is particularly true of its development of software tools for local data collection. BJS developed, and made freely available to state and local agencies, software for conduct- ing victimization surveys, using the NCVS (including the detailed Incident Report) as a template. The software has since been relabeled Justice Survey Software and is administered by SureCode Technologies, and remains avail- able at http://www.bjsjss.org; templates have been added for victimization surveys conducted in individual states, as well as for BJS's National Survey of Prosecutors, Police-Public Contact Survey supplement, and State Court Processing Statistics inventory. More recently, this survey-building software has been ported to the Web as "BJScvs" and made available to state and lo- cal agencies through the website http://www.bjscvs.org (Justice Research and Statistics Association, 2006b:5).

4–A.2 Assessment

Some of the basic benefits of a strong and active partnership between BJS and its state SACs can be listed in brief:

- SACs are familiar with state, local, and often regional justice issues and can provide context for federal initiatives.

- Federal-state cooperation promotes consistency in definitions and con- cepts across data collections, which in turn has the benefit of facili- tating more effective comparisons between jurisdictions and agencies. With partnerships, BJS is also in a position to provide guidance to states and localities on common standards for data quality, measure- ment, and analysis.

- The response to crime is predominantly local and state in nature, hence the need for familiarity with justice systems, data, and agencies at this level. The SACs are often able to facilitate access to key agencies, data systems, and collection mechanisms that benefit federal statistical system efforts.

- States benefit from a strong state-federal partnership in several key ways:
  - BJS is able to provide technical assistance that would not otherwise be available to states, or that might be cost-prohibitive for any one jurisdiction to obtain.
  - Although the financial benefit of the SAC program has not been large, states have been able to leverage BJS assistance (including NCHIP) for system, data quality, and analytical enhancements that might otherwise not be available from state resources alone.
  - A major benefit for BJS and the nation is improvement in justice data systems and a national perspective on crime and justice. Policy development in the states often requires benchmarking and an understanding of crime on the national scene; BJS is uniquely situated to provide this perspective.

- BJS benefits from a strong relationship with the SACs through the inventiveness of research performed at the state level and through the states' direct contact with issues of local interest; feedback from SAC partners and successes with state-level activities can inform the development of national-level data collections.

The capacity of the state partnerships to assist BJS to more nimbly meet new data needs was clearly illustrated by the data collection efforts set up to comply with the Deaths in Custody Reporting Act of 2000 (P.L. 106-297). As we also discuss in Box 3-3 and Section 3–B.3, the act tasked BJS with quarterly data collection on incidents involving the death of persons while in criminal justice system physical custody. BJS was able to use its ongoing relationship with state Departments of Correction to collect data on deaths occurring while in correctional custody. However, a data collection system for deaths occurring in law enforcement or other justice system custody proved more problematic, given significant variation in the ability to identify and capture data on incidents even within states. SACs assisted in developing data collection systems initially and in some cases continue to assist BJS in preparing reports for the Deaths in Custody project mandated by Congress; it remains one of the suggested program themes in the 2008 program solicitation.

In the panel's assessment, the BJS-state SAC network stands as a relatively low-cost activity on BJS's part with great dividends in terms of outreach and feedback, as well as dissemination of data and products to state policy makers. Consistent with other recommendations we make in Chapter 5 on

BJS developing mechanisms for securing external advice, BJS's good work in establishing state-level ties through the SACs, coordinated through JRSA, should continue to be a high-priority activity for the agency.

> **Finding 4.1:** BJS's state Statistical Analysis Center (SAC) program has cultivated a strong federal-state relationship, relative to other federal statistical agencies. Development of the SAC network—which provides points of contact across the justice system to facilitate research on individual data series, dissemination of BJS information, and coordination of activities—has involved forging unique relationships adapted to state environments (for instance, whether the SAC is part of a state law enforcement department or is housed at a university).

Implicit in this finding is the determination that BJS's SACs are appropriately positioned and that the heterogeneity of organizational structures across the SACs is a strength of the program. However, going forward, a challenge that would be useful for BJS and the SACs to consider (together with JRSA and BJS's data collection agents) is finding a stronger role for the SACs in facilitating data collection activities. The Deaths in Custody example is a good one, where the existence and expertise in the SAC network made it possible to establish a new data series (and respond to a legal mandate) in a short time frame; it would be beneficial to find other avenues where such efficiency can be achieved and where the SACs can serve as an active point of contact or a collaborator in gathering information. We do not suggest by this language that the BJS-SAC relationship be revamped to look more like the vital statistics (dominant state, coordinating federal roles) or the NASS (dual federal-state staffing) models. Such models are not viable in the justice case because the major state-level operations of interest—law enforcement, corrections, and judiciary—are generally not located within the same department.

Still, acknowledging the fact that the capacity to use the SACs for data collection will always be limited due to the range of types of SACs and the lack of a central justice information authority in many states, good statistical systems are ones in which states are active partners in data systems. BJS's state partnerships—not only the work of the SACs but also its role in administering grants such as NCHIP—give it distinctive possibilities relative to other federal statistical agencies. Through these works, BJS has the ability to subtly but directly affect the quality of the data that it receives as input to its ongoing series and the technical systems used to generate those data.

> *Recommendation 4.1:* **Through its Statistical Analysis Center and State Justice Statistics programs, BJS should continue to develop its ties with the states, and more fully exploit the potential for using states as partners in data collections.**

It is particularly essential that the state SAC perspective be brought to bear in addressing the points raised in Section 3–F.1 on emphasizing longitudinal structures within series. Tapping state expertise on available data and information systems would be highly beneficial in finding new ways to link existing data sets or to design panel surveys to follow cohorts of persons through the various steps of the justice system. Because states are the most likely immediate consumers and disseminators of small-area data, efforts by BJS to generate subnational measurements from the NCVS or other surveys should certainly be done with the active input from the states.

> *Recommendation 4.2:* **Developments toward longitudinal and small-area measurement systems should involve state partners who are active in data collection and knowledgeable about state justice systems.**

4–B NATIONAL CRIMINAL HISTORY IMPROVEMENT PROGRAM AND RELATED GRANT PROGRAMS

4–B.1 Background Checks and the Development of NCHIP

NCHIP makes grants to states for establishment or upgrading of information systems, in response to a provision in the Brady Handgun Violence Prevention Act of 1993 (P.L. 103-159):

> The Attorney General, through the Bureau of Justice Statistics, shall, subject to appropriations and with preference to States that as of the date of enactment of this Act have the lowest percent currency of case dispositions in computerized criminal history files, make a grant to each State to be used—
>
> (A) for the creation of a computerized criminal history record system or improvement of an existing system;
>
> (B) to improve accessibility to the national instant criminal background system; and
>
> (C) upon establishment of the national system, to assist the State in the transmittal of criminal records to the national system.

Shortly thereafter, the National Child Protection Act of 1993 (P.L. 103-209) added similar grant-making authority with specific reference to improving state computerization and transmittal of criminal history records involving child abuse.

The specific background check system created by the Justice Department in response to this mandate is the National Instant Criminal Background Check System, better known as NICS, which began operating in November 1998. The Brady Act requires federal firearm licensees to check potential firearm purchasers against the NICS database to determine whether the applicant is disqualified from making the purchase. NICS was developed by the FBI in consultation with federal, state, and local law enforcement agencies; like the UCR program, NICS is administratively housed in the FBI's Criminal Justice Information Services Division in Clarksburg, West Virginia.

A query against NICS is an instant check against three separate databases:

- The *Interstate Identification Index* ("Triple I" or III) is an index of criminal history records, including persons arrested for felonies and some serious misdemeanors. It is an index (including identifying information such as name, gender, race and ethnicity, and data of birth) of criminal histories rather than a full-fledged compilation of records. The basic "instant" query against the index takes only a few seconds and indicates whether arrest records exist for the target person in any state. If the instant query suggests a match, separate record requests (using either the FBI-assigned or state-issued identification numbers coded on the record) retrieve the specific, detailed records corresponding to the individual from state record repositories. Ramker and Adams (2008:9) note that "forty-nine states (Vermont is working toward participation) and the District of Columbia currently participate in III and the system now includes over 66 million criminal records."

- The *National Crime Information Center* (NCIC) database is a compilation of a wide variety of personal and property records; it includes sex offender and protection order registries, arrest warrant records, and parole and conviction records, as well as records of vehicle or property theft and existing firearm records.

- The *NICS index* culls from federal and state records to cover information on characteristics that are identified by law (18 USC § 922) as disqualifying a potential firearm purchaser but that are not covered by either the Triple I or the NCIC databases. Notably, these characteristics include immigration (alien) status and mental health history.

A NICS query results in one of three responses: "Proceed," "Denied," or "Unresolved." If the instant check against these databases suggests that the potential purchaser falls into a prohibited category, then the sale or transfer is "denied;" the query itself does not tell the federal firearm licensee (or the potential purchaser) the category or categories that resulted in the disqualification, though the individual has the right to request such information and appeal any inaccurate information.

Exactly why BJS was designated as the administrator of the grant program to support implementation of NICS and related criminal history databases—as opposed to the FBI (which housed the existing record systems) or a purely grant-based agency such as the Bureau of Justice Assistance—is not clear. One possible reason is simply that BJS, as part of OJP, has grant-making authority that the FBI lacks; another is that references to criminal record systems remained among BJS's legally mandated duties (Box 1-2) following its creation from the predecessor LEAA. However the authority came about, BJS made its first grants related to computerized record improvement as early as 1995, and its grantmaking program came to be known as NCHIP.

NCHIP was expanded in scope by the Crime Identification Technology Act of 1998 (codified as 42 USC § 14601(a)), which directed that grants be made to "the State[s], in conjunction with units of local government, State and local courts, other States, or combinations thereof." The intended purpose of these grant monies was to:

> establish or upgrade an integrated approach to develop information and identification technologies and systems to—
>
> > (1) upgrade criminal history and criminal justice record systems, including systems operated by law enforcement agencies and courts;
> >
> > (2) improve criminal justice identification;
> >
> > (3) promote compatibility and integration of national, State, and local systems for—
> >
> > > (A) criminal justice purposes;
> > >
> > > (B) firearms eligibility determinations;
> > >
> > > (C) identification of sexual offenders;
> > >
> > > (D) identification of domestic violence offenders; and
> > >
> > > (E) background checks for other authorized purposes unrelated to criminal justice; and
> >
> > (4) capture information for statistical and research purposes to improve the administration of criminal justice.

Authority for the issuance of these grants was specifically vested in "the Office of Justice Programs relying principally on the expertise of the Bureau of Justice Statistics." As indicated in Box 1-2, the description of BJS's duties under the law was subsequently modified in 2006 to specifically reference NCHIP-related activities.

The 1998 Crime Identification Technology Act served to dramatically increase the scope of information and identification systems eligible for improvement grants. As detailed in Box 4-3, the act specifically covered a wide array of information systems used by state and local law enforcement agencies, courts, and support agencies, ranging in content and data type from person-level attributes (e.g., sexual offender registration and criminal history records) to graphic images (e.g., scans of fingerprints and images of the toolmarks on ballistics evidence [spent bullets and cartridge casings]). The act also formally defined funds for improving systems for tracking domestic violence and stalking activity, including filed protective orders, as was authorized by amendments to the Violence Against Women Act.[3]

---

[3]Funds for these purposes are sometimes called by a separate name and acronym—the Stalking and Domestic Violence Records Improvement Program, or SDVRIP—but are administered as part of NCHIP.

Box 4-3   Information Systems Covered by the Crime Identification Technology Act of 1998 and National Criminal History Improvement Program

Grants under this section may be used for programs to establish, develop, update, or upgrade—

(1) State centralized, automated, adult and juvenile criminal history record information systems, including arrest and disposition reporting;

(2) automated fingerprint identification systems that are compatible with standards established by the National Institute of Standards and Technology and interoperable with the Integrated Automated Fingerprint Identification System (IAFIS) of the Federal Bureau of Investigation;

(3) finger imaging, live scan, and other automated systems to digitize fingerprints and to communicate prints in a manner that is compatible with standards established by the National Institute of Standards and Technology and interoperable with systems operated by States and by the Federal Bureau of Investigation;

(4) programs and systems to facilitate full participation in the Interstate Identification Index of the National Crime Information Center;

(5) systems to facilitate full participation in any compact relating to the Interstate Identification Index of the National Crime Information Center;

(6) systems to facilitate full participation in the national instant criminal background check system established under [the] Brady Handgun Violence Prevention [Act] for firearms eligibility determinations;

(7) integrated criminal justice information systems to manage and communicate criminal justice information among law enforcement agencies, courts, prosecutors, and corrections agencies;

(8) noncriminal history record information systems relevant to firearms eligibility determinations for availability and accessibility to the national instant criminal background check system established under [the] Brady Handgun Violence Prevention [Act];

(9) court-based criminal justice information systems that promote—

(A) reporting of dispositions to central State repositories and to the Federal Bureau of Investigation; and

(B) compatibility with, and integration of, court systems with other criminal justice information systems;

(10) ballistics identification and information programs that are compatible and integrated with the National Integrated Ballistics Network (NIBN);

(11) the capabilities of forensic science programs and medical examiner programs related to the administration of criminal justice, including programs leading to accreditation or certification of individuals or departments, agencies, or laboratories, and programs relating to the identification and analysis of deoxyribonucleic acid;

(12) sexual offender identification and registration systems;

(13) domestic violence offender identification and information systems;

(14) programs for fingerprint-supported background checks capability for noncriminal justice purposes, including youth service employees and volunteers and other individuals in positions of responsibility, if authorized by Federal or State law and administered by a government agency;

(continued)

Box 4-3   (continued)

(15) criminal justice information systems with a capacity to provide statistical and re-search products including incident-based reporting systems that are compatible with the National Incident-Based Reporting System (NIBRS) and uniform crime reports;

(16) multiagency, multijurisdictional communications systems among the States to share routine and emergency information among Federal, State, and local law enforcement agencies;

(17) the capability of the criminal justice system to deliver timely, accurate, and complete criminal history record information to child welfare agencies, organizations, and programs that are engaged in the assessment of risk and other activities related to the protection of children, including protection against child sexual abuse, and placement of children in foster care; and

(18) notwithstanding subsection (c) of this section, antiterrorism purposes as they relate to any other uses under this section or for other antiterrorism programs.

NOTE: The network referred to in point 10 is mislabeled in this legislative text; it should be the National Integrated Ballistic Information Network (NIBIN), which is operated by the Bureau of Alcohol, Tobacco, Firearms, and Explosives. See National Research Council (2008a) for additional description of NIBIN.
SOURCE: Excerpted from 42 USC § 14601(b).

The 1998 act established some basic eligibility criteria for the funds, as well as conditions and limitations on their use. To be eligible to receive these funds, states must demonstrate "the capability to contribute pertinent information to the national instant criminal background check systems" and have documented plans for developing integrated information technology systems (42 USC § 14601(c)). An important condition placed on the funds (42 USC § 14601(e)) is a 5 percent set-aside for BJS study and documentation purposes:

> Not more than 5 percent may be used for technical assistance, training and evaluations, and studies commissioned by Bureau of Justice Statistics of the Department of Justice (through discretionary grants or otherwise) in furtherance of the purposes of this section.

Every 2 years since 1989, BJS has sponsored a Survey of Criminal History Information Systems that, in part, serves to measure progress made through NCHIP grants. The survey is conducted by SEARCH, the National Consortium for Justice Information and Statistics. BJS has issued periodic updates on progress in criminal history record improvement (e.g., Bureau of Justice Statistics, 2001) and, in 2005, published a self-review of accomplishments of the NCHIP program (Brien, 2005). However, the only current data series that actually measures uses of and results from NCHIP-covered databases is the Firearm Inquiry Statistics program mandated by the original Brady Act. This program provides summaries of the number of handgun-purchase background checks completed (and failed) each year; see, e.g., Bureau of Justice Statistics (2008a).

In 2004, the General Accounting Office (GAO; later the Government Accountability Office) issued a review of the NCHIP program conducted at the request of the House Committee on the Judiciary. The study concluded that, "using their own funds, as well as NCHIP and other federal grants, states have made much progress in automating their records and making them accessible nationally" (U.S. General Accounting Office, 2004:35). However, the report also warned of both increasing demands for background check services and the costs of upgrading and replacing computer systems infrastructure. Four years later, GAO conducted a second audit of NCHIP, with specific mandates from congressional requesters to describe Department of Justice oversight of the funds. The second review again reported significant progress in automating criminal history records; replying to a review version of the report, BJS indicated technology reporting and better case disposition reporting from court information systems as particular priorities for NCHIP work (Larence, 2008). In addition to the two GAO reviews, the NCHIP program was formally submitted to OMB's Performance Assessment Rating Tool (PART) process in 2003 and deemed to be "moderately effective," the second-highest ranking in the PART framework.[4]

4–B.2   Recent Law and Developments

In January 2008, the NICS Improvement Amendments Act of 2007 was signed into law (P.L. 110-180). The legislation was developed in the aftermath of the April 2007 mass homicide at Virginia Polytechnic Institute and State University; that shooting is specifically cited in the initial findings section of the act as the act's motivation, along with a 2002 shooting in a church in Lynbrook, New York. The existing NICS index coverage of mental health history includes only formal determinations by a legal authority (such as a finding of insanity or incompetence to stand trial) or actual commitment to an institution. In the Virginia Tech incident, the perpetrator evidenced a history of mental illness but not the level of formal legal commitment that would be recorded in the NICS index; in the Lynbrook incident, the perpetrator did have a mental health commitment as well as a restraining order against him, but neither of those disqualifying factors was registered in the instant background check. The act seeks to improve the coverage of mental health adjudications and commitments in the NICS databases; it further

---

[4]See   http://www.whitehouse.gov/omb/expectmore/detail/10001094.2003.html   for   the PART summary. The PART evaluation was completed before BJS had developed a "Record Quality Index" to assess the quality of existing systems in individual states in order to better target resources—the sole substantive point on which the PART found fault in the NCHIP program.

requires states to provide records of convictions on misdemeanor domestic violence charges. To do so, the act authorizes the attorney general to make grants "in a manner consistent with the [NCHIP] program" to help states supply these records and generally improve submittal of records for NICS purposes. The act further explicitly directs the director of BJS to conduct ongoing evaluations of NICS[5] and to submit two annual reports to Congress, one on the general operations of the background check system and the other on specific practices by the states in assembling and providing the relevant records (identifying and recommending best practices for all states).

Although the new act supported an increased role for NCHIP-type grants, the level of funds appropriated by Congress for NCHIP has declined dramatically in recent years. In fiscal year 2003, BJS had funds to allocate about $47.5 million to states and territories; award totals dropped to about $26 million in fiscal year 2005 and $8.5 million in both fiscal years 2007 and 2008.[6]

Although NCHIP funding levels may have decreased, recent legislation has also created the possibility for BJS to actually use for research purposes the criminal history record data that its NCHIP grants have helped to improve. The most recent reauthorization of the Department of Justice (P.L. 109-162, which became law in 2006) did three specific things to put BJS in a position where it can actually utilize criminal history record databases for research. First, it added specific detail to the 19th listed duty of BJS (see Box 1-2), in particular authorizing "statistical research for critical analysis of the improvement and utilization of criminal history records." Second, it vested the director of BJS with responsibility for maintaining the integrity and confidentiality of data in BJS hands: the director "shall be responsible for the integrity of data and statistics and shall protect against improper or illegal use or disclosure" (42 USC § 3732(b)). Third, it expanded existing authority for BJS to request information from federal, state, and local agencies by authorizing BJS to enter into data-sharing agreements: the BJS director shall "confer and cooperate with Federal statistical agencies as needed to carry out the purposes of this subchapter, including by entering into cooperative data sharing agreements in conformity with all laws and regulations applicable to the disclosure and use of data" (42 USC § 3732(d)(6)).

Armed with these new legal authorities, BJS began the process of negotiating access to criminal history records with the FBI. One important, and somewhat complex, step in this process involved arranging for the FBI to issue BJS an Originating Agency Identification number—codes that are normally issued to law enforcement agencies—for the purpose of issuing Triple

---

[5]A separate title of the act obligates GAO to audit the funds allocated under the act and report to Congress on how they were spent.

[6]See http://www.ojp.usdoj.gov/bjs/stfunds.htm, which summarizes the amount of NCHIP awards by fiscal year and by state.

I requests for research purposes. In August 2008, BJS entered into a cooperative agreement with Nlets—the International Justice and Public Safety Information Sharing Network—to develop the information technology for BJS to work with Triple I records.[7] Specific tasks to be completed by Nlets—within approximately 1 year—include "provid[ing] BJS with the capability to request/obtain multiple electronic criminal history records at one time" and to "develop and implement a simplified uniform criminal history record format to facilitate BJS's statistical analysis of the criminal history record information" (Ramker and Adams, 2008:9–10).

4–B.3 Assessment

On one hand, the NCHIP and related grant programs are the easiest targets for criticism in a review of the programs of BJS as a statistical agency, precisely because they are not statistical data collection programs. BJS's role in the grantmaking programs is generally limited to award and administration of the grant funds and it does not acquire significant series of data as a direct result of the funds (save for the firearm inquiry counts). Significantly, BJS lacked any access whatsoever to the data systems that the grant funds sought to improve, and the research on quality and content of the resulting record databases has not been commensurate with what one would expect given the 5 percent set-aside for evaluation purposes.

> **Finding 4.2:** The National Criminal History Improvement Program (NCHIP) is a grantmaking program but not directly a statistical collection, even though it is administered by BJS. However, improved criminal history records are important for the prospects of longitudinal analysis of the criminal justice system. Analysis of the National Instant Background Check System serves as one approach to provide the data necessary to evaluate national policy on regulation of firearms purchases.

However, the caveat we noted above—that BJS can signal particular priorities and interests in its solicitation announcements, and so can subtly influence the quality of the information systems that will ultimately be used to generate data on justice system operations—is a real and significant one. BJS's capacity to let systems improvement grants through NCHIP directly affects the level of entry of criminal history records into a central repository but, over time, also affects the input streams from court processing systems,

---

[7]As summarized by Ramker and Adams (2008:9), "Nlets is responsible for all interstate exchange of federal and state criminal history records, and operates a national telecommunications infrastructure for this purpose. Nlets is also a member of the FBI's Joint Task Force for Rap Sheet Standardization and serves as a custodian of the standardized rap sheet layout and national standard format."

law enforcement booking and case management files, and correctional supervision records. Although the effects may not be as immediate or massive as might be hoped, BJS's grantmaking authority does put it in a distinctive position relative to other statistical agencies of being able to do something about the quality of source-level data rather than just bemoaning or adjusting for shortcomings in the data.

The developments in 2008 that will, apparently, put BJS in a position to harness criminal history records for research purposes are extremely promising. BJS's new recidivism, reentry, and special projects unit should be encouraged to be wide ranging in considering the ways in which access to these data can inform studies of histories and "careers" in crime. The use of the records to support an ongoing measure of recidivism and recurrence of criminal behavior—as a relatively low-cost complement to formal panel studies of persons released from correctional supervision—is a solid first step. More generally, access to criminal history records is a linchpin to improving BJS's collections on longitudinal flows within the justice system, as we discussed in Section 3–F.1.

> *Recommendation 4.3:* **BJS should actively utilize the NCHIP program to improve criminal history records necessary for longitudinal studies of crime.**

It is appropriate, in this chapter on federal-state partnerships, to observe that BJS can learn from and build on work done in several of its SAC affiliates, some of which (being parts of law enforcement departments) have been able to utilize electronic records in their work. In particular, Burton et al. (2004) provide a good example of the type of analysis that could be done through actual analysis of computerized criminal history records, developing and assessing a measure of "seriousness" of criminal career trajectories.

4–C BJS AND THE UNIFORM CRIME REPORTING PROGRAM

A federal criminal investigative agency within the U.S. Department of Justice, the FBI was formally founded by executive order in 1933, expanding the authority vested in a substantially smaller Bureau of Investigation (founded in 1908) by incorporating key functions from the Bureau of Prohibition. Legislation in 1935 dubbed the agency the "Federal Bureau of Investigation," the name it has retained since. Significantly for the purposes of this report, the FBI has been engaged in the collection of crime statistics since its earliest days, including through direct collection from state and local law enforcement officials, and so is a natural point of comparison for BJS's programs.

The general functions of the director of the FBI (and, hence, of the agency) are articulated in Title 28, Section 0.85 of the Code of Federal Regulations. Chief among these is its basic criminal justice role: to "investigate violations of the laws, including the criminal drug laws, of the United States and collect evidence in cases in which the United States is or may be a party in interest." However, the FBI's duties also include a provision that creates overlaps with BJS's responsibilities in several respects: "operate a central clearinghouse for police statistics under the Uniform Crime Reporting Program, and a computerized nationwide index of law enforcement information under the National Crime Information Center" (28 CFR § 0.85(f)). Specifically, the reasons why it is useful to consider the relationship between the FBI and BJS are:

- The summary records from the FBI's UCR program are published annually as *Crime in the United States* and are frequently used as a national indicator of crime. In this function as national indicator of the incidence of crime, the UCR is a counterpart to BJS's NCVS. The two measures differ conceptually and so provide the benefit of offering multiple vantages on the same underlying phenomenon of crime in the United States. However, since the existence of two measures may also commonly be seen as redundant or wasteful, it is important that the relative strengths and weaknesses of the two data sources be well documented and conveyed to the public.

- A component of the UCR summary reporting program also generates administrative information on law enforcement personnel, a point of overlap with BJS's Law Enforcement Management and Administrative Statistics (LEMAS) series. Specific aspects of coverage of law enforcement in the UCR program are described in Box 4-4. Table 4-1 summarizes further similarities and differences in coverage and content between the UCR and the NCVS.

- As discussed in the preceding section, the FBI maintains and administers national-level criminal history record databases, such as the instant background check database used to screen potential firearm purchasers, through its NCIC. BJS supports state and local law enforcement departments and their capacity to populate these FBI databases through NCHIP and other grants but it has no "ownership" of—or, until recently, access to, for statistical purposes—the resulting data compiled by the FBI.

4–C.1   Overview of the UCR Program

In 1930, the Justice Department was authorized to "acquire, collect, classify, and preserve identification, criminal identification, crime, and other records" (28 USC § 534(a)(1)). In turn, the attorney general delegated authority for collecting crime information via the UCR program to the FBI.

Box 4-4   Law Enforcement Coverage in the Uniform Crime Reporting Program

As part of the Summary Reporting System of the Uniform Crime Reporting (UCR) program, local law enforcement agencies report summary counts of the number of offenses reported to police, the number and basic characteristics (age, sex, race) of arrestees, and the number of "clearances" for each major (Type I) crime. A clearance is an offense-level attribute, not a person-level count; hence, "Several crimes may be cleared by the arrest of one person, or the arrest of many persons may clear only one crime" (Federal Bureau of Investigation, 2004b:79). Under FBI definitions, an offense can be cleared in only one of two ways:

- *Clearance through arrest,* or "solved for crime reporting purposes," occurs when "at least one person is (1) arrested, (2) charged with the commission of the offense, and (3) turned over to the court for prosecution" (Federal Bureau of Investigation, 2004b:79); or

- *Clearance through exceptional means,* such as when the offender is killed or when a confession is obtained from a person already serving a sentence for another crime. Technically, a crime may be cleared exceptionally if an agency "can answer all of the following questions in the affirmative" (Federal Bureau of Investigation, 2004b:80–81):

    1. Has the investigation definitely established the identity of the offender?
    2. Is there enough information to support an arrest, charge, and turning over to the court for prosecution?
    3. Is the exact location of the offender known so that the subject could be taken into custody now?
    4. Is there some reason outside law enforcement control that precludes arresting, charging, and prosecuting the offender?

Even absent full compliance with National Incident-Based Reporting System reporting, law enforcement officials are also asked to supply detailed incident-level information on homicides on monthly Supplementary Homicide Reports. These data, which are separately tabulated and made available for analysis, include detail on the circumstance of the incident, the type of weapon used in the murder, and what information is known about the relationship between the victim and the offender. In recent years, police departments have also been required to submit quarterly reports of hate crime incidents; these reports, too, include incident-level characteristics such as the type of bias motivation.

On an annual basis, agencies are asked to provide counts of the number of personnel in their employ (total and sworn officers, specifically) as of October 31; these are tabulated and published in the annual *Crime in the United States* report. In addition, a UCR subprogram asks agencies to submit information on incidents in which officers are killed (feloniously or accidentally) or assaulted in the line of duty. (A record is supposed to be made in cases in which an officer is off-duty but is "acting in an official capacity, that is, reacting to a situation that would ordinarily fall within the scope of his or official duties as a law enforcement officer" [Federal Bureau of Investigation, 2004b:109].) These data are labeled as Law Enforcement Officers Killed or Assaulted, and are tabulated in a separate publication by the FBI.

The FBI's authority for operating the UCR is currently assigned by regulation (28 CFR § 0.85(f), tasking the director of the FBI to "operate a central clearinghouse for police statistics under the Uniform Crime Reporting Program"). Authority for that designation was further affirmed by the Uniform Federal Crime Reporting Act of 1988 (P.L. 100-690 § 7332), which authorized the attorney general to "designate the Federal Bureau of Investigation as the lead agency" for UCR purposes and to establish "such advisory and oversight boards as may be necessary."[8]

The current UCR program is a cooperative program of law enforcement agencies that produces aggregate data on crimes reported to police. When data collection began in January 1930, about 400 cities contributed information; indeed, national-level estimates of crime rates from the UCR were not issued until 1958 because of incomplete coverage (Maltz, 1999:4). As of 2004, 17,000 law enforcement agencies were participating in UCR (Federal Bureau of Investigation, 2004b:Foreword). Like the NCIC that administers the FBI's criminal history databases, the UCR program is administratively housed in the FBI's Criminal Justice Information Services Division in Clarksburg, West Virginia.

For years, the FBI has been in the process of trying to transition from collection of UCR data as has been done for decades—what is now known as the Summary Reporting System (SRS)—to a newer and more detailed system. The National Incident-Based Reporting System (NIBRS) is poised to eventually supplant the SRS but has been slow to develop.

Summary Reporting System

The core content of the SRS inherits directly from the work of a Committee on Uniform Crime Records convened by the International Association of Chiefs of Police (IACP) in 1927. "Recognizing a need for national crime statistics," that committee "evaluated various crimes on the basis of their seriousness, frequency of occurrence, pervasiveness in all geographic areas, and likelihood of being reported to law enforcement" (Federal Bureau of Investigation, 2004b:2). Although the labels have changed slightly, the seven crimes identified by the 1927 IACP committee remain the focus of today's Uniform Crime Reports and are known as "Part I offenses." Three of these are crimes against persons—criminal homicide, forcible rape, and aggravated assault—and four are crimes against property: robbery, burglary, larceny-theft, and motor vehicle theft. The only substantive change to this list of Part I offenses was made in 1978, when legislation directed that arson be designated a Part I offense; however, arson continues to be reported on a separate form rather than the standard "Return A" used to report the

---

[8] Provisions of the 1988 act also required the UCR to collect data on federal criminal offenses and to "classify offenses involving illegal drugs and drug trafficking as a part I crime."

other Part I offenses. The Part I offenses are also known as "index crimes" because they are used to derive a general, national indicator of criminality—the national Crime Index. The index, first computed and reported in 1958, consists of the sum of the seven original Part I offenses, except that larceny is restricted to thefts of over $50.

The FBI instructs agencies to follow a specified "hierarchy rule" in coding offenses for generation of the monthly summary counts. The FBI directs that multiple-offense situations—incidents in which more than one crime is committed simultaneously—are to be handled by "locat[ing] the offense that is highest on the hierarchy list and scor[ing] that offense involved and not the other offense(s)" (Federal Bureau of Investigation, 2004b:10). This rule is described in Box 4-5. BJS uses a similar "seriousness hierarchy" for classification of events in some of its work on the NCVS, with the important distinction that the hierarchy is applied after data collection for generation of some incident count tables. Thus, "the NCVS collects and preserves information for each crime occurring in the incident, which enables researchers to create their own classification scheme." In comparison, the application of the UCR hierarchy rule at the point of data collection collapses incidents involving several crime types to record just one type, losing the full incident detail (Addington, 2007b:229).

UCR participants are asked to provide monthly reports under the SRS. The basic form tallying the monthly counts of Part I offenses known to law enforcement is known as Return A (arson incidents are reported on a separate monthly return). However, Return A is not the only data collection requested by the FBI and UCR. The SRS also asks participating agencies to complete additional forms. Unless otherwise noted, all of these supplemental forms are also expected to be completed on a monthly basis by reporting agencies:

- *Type and value of property stolen and recovered:* A monthly supplement to Return A queries departments for estimates of the value of stolen property in their jurisdictions; in most instances, this is taken to be the reporting victim's evaluation of the value of the property but may also include estimates researched by the police. Estimates are requested for each of 11 property types (e.g., jewelry, office equipment). A separate table in the supplement asks departments to further classify and total these stolen property incidents (and corresponding values) by the type of crime involved (e.g., if the theft occurred as part of a murder) or the location of the crime (e.g., convenience store, residence).

- *Supplementary Homicide Reports (SHRs):* Since 1962, reporting agencies have also been asked to complete SHRs for every incidence of murder and nonnegligent manslaughter. SHR data provide a wealth of detail about the particular crime of homicide, including what is

---

Box 4-5   Hierarchy Rule for Part I Offenses and Suboffenses, Uniform Crime Reporting Program

1. Criminal homicide
   a. Murder and nonnegligent manslaughter
   b. Manslaughter by negligence
2. Forcible rape
   a. Rape by force
   b. Attempts to commit forcible rape
3. Robbery
   a. Firearm
   b. Knife or cutting instrument
   c. Other dangerous weapon
   d. Strong-arm (hands, fists, feet, etc.)
4. Aggravated assault
   a. Firearm
   b. Knife or cutting instrument
   c. Other dangerous weapon
   d. Strong-arm (hands, fists, feet, etc.)
5. Burglary
   a. Forcible entry
   b. Unlawful entry (no force)
   c. Attempted forcible entry
6. Larceny-theft (except motor vehicle theft)
7. Motor vehicle theft
   a. Autos
   b. Trucks and buses
   c. Other vehicles
8. Arson
   a.–g. Structural
   h.–i. Mobile
   j. Other

The *Uniform Crime Reporting Program Handbook* (Federal Bureau of Investigation, 2004b) defines three major exceptions to use of this hierarchy rule for crime reporting. First, motor vehicle theft—as a special class of larceny, generally—can outrank larceny; hence, the theft of a car with valuables inside it would be coded as a motor vehicle theft (trumping the classification as larceny) even if the vehicle is subsequently recovered but the valuables are not. Arson is also a special case because it is reported on a separate form from the other Part I offense: multiple-offense crimes involving arson can include *two* reported Part I offenses, the arson tally on the separate schedule and the highest-ranking Part I offense under the usual rule reported on Return A. The third exception to the hierarchy rule is justifiable homicide, "defined as and limited to the killing of a felon by a police officer in the line of duty [or] the killing of a felon, during the commission of a felony, by a private citizen." By this definition, justifiable homicide necessarily "occurs in conjunction with other offenses"; those offenses are the ones to be considered in classifying the incident.

known about the relationship between the victim(s) and offender(s), the circumstances of the killing, and the use of weapons. Similar information is also requested for incidents of negligent manslaughter, though traffic fatalities and accidental deaths are not included in this accounting.

- *Age, race, and sex arrest data:* On a monthly basis, agencies are asked to provide counts of completed arrests by the age, race, and sex of the arrestee(s). These data are counts aggregated by type of crime, not individual records per crime incident; moreover, separate counts are requested by age group (16 groups; individual ages 18–24, 5-year groups from 25 to 64, and 65 and over) crossed with sex, but not by race. Totals by race group, four categories, are tallied separately. The age, race, and sex data are requested for Part II offenses as well as the Part I offenses considered in Return A, making these data the UCR's only systematic source of information on these offenses as well as the only source of offender attributes.[9] A separate schedule is used to count crime totals for juveniles (under 18 years of age).

- *Law Enforcement Officers Killed and Assaulted (LEOKA):* Collected on monthly forms and published annually since 1972, the LEOKA data are intended to count line-of-duty deaths or assaults of law enforcement officers. "Line-of-duty" does not mean "on duty" but rather that the officer is acting in an official capacity, responding to a situation that would normally fall within official duties. An eight-page follow-up questionnaire is used to provide additional information on LEOKA incidents in which a firearm or a knife (or other cutting instrument) was used against the officer.

- *Hate crime statistics:* The Hate Crime Statistics Act of 1990 led to the collection of a variable on "bias motivation in incidents in which the offense resulted in whole or in part because of the offender's prejudice against a race, religion, sexual orientation, or ethnicity/national origin" (Federal Bureau of Investigation, 2004b:3). The scope of hate crimes reported in this series was expanded in 1994 to include crimes motivated by victims' physical or mental disability. Aggregate counts of hate crime incidents are reported on a quarterly basis; a one-page report is also requested for every specific incident, recording the of-

---

[9]The 21 offenses currently tallied as Part II offenses are other assaults; forgery and counterfeiting; fraud; embezzlement; stolen property (buying, receiving, possessing); vandalism; weapons (carrying, possessing, etc.); prostitution and commercialized vice; sex offenses; drug abuse violations; gambling; offenses against the family and children; driving under the influence; liquor laws; drunkenness; disorderly conduct; vagrancy; all other offenses; suspicion; curfew and loitering laws (persons under 18); and runaways (persons under 18) (Federal Bureau of Investigation, 2004b:8).

fense type, location, type of bias motivation, victim type, and number and race of known offenders.

- *Law enforcement employees report:* On an annual basis, UCR participant agencies are sent a form to provide a count of full-time sworn and civilian personnel on the payroll as of October 31. Though the *Uniform Crime Reporting Handbook* provides considerable detail on who should and should not be included in the count (Federal Bureau of Investigation, 2004b:124), only the aggregate employee count is requested.

The UCR is, ultimately, a program in which participation in voluntary; consequently, UCR coverage by reporting law enforcement agency can be spotty. Complete nonresponse to the UCR program, for individual years or for long stretches of crime, occurs and is sometimes pervasive for some states and large states and localities. Gaps in UCR coverage have been described most thoroughly by Maltz (1999, 2007). The FBI uses imputation to bridge some of these gaps for deriving national-level estimates.

The status of UCR estimates as crimes officially reported to police imparts a veneer of legitimacy in some respects. For example, government grant programs that use crime information in scoring areas or allocating funds typically use UCR numbers; these include "renewal community" funds from the U.S. Department of Housing and Urban Development (24 CFR § 599.303) and even Justice Department grants to correctional facilities (28 CFR Part 91). In the latter example, the measure of "Part 1 violent crimes" required in grant submissions is defined by 28 CFR § 91.2(c) as those "reported to the Federal Bureau of Investigation for purposes of the Uniform Crime Reports. If such data [are] unavailable, Bureau of Justice Statistics (BJS) publications may be utilized." It is also worth noting that the FBI's duties under current regulation also include "carry[ing] out the Department's responsibilities under the Hate Crime Statistics Act" (28 CFR § 0.85(m)). That is, it is the UCR measure of hate crimes—and not any product of BJS—that is used as the official, legally mandated measure of hate crime prevalence.[10]

---

[10]A BJS report by Harlow (2005) acknowledges the legal distinction, noting that "the Attorney General delegated data collection of hate crimes principally to the FBI." The report compares the NCVS and UCR measures of hate crimes, using the NCVS to provide detail on the circumstances and characteristics of such attacks; the analysis suggests that less than half (44 percent) of hate crimes are reported to police, with no significant difference in reporting of hate- and nonhate-related crimes of the same violent crime type (Harlow, 2005:4). Direct comparison of the NCVS and UCR measures suggest general similarity on some characteristics (categories of offenses, reported motivation) but some differences in demographics (age and race of victim) and some contextual factors (use of weaponry in the incident).

National Incident-Based Reporting System

Development of NIBRS dates to the publication of a joint BJS-FBI task force study (Poggio et al., 1985). Recommendations in this *Blueprint for the Future of the Uniform Crime Reporting Program* led to pilot work with several law enforcement agencies in South Carolina in 1987 and the presentation of the new NIBRS concepts at a national UCR conference in March 1988; the first NIBRS data were received by the FBI in January 1989, and the first NIBRS data for public use were released in 1998 (for 1996 data).

The *Uniform Crime Reporting Program Handbook* (Federal Bureau of Investigation, 2004b:3) notes:

> The intent of NIBRS is to take advantage of available crime data maintained in modern law enforcement records systems. Providing considerably more detail, NIBRS yields richer and more meaningful data than those produced by the traditional summary UCR system. The conference attendees recommended that the implementation of national incident-based reporting proceed at a pace commensurate with the resources and limitations of contributing law enforcement agencies.

The handbook also summarizes the basic content of NIBRS as follows:

> NIBRS collects data on each incident and arrest within 22 offense categories made up of 46 specific crimes called Group A offenses. For each incident known to police within these categories, law enforcement collects administrative, offense, victim, property, offender, and arrestee information. In addition to the Group A offenses, there are 11 Group B offenses for which only arrest data are collected.

The NIBRS incident report is quite intricate and allows for great flexibility in the coding of individual events: Each report can include up to 10 offenses, 3 weapons, 10 relationships to victim, and 2 circumstance codes.

Work on NIBRS has generally concentrated on implementation and in getting additional agencies to use the new format, rather than refinement of the NIBRS instruments themselves. An explanatory webpage published by the FBI observes that (http://www.fbi.gov/ucr/cius2006/about/about_ucr.html):

> In the late 1980s, the FBI committed to hold all changes to the NIBRS in abeyance until a substantial amount of contributors implemented the system. [However,] three modifications have been necessary. To meet growing challenges in the fight against crime, the system's flexibility has permitted the addition of a new data element to capture bias-motivated offenses (1990), the expansion of an existing data element to indicate the presence of gang activity (1997), and the addition of three new data elements to collect data for law enforcement officers killed and assaulted (2003).

Coverage continues to be a problem for NIBRS usage; "in the official 2005 NIBRS data released through [the Inter-university Consortium for Political and Social Research], only about 0.3% of all of the NIBRS reporting jurisdictions for 2005 fall into the 500,000 and 999,999 population category, reporting slightly over 12% of the NIBRS incidents. There are no cities with populations over 1,000,000 reporting data through NIBRS" (Faggiani, 2007:3). As of September 2007, JRSA estimated that only about 25 percent of the nation's population is included in NIBRS-compliant jurisdictions; see http://www.jrsa.org/ibrrc/background-status/nibrs_states.shtml [12/1/07]. In all, about 26 percent of agencies that supply data to the UCR do so using the NIBRS format. Among the states that have not yet implemented NIBRS are California, New York, and Pennsylvania; in Illinois, the only NIBRS participant to date is the Rockford Police Department. Five states—Alaska, Florida, Georgia, Nevada, and Wyoming—have not yet specified any formal plan for participation in NIBRS.

4–C.2   BJS Role in UCR and NIBRS

BJS plays no role in the collection or dissemination of UCR or NIBRS data. However, it has issued grants over the past decade to promote the transition to NIBRS reporting by law enforcement agencies (particularly those in larger states and metropolitan areas). In part, it was BJS's work in the area of administering grant monies by block grant—a duty assigned to it by a 1994 law—that focused some attention on the limitations of UCR data; this led to a period in which BJS issued grants specifically for NIBRS improvement. As Maltz (1999:7, 9) recounts:

> In 1994, in reauthorizing the Omnibus Crime Control and Safe Streets Act of 1968, the U.S. Congress appropriated anticrime funding for jurisdictions under the Local Law Enforcement Block Grant Program. The amount of funds received by a jurisdiction was to be based on the number of violent crimes they had experienced in the 3 most recent years (1992–94). According to the statute, the UCR was to be the source of the crime data. [This action was significant because it] marked the first time that funding decisions were to be made on the basis of the data in the UCR.

BJS was called upon to develop the allocation formula, based on UCR data, to divide funds among localities. As intended by the law,

> BJS used the actual raw crime data as reported by each police agency to the FBI, rather than the imputed data, in the allocation formula. But in reviewing the raw UCR data, BJS immediately recognized their limitations: Of the 18,413 police agencies that reported to the FBI in 1992–94, 3,516 (19%) did not provide crime data for *any* month during the 36-month period used in the formula and another 3,197 (17%) reported between 1 and 35 months.

Hence, BJS worked with the FBI on improving its imputation procedures, including convening an expert conference on the topic.[11] As states struggled with implementation of NIBRS reporting standards, BJS's experience in information technology improvement grants and its basic authority to let local assistance grants led it to administer specific grants to states in the late 1990s to improve their crime reporting capabilities to conform to NIBRS standards.

The "Data Online" and "Data for Analysis" sections of BJS's website provide users with the ability to generate custom tables from UCR data from 1985 though the most recent year of release, and a separate tool generates tables from the UCR SHRs. Curiously, the BJS site provides no such direct tabulations or estimates based on its own NCVS data.

4–C.3   The UCR Program and the FBI's Strategic Priorities

Since September 11, 2001, the FBI has largely recast its mission as one of deterring and preventing acts of terrorism. The bureau's current strategic plan (Federal Bureau of Investigation, 2004a:9) identifies eight priorities, in descending order:

1. Protect the United States from terrorist attack;

2. Protect the United States against foreign intelligence operations and espionage;

3. Protect the United States against cyber-based attacks and high-technology crimes;

4. Combat public corruption at all levels;

5. Protect civil rights;

6. Combat transnational and national criminal organizations and enterprises;

7. Combat major white-collar crime; [and]

8. Combat significant violent crime.

The plan further identifies two "key enabling functions that are of such importance they merit inclusion:" "Support federal, state, local, and international partners" and "Upgrade technology to successfully perform the FBI's mission."

In light of these evolving priorities, the question can be raised as to whether the UCR program receives due resources and attention or whether administration of UCR detracts from the FBI's overall strategic objectives. The UCR program is mentioned briefly in the FBI's most recent strategic

---

[11]In 2004, BJS issued a technical report (Bauer, 2004) summarizing BJS's final allocation formula and the amounts of money dispersed to the statements under the Local Law Enforcement Block Grant Program from 1996 to 2004. BJS was formally tasked with deriving the formula for the Edward Byrne Memorial Justice Assistance Grant program, the successor to the block grant program (Hickman, 2005a).

plan, under the strategic goal of the FBI's Criminal Justice Information Services (CJIS) Division. The discussion of strategic objective IVD.1, "Expand information sharing capabilities to support customer needs," notes (Federal Bureau of Investigation, 2004a:98):

> An array of state-of-the-art technology in the Uniform Crime Report (UCR) Program is needed to provide more efficient, optimum quality, and timely products and services to law enforcement and other consumers of UCR crime data. The new system will optimize the production capabilities of the existing CJIS information systems by leveraging the immense amount of data already regularly contained in each system repository.

This cursory mention of the program appears to concentrate on the production of estimates and products, rather than any ongoing assessment of the quality and timeliness of the actual data from local agencies. The strategic plan also identifies an expansion of NIBRS content as an objective (Federal Bureau of Investigation, 2004a:99):

> [The "enhanced NIBRS"] data set combines the current 53 NIBRS crime descriptors with the specific personal and event identifiers which form the core of most police department incident reports. Many states collect more information than the current 53 NIBRS data elements describing incidents for their own in-house purposes. To realize the full potential of the information sharing capabilities of NIBRS data, additional identifying data (e.g., victim, offender, and suspect) must be included, which will provide law enforcement additional investigative leads.

However, this topic is only identified as an area for which an implementation plan should be developed.

4–C.4   Assessment

Managing the BJS-UCR Relationship

As we argued in more detail in our first report (National Research Council, 2008b:Sec. 3–F), we do not see the relationship between the FBI's UCR program and BJS's NCVS (and related data series) as an either-or proposition. Although the two programs overlap in that they cover a similar set of crimes and are both used to generate national-level estimates of violent crime, their major differences in scope and methodology make each a valuable source of information on crime and violence. The major features of the UCR compared with the NCVS are summarized in Table 4-1.

The intended successor to the UCR summary reporting program, NIBRS, has fallen short of its promise to date because of the slow adoption of the more detailed reporting scheme by local departments. In turn, the low coverage in NIBRS has also affected the extent of available research that, by

Table 4-1 National Data Sources Related to Crime Victimization in the United States

Data Characteristics	NCVS	UCR	
		Summary	NIBRS
Target population	Noninstitutionalized persons age 12 and older in the United States	Crime incidents occurring in the United States	Crime incidents occurring in the United States
Unit of observation	Individual	Law enforcement agency	Crime incident
Estimated coverage	Nationally representative sample	94.2% of U.S. population covered by agencies active in UCR reporting	Approximately 25% of U.S. population covered by agencies reporting in NIBRS format
Types of victimization covered			
Criminal homicide	No	Yes	Yes
Other index crimes	Yes	Yes	Yes
Geographic areas identified			
Region	Yes	Yes	Yes
State	Yes*	Yes	Yes
County	Yes*	Yes	Yes
Census tract	Yes*	No	No
Demographic coverage			
Age	Yes	No	No
Race	Yes	No	Yes
Sex	Yes	No	Yes
Ethnicity	Yes	No	Yes
Vulnerable groups			
Children	12 & older	No	Yes
Immigrants (native born)	No	No	No
Disabled (learning disability only)	No	No	No
Elderly	Yes	No	No
Timeliness of data availability			
Pre-announced schedule	Yes	Yes	Yes
Fixed schedule	Yes	Yes	Yes
Accuracy and quality			
Sampling error	Routinely estimated	Unmeasured	Unmeasured
Other errors (nonsampling)	No ongoing evaluation	Unknown	Unknown

* These areas are not identified in the public use files currently available from the National Archive of Criminal Justice Data, but are on area-identified files maintained by the Census Bureau. It should be noted that the NCVS sample is not drawn with the intent of producing estimates at these geographic levels.

generating particularly interesting or useful findings, could spur greater interest and participation in the series. As Faggiani (2007:2) summarizes:

> The implementation of NIBRS by local law enforcement agencies is an evolving but slow moving process and this has had an impact on its use for research. The early NIBRS data releases (1996 through 1998) were mostly limited to small and medium-sized law enforcement agencies representing primarily rural states. For example, the 1996 data covered only nine cities with populations in excess of 100,000 and no cities with populations over 250,000. Researchers examining the utility of NIBRS for scientific research began to raise serious questions about the overall representativeness of this supposed national data system.

For the purpose of studying the occurrence of crime in the United States, a healthy UCR program and, particularly, a full-fledged NIBRS are both critical data systems. A fully featured NIBRS with high participation has the potential to shed light on some dynamics of law enforcement operations that are not visible in current data (and not even envisioned within the tight management and administrative focus of the LEMAS series). One such example is the potential for NIBRS to provide detail on police clearance rates, which are currently reported as a gross indicator of departmental success. However, the aggregate rates, minus the type of contextual information on incidents and the extent of police contact with victims and offenders, mask a great deal of potentially useful information. On an explanatory basis, Addington (2007a) used the incident and police clearance date recorded in NIBRS data to test (and confirm) the conventional wisdom that those murder cases that are cleared by police tend to be cleared early and that there is a major drop in the clearance rate after more than a week has passed since the homicide. With fuller data resources, and study of different crime types, NIBRS could provide a useful platform to study the factors that influence the successful clearance of crimes by police.

> **Finding 4.3:** A full-fledged NIBRS would be a source of basic information on police responses to public complaints (911 calls), including whether or not a case is "cleared" by police through an arrest.

Having concluded that the UCR and NIBRS programs have their merits—and that the nation benefits from having multiple data systems (UCR/NIBRS and the NCVS) to measure the incidence and circumstances of crime from different perspective—the question that remains is whether they should be managed by separate parts of the Justice Department. Put more bluntly, the question is whether it makes sense for BJS, the principal statistical agency in the Justice Department, to lack authority for what is arguably the most prominent statistical data series produced by the department, and whether it would be preferable for BJS to "take over" UCR operations from the FBI.

Rosenfeld (2007:830) argues that "the FBI is no longer the appropriate institutional home for the UCR program, if it ever was." The principal argument for the transfer is that BJS is more likely to be able to provide the technical support and capability for ongoing improvement of the UCR than the FBI, given that "tracking conventional crime is not a high priority in the [FBI's] post 9/11 focus." The "appropriate focus and necessary human and technical resources" to best monitor locally recorded crimes reside in BJS rather than the FBI and its administrative placement is more historical artifact than organizational efficiency: "had the BJS existed 75 years ago, the responsibility to compile local crime statistics would have been placed there," but by the time BJS was founded in 1979, the UCR was well entrenched as part of the FBI and little incentive existed to transfer authority for the UCR program (Rosenfeld, 2007:831). To be sure, Rosenfeld (2007:831) argues that transfer of the UCR program "is a necessary but not sufficient condition to upgrade the nation's crime monitoring capabilities," and that BJS would require additional resources to make substantial improvements in the timeliness and quality of UCR estimates.

Rosenfeld (2007) further cites a recent example of both UCR and NCVS being bested by another data source—in terms of timeliness of information—as motivation for improving both benchmark measures of crime through an organizational realignment. By August 2006, "local police chiefs had been complaining for months . . . that violent crime was on the rise and that they lacked the resources to combat it." However, such a shift in crime rate (reversing several years of declining trends) could not be measured by UCR: "the FBI report [of UCR results for 2006] was not released until September 2006 and covered only the period through the end of 2005." The FBI released a preliminary report covering data from the first half of 2006 in December 2006—rapid dissemination by UCR standards, but in a sense the information was still too late. Any reading from the NCVS lagged behind the FBI's figures, meaning that "no single source of [publicly available] systematic data" existed to refute or corroborate the chiefs' claims. That August, the Police Executive Research Forum (PERF) convened a Violent Crime Forum of police chiefs, the result of which was a compilation of current crime data from several of the participating police departments. PERF and the chiefs used these data to describe apparent crime increases in its report *A Gathering Storm: Violent Crime in America* (Police Executive Research Forum, 2006)—a report that went public in October 2006. An update of that report, published in April 2007, pointedly reminded readers that PERF encourages police agencies "not to wait" for the FBI to release its UCR crime figures and to send their data directly to PERF for compilation and early release (Rosenfeld, 2007:826).

The basic arguments for BJS acquiring authority for the UCR and NIBRS programs include, in brief:

- The operational transfer would solidify BJS's position as the preeminent statistical agency within the Department of Justice and the preeminent governmental source for justice statistics, generally.

- The transfer would permit the FBI to sharpen its new organizational focus on antiterrorism efforts.

- Placing authority for the UCR and NIBRS within a true statistical agency would facilitate attention to methodological problems in the series, including adjustments for nonresponse and imputation routines.

The strongest argument *against* BJS acquiring control of the UCR is the potential disruption of the relationships that have built up over the decades of FBI administration of the UCR, brokered through outlets such as the IACP. A great strength of BJS's correctional data series and a key to their quality is the network of ties that BJS has built with state departments of corrections and individual facilities; likewise, BJS continues to develop ties to state court systems. Just as it is reasonable to expect that the quality of resulting data would be impaired by a sudden change in reporting structures, shifting lines of data reporting that have existed, in some cases, since the 1930s is not something that should be taken lightly. There is, moreover, a trust that may be implicit in UCR reporting relationships—law enforcement agencies providing data to a fellow law enforcement agency—that is nontrivial. It is certainly possible that, with strong endorsement and assistance from collaboratives such as the IACP and PERF, a transition from FBI to BJS could be successful in time, but the short-term impact on response rates could be significant. In our assessment, the prospect of BJS "taking over" the UCR picks unnecessary turf fights with both the police community and the FBI, both of which have historically been protective of the program.

Although the organizational transfer of UCR from one Justice Department agency to another would seem to be a fairly easy task, we think that this appearance is deceptive. The suggestion severely underestimates the level of energy and expense that would be necessary to get the UCR SRS (much less a full-fledged NIBRS) to function efficiently and effectively under a new administrative parent and as a part of the statistical system. In our interim report, we noted the inherent rigidity of the UCR—that, for instance, the core set of crime types covered by the UCR has remained the same since the UCR's creation in 1929 (save for the addition of arson as a top-tier "Part I" crime. We commended the value of the NCVS as an independent check on the UCR (and vice versa), but urged that "the utility of an UCR-independent measure of crime should not prevent consideration of [NCVS] design options that reduce lockstep similarity between the UCR and the NCVS" (National Research Council, 2008b:77). To make clear the tacit criticism in that statement, the UCR is in some respects an antiquated and inefficient system for collecting and disseminating annual estimates of level and changes in crime reported to the police at the national level. As a census-type measure of all jurisdictions, the UCR has and will continue to have essential roles, including such purposes as allocating funding across all jurisdictions based on crime counts. However, our concern is that the products of a principal statistical agency are held to high standards (as we describe in great detail throughout Chapter 5). In particular, recast as a core statistical collection within a principal statistical agency, a BJS-led UCR would require much more intensive—and expensive—attention to issues of data quality, response, data collection instrumentation, and documentation than has previously been brought to bear on the UCR.

Clearly, as we have indicated, it is in the national interest to have a high-quality UCR program. It follows that if the FBI's strategic goals shift even more heavily toward its expanded portfolio in terrorism surveillance, and hence that attention to and resources for administration of the UCR become so scarce that the UCR and NIBRS programs will atrophy, then an administrative transfer of authority of UCR to BJS would be sensible (short-term effects on response notwithstanding). Barring these conditions, we find no compelling reason, other than the organizational neatness of consolidating statistical functions in one agency, for UCR to shift away from the FBI.

Timely Records-Based Collection from Local Law Enforcement Files

Rather than "take over" UCR, as the option might be bluntly described, our recommendation is that BJS explore the possibility of doing something that is different from either the UCR or NCVS and that we think is better than the UCR in some respects. Specifically, we suggest that BJS work with local law enforcement agencies to develop a system under which BJS could regularly extract records from individual departments' own computer systems—data that many departments regularly compile on their own and some departments post on their websites—for a sample of jurisdictions. This system would shift much of the burden of response and data gathering from the local authorities (i.e., filling out UCR summary forms) to BJS (i.e., sample design, data editing, and inference). Such a system would be capable of providing more timely (if less detailed) glimpses of crime trends than is possible under UCR, NIBRS, or the NCVS.

The strategy we propose is consistent with one commonly used by statistical agencies, pairing sample- and census-based methods. Among U.S. federal statistical agencies, as well as in government statistical systems around the world, it is common to conduct sample-based measurements in parallel with more exhaustive, census-type measurements of the same basic phenomena. For example, BLS interviews a large sample of employers (covering about 390,000 worksites) each month as part of its Current Employment Statistics (CES) program. Each month, the CES data are used to produce a count of changes in the number of jobs in the country; this monthly "Employment Situation" report is a familiar and highly publicized barometer of economic conditions. The CES also supports production of employment figures for states and metropolitan areas.[12] This monthly measure of national, state, and local employment conditions is complemented by the Quarterly Census of Employment and Wages (QCEW), which taps into data from the state-based unemployment insurance systems. Thus, the QCEW provides a more comprehensive census-type count in the change of jobs at the national level (with QCEW coverage estimated to include 98 percent of U.S. jobs), while yielding detailed estimates down to the county level.[13] Another example is the national Vital Statistics program administered by NCHS, which we described in Section 4–A.1. Drawing from reports from state health departments and reporting authorities, the vital statistics estimates represent a census-type measure of births in the United States. However, NCHS also fields the National Survey of Family Growth (NSFG; see http://www.cdc.gov/nchs/NSFG.htm), which measures self-reported births among probability samples of females, providing independent measures of births and fertility. The NSFG is also capable of generating more detail about the pregnancies and related births and yields national estimates of marriage and divorce, topics that are no longer covered by the Vital Statistics program. The UCR and the NCVS do not fit this paired census- and sample-based measurement approach because of the much more extensive scope of the NCVS, including crimes and general incidents of victimization that are not reported to the police.

There are three reasons that such paired census-based and sample-based measures are useful to policy makers and professionals. First, the sample-based measurements can be constructed to be more timely than the census-based measurements. Individual attention can be paid to each sample unit; efforts of interviewers and other agents can raise the level of quick response to the survey request. Second, by focusing survey research resources on a small number of sample units, the quality of reporting can be raised, over that expected from administrative systems. Often this higher quality is obtained through increased standardization in reporting across the various sample units. Because participation is assisted through the survey data collectors, estimates can be published more quickly, to the benefit of the country. However, and the third reason for parallel samples and censuses,

---

[12]Because of cuts in BLS's funding for fiscal year 2008, BLS has had to eliminate the production of some CES estimates for all metropolitan areas and completely eliminate the generation of estimates for the smallest metropolitan areas; see http://www.bls.gov/sae/msareductions.htm.

[13]See http://www.bls.gov/ces/ and http://www.bls.gov/sae/ for additional information on the CES program and http://www.bls.gov/cew/ regarding the QCEW.

those benefits come at a price: the sample sizes are generally inadequate to offer stable estimates of the prevalence of rare events or estimates of differences of small subsets of the population. Only with full census-based measurement can analysis at such levels be accomplished.

In short, the sample-based methods are appealing because they can offer higher-quality responses, obtained in a timelier fashion. In contrast, census-based methods are necessary to understand small subpopulations or rare events related to the phenomenon of interest. Facilitating both of these systems makes sense, especially when the census-based system is already in place for administrative and management reasons.

The panel believes that the country would be well served by a similar parallel structure applied to crimes reported to police. The new system would rely on the UCR as the census-based vehicle that would permit fine-grade comparison of areal patterns of crime, while a sample-based measurement drawn from the crime information compiled and disseminated in real time by many police departments would provide a more timely indicator of general crime trends. The NCVS, of course, would be a third component of this new structure, adding benefit through its rich contextual information and coverage of incidents not reported to police.

> **Recommendation 4.4:** **To improve the timeliness of crime statistics, BJS should explore the development of a crime reporting system based on a probability sample of police administrative records. The goals of such a system would be national representativeness, high response, high data quality, timeliness and flexibility in terms of crime classification and analysis, and national statistics for the monitoring of crime trends.**

Such a system has two notable precursors. First, it hearkens back to the original purpose of the NIBRS program, "to take advantage of available crime data maintained in modern law enforcement records systems" (Federal Bureau of Investigation, 2004b:3). Where NIBRS has endured slow implementation because of the need to develop systems to provide rich incident detail, the hope of a timely records-based summary system would be to tap the more aggregate crime statistics used by departments to measure their own progress. The second key precursor is the one that makes such a records-based system feasible at this time, when it has not in the past. The New York City Police Department, under then-commissioner William Bratton, is credited with developing the COMPSTAT approach to measuring and planning responses to crime in 1994. Alternatively described as a contraction for "computational statistics" or "comparative statistics," COMPSTAT combined a managerial focus emphasizing internal accountability among district and regional commanders for crime activity in their areas with a technical basis in the use of timely crime incident data to identify problems and tailor responses. The COMPSTAT program was popularly credited, at least in part, for major crime rate decreases in New York, which in turns spurred the adoption of similar programs in other cities. COMP-STAT implementation and use in a variety of sites, including detailed case studies, have been reviewed by Willis et al. (2003), Weisburd et al. (2003), and Weisburd et al. (2004).

The system we suggest is akin to, but not as extensive as, the "national COMPSTAT" that has been suggested by PERF in the wake of PERF's initial work in combining local-agency records (and hence monitoring crime-rate shifts earlier than data provided by either the UCR or the NCVS). In his introduction to PERF's report on its second Violent Crime Summit, PERF Executive Director Chuck Wexler described the "national COMPSTAT" idea (Police Executive Research Forum, 2007:iv; emphases in original):

> *We want to change the way that people view crime. In the past, crim-inologists waited several years to make conclusions about crime trends. They were cautious about drawing conclusions and waited until they could state with scientific certainty that there was a changing pattern. The problem with that approach is that by the time a crime trend has been identified, the information is so old as to make it useless, because new trends, new crime patterns, and new causes of crime have taken hold. Programs and policies that we undertake today, to respond to the crime problems of last year, are not likely to succeed.*
>
> [A "national COMPSTAT" approach would use] accurate, timely in-formation to track crime as it happens, to search for pockets of violence wherever and whenever they occur, and to react quickly. **In a sense, we believe that police leaders should act more like public health epidemi-ologists, who don't wait for a pandemic to overtake the nation, with hundreds or thousands of people dead, before they sound an alarm and start implementing countermeasures.**

We say that our proposal is akin to but not identical to a "national COMPSTAT" in that our proposal is to generate a timely summary or index value of crime, and not a direct analysis of the geographically coded incident data that can be used to inform COMPSTAT managers' decisions to allocate personnel and resources. Valuable though such data would be, there is an in-herent tension between the statistical information that an agency such as BJS can provide and the fine-grained, "tactical" assessments—informing specific interventions and used to hold line officers and commanders accountable—that are of particular interest to law enforcement agencies. Rather, our in-tent is to provide policy makers with crime data that provide information that is both *contemporaneous* (immediately of interest and relevance to con-sumers in law enforcement) and *timely* (short time between collection and dissemination).

Our intent in proposing this system is not to be duplicative of effort in the existing UCR data collection. Rather, the idea is to maximize the use of those data systems that local police departments prepare and maintain on their own, as a course of doing business and, in several cases, to promote public transparency by providing up-to-date crime information for public view on websites. Rather than put the burden for processing, coding, and interpreting results on the local departments (through completion of summary report questionnaires), the intent of this system would be to create means by which available electronic data could be transmitted to BJS with minimal effort on the part of local departments (save, perhaps, for stripping personal identifiers). The burden of sample design, data processing, and compilation would be on BJS. However, the availability of the data in electronic form and the development of routines for handling specific varieties of local data types would, ideally, yield a system in which some summary measure or index could be generated and disseminated quickly.

The value of BJS involvement in compiling crime data from local departments (and, in some cases, publicly posted online as well as used in internal meetings and assessments) would be rigor in design (documenting that index measures are representative of some larger whole, if not the entire nation then something like urban areas of a particular size), consistency in definition and coding, and attention to data quality. Some further comments along these lines may be useful:

- Though the objective of such a system would be to be minimally invasive—making use of data that departments already have, and have in electronic format—success in implementation will depend on building ties with and commitments from individual departments. Making use of consortia such as PERF (given its initial work in the area), the Major City Chiefs Association, and IACP would be instrumental.

- A necessary first step in constructing such a system is to assess its basic feasibility: studying individual departments' technical capability and the availability of suitable data resources. Short of a complete inventory of technical systems status (such as BJS performed for state corrections departments and individual facilities; Bureau of Justice Statistics, 1998b) or its SEARCH-conducted Survey of Criminal History Information Systems (Section 4–B), a module of questions on the LEMAS survey would be a useful start.

- The sample panel of departments would likely emphasize the largest cities, given that they are most likely to have amenable record systems. The preliminary work we just described would be useful in determining whether the content must be limited to such large cities (say, above 250,000 population) or whether a more nationally representative sample including smaller agencies is feasible. In constructing the sample, BJS should take into account the distribution in crime with respect to population size and changes in that distribution. Crime is not as highly concentrated in the most populous of cities as in past decades, with historically high-rate large cities having been some of the beneficiaries of major declines in the 1990s and 2000s.

- For jurisdictions that organize their collections by their relevant state criminal codes, and whose definitions of standard crimes are not consistent with the FBI's UCR definitions, BJS will need to learn and apply the techniques used by the localities in "converting" their data in order to report UCR summaries. A crime type such as aggravated assault, for instance, can vary substantially in scope across the states, and so a common metric would need to be defined.

- As a sample-based equivalent to the UCR, and for keeping collection tractable, an initial focus on UCR Part I crimes is sensible. Over time, an interesting question is whether to expand beyond that scope. Data on arrests with no "crimes known to police" equivalent in the UCR, such as drug, weapons, and DWI arrests would likely have strong policy interest.

- Using a panel of reporting departments should make it possible to develop quarterly estimates. However, at least two technical challenges would need to be resolved in order to smooth the introduction of a new data collection effort. The first is the exact timing and coordination of collection from "live" incident files. Department incident files may be subject to revision after initial entry, "closing" for reporting purposes at some defined data period (such as a month or quarter). Protocols for timing access to and collection of data would need to be developed, as well as mechanisms for updates as necessary (e.g., when assault turns into homicide when a victim survives a shooting for a long period). The second technical challenge is the resolution of the data that the local departments could, and could most feasibly, give to BJS: whether incident-level files as in existing NIBRS or aggregate summary reports like the agencies now send to the FBI. In some ways, aggregated summary figures would be the least demanding on both suppliers and BJS, but may require additional care (and timing) in coding to reflect federal definitions. Raw incident-level files would obviously be analytically useful, and BJS could request them geocoded, but this would be substantially harder to negotiate and would impose a greater burden on the data collector and on BJS as aggregator. BJS and its data collectors would have to anticipate a two-track plan, working with data of either type, until one became more universally preferable.

The implementation of such a sample-based analogue to the UCR would enable great strides in making BJS's law enforcement data more timely and relevant to data users, and would well complement a reengineered core-supplement design for LEMAS. In time, success in building two relatively nimble systems could prove useful for law enforcement agencies and policy makers alike. For instance, a LEMAS supplement could provide a "quick response" capability for documenting local agencies' experience with some new or emerging crime problem; the records-based selection could begin to show continued growth of the problem (or hint at signs of resolution); and the UCR and, particularly, the NCVS would be poised to provide richness of information on contexts and possible causes that are not possible with the interim indicators. All of this—well-designed data systems working at multiple resolutions—would contribute greatly to BJS's core mission to assist state and local governments and agencies.

– 5 –

Principles and Practices: BJS as a Principal U.S. Federal Statistical Agency

OUR CHARGE DIRECTS US to provide guidance to the Bureau of Justice Statistics (BJS) regarding its strategic priorities and goals. Before doing this, it is important to consider the functions—and expectations—of BJS from a higher, agency-level perspective.

One important filter through which to view the priorities and operations of BJS is its role as one of the principal statistical agencies in the U.S. federal statistical system. Relative to other countries, the U.S. federal statistical system is highly decentralized. Whereas other countries vest the primary authority for collection and dissemination of statistical data in a single agency—the Australian Bureau of Statistics, Statistics Canada, and Statistics Netherlands, for example—authority for production of official statistics in the United States is divided across numerous agencies.[1] These agencies are by no means equal in terms of their staffing levels and budgetary re-

---

[1] The statistical system of the United Kingdom is also frequently cited as an example of centralization; it is currently in a state of change. Effective as of April 2008, a new Statistics and Registration Service Act formally abolished the legal role of the Office for National Statistics (ONS), previously the United Kingdom's dominant statistical agency. ONS functions continue, but the office is now a subsidiary of the Statistics Board, created as an independent corporate body as the arbiter and producer of official statistics in the country. As discussed in our panel's interim report (National Research Council, 2008b), the British Crime Survey is an example of a United Kingdom data collection that is *not* collected by ONS or the Statistics Board; it is administered by the Home Office.

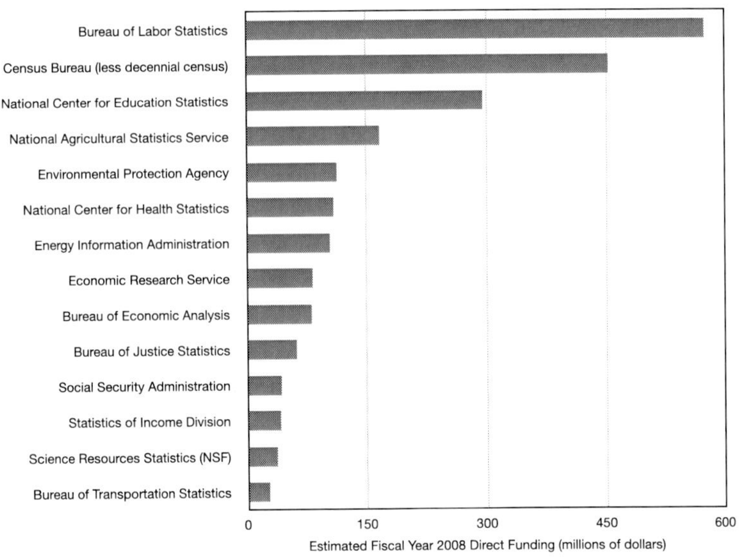

Figure 5-1   Estimated direct funding levels for principal federal statistical agencies, fiscal year 2008

NOTES: NSF, National Science Foundation. Including the costs associated with the decennial census would add $797.1 million to the Census Bureau total.

SOURCE: U.S. Office of Management and Budget (2007:Table 1).

sources. As shown in Figure 5-1, the three largest statistical agencies—the Census Bureau, the Bureau of Labor Statistics, and the National Center for Education Statistics—dominate the others in terms of resources even though the subject-matter portfolios of the smaller agencies—justice, transportation, agriculture, and so forth—are undeniably important.

It is appropriate, in the panel's judgment, to evaluate BJS in the context of the larger federal statistical system, especially the principal statistical agencies whose primary mission is the collection and dissemination of statistical information. (There are 60–70 other federal agencies that spend more than $500,000 per year on statistical information dissemination, but whose program duties outweigh their statistical focus.)

The panel benefited from a preexisting, fully vetted set of evaluative criteria for a federal statistical agency. The observations that we make in this chapter are generally structured around the *Principles and Practices for a Federal Statistical Agency*, a white paper of the Committee on National Statistics (CNSTAT) (National Research Council, 2005b). *Principles and Practices* articulates the basic functions that are expected of a unit of the U.S. federal statistical system; it also outlines ideals for the relationship between individual statistical agencies and their parent departments. As such, it has been widely used by various statistical agencies in their interactions with Congress and with officials in their departments. Indeed, BJS has already embraced such evaluative criteria: a summary version of the *Principles and Practices* is featured prominently on the front page of BJS's current strategic plan (Bureau of Justice Statistics, 2005a) and as a top-level link on BJS's website (under "BJS Statistical Principles and Practices").

The two sections of this chapter assess BJS, its products, and its performance relative to the *Principles and Practices of a Federal Statistical Agency*; each subsection begins with a précis of the relevant descriptive text from the fourth edition of *Principles and Practices* (National Research Council, 2009). In the course of this review, we provide extended discussion of two recent "flashpoints" in recent BJS experience—the circumstances surrounding release of data from the 2002 Police-Public Contact Survey (PPCS) that led to the dismissal of a BJS director and the reporting requirements imposed by the Prison Rape Elimination Act of 2003—that are particularly relevant to examination of the major principles of a statistical agency. We defer conclusions and assessments based on the chapter as a whole to Chapter 6, a more comprehensive statement on strategic goals for BJS.

5–A PRINCIPLES OF A FEDERAL STATISTICAL AGENCY

5–A.1 Trust Among Data Providers

A federal statistical agency must have the trust of those whose information it obtains. Data providers, such as respondents to surveys and custodians of administrative records, must be able to rely on the word of a statistical agency that the information they provide about themselves or others will be used only for statistical purposes. An agency earns the trust of its data providers by appropriately protecting the confidentiality of responses. Such protection, in particular, precludes the use of individually identifiable information maintained by a statistical agency—whether derived from survey responses or another agency's administrative records—for any administrative, regulatory, or law enforcement purpose. (National Research Council, 2009:5–6)

In a democracy, government statistical agencies depend on the willing cooperation of resident respondents to provide information about themselves and their activities. This willingness requires assurance that their information will not be used to intervene in their lives in any way. When respondents to BJS data collections provide data to the agency (or contractors representing the agency) they are told that their individual data will never be used to harm them; they are told that the purposes of the data collection will be fulfilled by publicly available statistical results; they are informed that the agency is an independent statistical organization transcending the current administration in power; they are informed that their data will be kept confidential. Agencies that fulfill such pledges build over time with their data providers a sense of trust that the agency's intentions are benign and that the agency respects their rights as data providers. When political interference is suspected by data providers, the trust that their reports are being used appropriately—solely to create statistical information—can be shaken, and restoring that trust can be a much slower process than its destruction.

BJS has many different target populations of data providers. Some are very large (e.g., the entire U.S. household population for the National Crime Victimization Survey (NCVS) or the full set of state courts of general jurisdiction) whereas others are quite small (e.g., state-level departments of correction or federal prisons). Small populations of data providers generally are repeatedly asked for data in ongoing BJS series. In turn these data providers are often more interested in the outcome of the data collections and may use the statistical information for their own purposes.

From its review of BJS documents, knowledge of its data sets, and interactions with its respondents, the panel concludes that BJS and its data collection agents are generally very diligent in preserving the confidentiality of responses from its respondents. This is particularly true for the NCVS, the effectiveness of which is wholly predicated on building trust and rapport between interviewer and respondent in order to obtain full and accurate accounts of victimization incidents. The use of respondents' data solely for statistical purposes is generally well known and presented.

However, we note that this is only "generally" true in that there exists a flagrant exception. In the judgment of the panel, the reporting requirements of the Prison Rape Elimination Act of 2003 (PREA) oblige BJS to violate the principle of trust among its institutional data providers. Specifically, the provision of information to a statistical agency is fundamentally different from the provision of information to a regulatory or enforcement agency. Regulatory agencies, by their very nature, have the goal of intervention in individual activities when they are found to violate some prescriptive actions sanctioned by the government. The crux of the problem is that the PREA reporting requirements assign to BJS a quasi-regulatory role, directly using data collected from responding institutions to impose sanctions. In the remainder of this section, we describe this breach of principle by describing the history and the implementation of the PREA reporting requirement.

Historical Development of the Prison Rape Elimination Act

In *Farmer v. Brennan* (511 U.S. 825 [1994]), the U.S. Supreme Court ruled that "deliberate indifference" to serious health and safety risks by prison officials constitutes a violation of the Eighth Amendment protection against cruel and unusual punishment. The particular case in *Farmer* involved a preoperative transsexual prisoner who was raped and beaten shortly after transfer to a federal penitentiary; the Court's ruling vacated lower court rulings that rejected the plaintiff's argument on the grounds that prison officials had not been demonstrated to be criminally reckless.

By 2002–2003, the general problem of sexual assault in prison drew legislative interest in Congress. Ultimately, the legislative initiative produced PREA. The final act is lengthy, including specification of grant monies targeted at reduction strategies, the establishment of a national commission, and adoption of national standards. However, in this section, we focus on the specific demands put on BJS by a section of the act covering "national prison rape statistics, data, and research"—reporting requirements that, in certain respects, run counter to the proper and accepted role of a federal statistical agency.

In the 107th Congress, identical versions of a proposed "Prison Rape Reduction Act" were introduced in both houses (H.R. 4943 and S. 2619). On the occasion of the introduction of the measure in the Senate, cosponsor Sen. Edward Kennedy (D-Mass.) described what little was known quantitatively about the extent of sexual assault in U.S. prisons:[2]

> Prison rape is a serious problem in our Nation's prisons, jails, and detention facilities. Of the two million prisoners in the United States, it is conservatively estimated that one in ten has been raped. According to a 1996 study, 22 percent of prisoners in Nebraska had been pressured or forced to have sex against their will while incarcerated [(Struckman-Johnson et al., 1996)].[3] Human Rights Watch recently reported, "shockingly high rates of sexual abuse" in U.S. prisons [(Human Rights Watch, 2001)].[4]

Cosponsor Sen. Jeff Sessions (R-Ala.) concurred, and briefly described the statistical analysis section of the bill:[5]

> Some studies have estimated that over 10 percent of the inmates in certain prisons are subject to rape. I hope that this statistic is an exaggeration. . . .

---

[2]*Congressional Record*, June 13, 2002, p. S5337.

[3]Struckman-Johnson et al. (1996:69–70) distributed questionnaires (for response by mail) to all inmates and staff at two maximum security men's prisons, one minimum security men's prison, and one women's facility, all of which are "in the state prison system of a rural Midwestern state." The state is not explicitly identified, but later discussions of the results included the acknowledgment of Nebraska as the survey site. In all, 1,801 prisoners and 714 staff members at these facilities were eligible to participate; 528 inmates and 264 staff members responded.

[4]No formal survey or statistical data collection was used by Human Rights Watch (2001); instead, the report's observations were based on written reports from about 200 prisoners, responding to announcements in publications and leaflets.

[5]*Congressional Record*, June 13, 2002, pp. S5337, S5338.

> [This] bill will require the Department of Justice to conduct statis-
> tical surveys on prison rape for Federal, State, and local prisons and
> jails. Further, the Department of Justice will select officials in charge of
> certain prisons with an incidence of prison rape exceeding the national
> average by 30 percent to come to Washington and testify to the Depart-
> ment about the prison rape problem in their institution. If they refuse
> to testify, the prison will lose 20 percent of certain Federal funds.

In both chambers, the legislation was referred to Judiciary subcommittees
and no further action was taken (save that the Senate Judiciary Committee
held a hearing on the bill on July 31, 2002).

In the 108th Congress, legislation identical to the previous bill was introduced by Rep. Frank Wolf (R-Va.) and Rep. Bobby Scott (D-Va.) in the
House as H.R. 1707 on April 9, 2003.[6] However, deliberations between
members and staff in both chambers were progressing toward a revised, bi-
partisan proposal, and these deliberations resulted in rapid passage of the
bill. On June 11, 2003, the House Subcommittee on Crime, Terrorism, and
Homeland Security replaced the existing text of H.R. 1707 with substitute
language and favorably reported it to the full Judiciary Committee. In turn,
the Judiciary Committee approved the revised bill on July 9. The Judiciary
Committee's report on the bill, H.Rept. 108-219, offers no explanation for
the revised wording in the BJS data collection section of the act. On July
21, Sen. Sessions introduced S. 1435—consistent with[7] the revised House
language, but now bearing the name "Prison Rape Elimination Act." Upon
introduction, the bill was immediately passed by unanimous consent without debate or amendment; the House took up the Senate bill on July 25
and passed it without objection; and the bill was signed on September 4,
becoming Public Law 108-79.

Text of the Act and Reporting Requirements

Box 5-1 shows the alterations to the section of PREA concerning BJS
data collection between its original introduction in the 107th Congress and
final passage. Both the original and final versions of the bill establish a Review Panel on Prison Rape; the original would have administratively housed
the Review Panel in BJS while the final version makes it an organ of the Justice Department. To be clear, it is important to note that the Review Panel
is more limited in scope than the National Prison Rape Elimination Commission created by other sections of the act. The Review Panel's work is
structured around the BJS work, while the formally appointed Commission

---

[6]A variant on the same bill, with the same reporting requirements on BJS, was introduced
on April 10, 2003, as H.R. 1765 but progressed no further than referral to committee.

[7]Judiciary Committee Chairman James Sensenbrenner (R-Wisc.) described the Senate bill as
"substantively identical to H.R. 1707" in his floor remarks on passage of the act (*Congressional
Record*, July 25, 2003, p. 7765).

Box 5-1 Statistical Reporting Provisions of Original and Final Versions of the Prison Rape Elimination Act

The following excerpt compares text from Section 2 of H.R. 4943 (107th Congress) and Section 4 of S. 1435 (108th Congress), the latter of which was enacted as Public Law 108-79. Subsections (d) and (e) on contracts and authorization of appropriations are omitted. Deletions from the earlier version are marked in ~~strikethrough text~~; additions in the newer version are shown in *italic type*.

NATIONAL PRISON RAPE STATISTICS, DATA, AND RESEARCH.

   (a) ANNUAL COMPREHENSIVE STATISTICAL REVIEW-

     (1) IN GENERAL- The Bureau of Justice Statistics of the Department of Justice (in this section referred to as the 'Bureau') shall carry out, for each calendar year, a comprehensive statistical review and analysis of the incidence and effects of prison rape. The statistical review and analysis shall include, but not be limited to the identification of the common characteristics of—

     (A) ~~inmates who have been involved with prison rape, both victims and perpetrators~~ *both victims and perpetrators of prison rape*; and

     (B) prisons and prison systems with a high incidence of prison rape.

     (2) *CONSIDERATIONS- In carrying out paragraph (1), the Bureau shall consider—*

     (A) how rape should be defined for the purposes of the statistical review and analysis;

     (B) how the Bureau should collect information about staff-on-inmate sexual assault;

     (C) how the Bureau should collect information beyond inmate self-reports of prison rape;

     (D) how the Bureau should adjust the data in order to account for differences among prisons as required by subsection (c)(3);

     (E) the categorization of prisons as required by subsection (c)(4); and

     (F) whether a preliminary study of prison rape should be conducted to inform the methodology of the comprehensive statistical review.

     (3) *SOLICITATION OF VIEWS- The Bureau of Justice Statistics shall solicit views from representatives of the following: State departments of correction; county and municipal jails; juvenile correctional facilities; former inmates; victim advocates; researchers; and other experts in the area of sexual assault.*

     ~~(2)~~*(4)* SAMPLING TECHNIQUES- The analysis under paragraph (1) shall be based on a random sample, or other scientifically appropriate sample, of not less than 10 percent of all Federal, State, and county prisons, and a representative sample of municipal prisons. *The selection shall include at least one prison from each State.* The selection of facilities for sampling shall be made at the latest practicable date prior to conducting the surveys and shall not be disclosed to any facility or prison system official prior to the time period studied in the survey. Selection of a facility for sampling during any year shall not preclude its selection for sampling in any subsequent year.

     ~~(3)~~*(5)* SURVEYS- In carrying out the review ~~required by this subsection~~ *and analysis under paragraph (1)*, the Bureau shall, in addition to such other methods as the Bureau considers appropriate, use surveys and other statistical studies of current and former inmates from a sample of Federal, State, county, and municipal prisons. The Bureau shall ensure the confidentiality of each survey participant.

(continued)

Box 5-1 (continued)

(6) PARTICIPATION IN SURVEY- Federal, State, or local officials or facility adminis-trators that receive a request from the Bureau under subsection (a)(4) or (5) will be required to participate in the national survey and provide access to any inmates under their legal custody.

(b) REVIEW PANEL ON PRISON RAPE-

   (1) ESTABLISHMENT- To assist the Bureau in carrying out the review and analysis under subsection (a), there is established, within the ~~Bureau~~ *Department of Justice*, the Review Panel on Prison Rape (in this section referred to as the 'Panel').

   (2) MEMBERSHIP-

      (A) COMPOSITION- The Panel shall be composed of 3 members, each of whom shall be appointed by the Attorney General, in consultation with the Secretary of Health and Human Services.

      (B) QUALIFICATIONS- Members of the Panel shall be selected from among individuals with knowledge or expertise in matters to be studied by the Panel.

   (3) PUBLIC HEARINGS-

      (A) IN GENERAL- The duty of the Panel shall be to carry out, for each calendar year, public hearings concerning the operation of ~~each entity identified in a report under clause (ii) or (iii) of subsection (c)(2)(B)~~ *the three prisons with the highest incidence of prison rape and the two prisons with the lowest incidence of prison rape in each category of facilities identified under subsection (c)(4). The Panel shall hold a separate hearing regarding the three Federal or State prisons with the highest incidence of prison rape.* The purpose of these hearings shall be to collect evidence to aid in the identification of common characteristics of ~~inmates who have been involved in prison rape, both victims and perpetrators~~ *both victims and perpetrators of prison rape*, and the identification of common characteristics of prisons and prison systems ~~with a high incidence of prison rape~~ *that appear to have been successful in deterring prison rape.*

      (B) TESTIMONY AT HEARINGS-

         (i) PUBLIC OFFICIALS- In carrying out the hearings required under subparagraph (A), the Panel shall request the public testimony of Federal, State, and local officials (and organizations that represent such officials), including the warden or director of each prison, *who bears responsibility for the prevention, detection, and punishment of prison rape at each entity,* and the head of the prison system encompassing such prison~~, who bear responsibility for the prevention, detection, and punishment of prison rape at each entity.~~

         (ii) VICTIMS- The Panel may request the testimony of prison rape victims, organizations representing such victims, and other appropriate individuals and organizations.

      ~~(C) FAILURE TO TESTIFY- If, after receiving a request by the Panel under subparagraph (B)(i), a State or local official declines to testify at a reasonably designated time, the Federal funds provided to the entity represented by that official pursuant to the grant programs designated by the Attorney General under section 9 shall be reduced by 20 percent and reallocated to other entities. This reduction shall be in addition to any other reduction provided under this Act.~~

(continued)

Box 5-1 (continued)

(C) SUBPOENAS-

   (i) ISSUANCE- The Panel may issue subpoenas for the attendance of witnesses and the production of written or other matter.

   (ii) ENFORCEMENT- In the case of contumacy or refusal to obey a subpoena, the Attorney General may in a Federal court of appropriate jurisdiction obtain an appropriate order to enforce the subpoena.

(c) REPORTS-

   (1) IN GENERAL- Not later than ~~March~~ *June* 30 of each year, the ~~Bureau~~ *Attorney General* shall submit a report on the activities of the Bureau ~~(including the Review Panel)~~ *and the Review Panel*, with respect to prison rape, for the preceding calendar year to–

   (A) Congress; *and* ~~(B) the Attorney General; and~~

   ~~(C)~~*(B)* the Secretary of Health and Human Services.

   (2) CONTENTS- The report required under paragraph (1) shall include—

   (A) with respect to the effects of prison rape, statistical, sociological, and psychological data; and

   (B) with respect to the incidence of prison rape—

      (i) statistical data aggregated at the Federal, State, prison system, and prison levels;

      ~~(ii) an identification of the Federal Government, if applicable, and each State and local government (and each prison system and institution in the representative sample) where the incidence of prison rape exceeds the national median level by not less than 30 percent; and~~

      ~~(iii) an identification of jail and police lockup systems in the representative sample where the incidence of prison rape is significantly avoidable.~~

      (ii) a listing of those institutions in the representative sample, separated into each category identified under subsection (c)(4) and ranked according to the incidence of prison rape in each institution; and

      (iii) an identification of those institutions in the representative sample that appear to have been successful in deterring prison rape; and

   (C) a listing of any prisons in the representative sample that did not cooperate with the survey conducted pursuant to section 4.

   (3) DATA ADJUSTMENTS- In preparing the information specified in paragraph (2), the ~~Bureau shall, not later than the second year in which surveys are conducted under this Act,~~ *Attorney General* shall use established statistical methods to adjust the data as necessary to account for ~~exogenous factors, outside of the control of the State, prison system, or prison, which have demonstrably contributed to the incidence of prison rape~~ *differences among institutions in the representative sample, which are not related to the detection, prevention, reduction and punishment of prison rape, or which are outside the control of the State, prison, or prison system, in order to provide an accurate comparison among prisons. Such differences may include the mission, security level, size, and jurisdiction under which the prison operates.* For each such adjustment made, the ~~Bureau~~ *Attorney General* shall identify and explain such adjustment in the report.

   (4) CATEGORIZATION OF PRISONS- The report shall divide the prisons surveyed into three categories. One category shall be composed of all Federal and State prisons. The other two categories shall be defined by the Attorney General in order to compare similar institutions.

has a broader charge to develop national standards for the detection and prevention of sexual violence in correctional facilities. It also appears that one of the intended roles of the Review Panel was to "assist" BJS in its data collection efforts (as is explicitly stated in both versions of the bill). This assistance function is consistent with concerns expressed at a congressional hearing on the bill, arguing that BJS should have an advisory group to work out definitional issues in measuring prison rape (U.S. House of Representatives, Committee on the Judiciary, 2003:19).

The critical difference in the legislative texts in Box 5-1 lies in the reporting requirements to support public hearings by the Review Panel. The original proposal called for public hearings with officials from institutions with high and low incidences of prison rape (facilities "where the incidence of prison rape exceeds the national median level by not less than 30 percent" and facilities "where the incidence of prison rape is significantly avoidable"). However, the final law directs that—each year, for different facility types—the facilities with the three highest and two lowest incidence rates be summoned to appear at hearings. Comparing the different versions of section (b)(3)(C) in Box 5-1, the original version of the act threatened institutions that refused to testify before the Review Panel with a 20 percent reduction in federal grant monies. The final version of the bill removed that threat but granted the Review Panel full subpoena power.[8] In addition to identifying the highest- and lowest-ranked institutions, the final legislative text also required the Review Panel (presumably using BJS's work) to provide a *complete* listing of all the facilities in the sample, "ranked according to the incidence of prison rape.

The original designation of "high" prison rape incidence—a value more than 30 percent greater than the national median—was a curious and intriguing one. Depending on the distribution of incidence rates across facilities, the criterion might have obliged the Review Panel to hear from an unworkably high number of parties, and perhaps that consideration drove the revision. Alternatively, singling out "the" highest-rate facilities may have been viewed by legislators as more consistent with the themes of accountability and action (as with the change in nomenclature from a "Prison Rape Reduction" to a "Prison Rape Elimination" Act). From the record, it is unclear exactly how and why the change came about. Indeed, both Rep. Scott's prepared statement for the Judiciary Committee markup of the bill on July 9, 2003 (H.Rept. 108-219, p. 114), and floor statement on the Senate bill

---

[8]Although the text does not appear in H.R. 3493 in Box 5-1, Corlew (2006) notes that the bill as originally proposed "would have granted a ten percent funding increase to prison systems that, because of their high percentages of prison rape, were required to provide testimony to the Review Panel." Both the American Correctional Association and the Association of State Correctional Administrators objected that this provision "appeared to reward undeserving systems"—another reason for the change to subpoena authority in the final bill text.

on July 25 (*Congressional Record*, p. H7764), refer to "conduct[ing] public reviews of institutions where the rate of prison rape is 30% above the national average rate"—even though that provision no longer existed in the revised language.

The principal congressional hearing on the bill was held before the House Judiciary Subcommittee on Crime, Terrorism, and Homeland Security on April 29, 2003. Being a House hearing, the bill referred to at the hearing was the original version of the legislation with the 30-percent-above-median reporting requirement. At that hearing, the only discussion of BJS's reporting role was raised by then-principal deputy attorney general for the Office of Justice Programs (OJP) Tracy Henke, and that concern came as a brief ending to her opening statement. Although Henke's remarks hinted at the inappropriateness of BJS's use of data for administrative and regulatory purposes, the specific objection was raised to the original bill's vague definition of low-prevalence facilities (U.S. House of Representatives, Committee on the Judiciary, 2003:13–14):

> I know my time is up, but real quickly, sir, another concern to the Department is that the Department believes that [it is of the utmost importance that[9]] the integrity of the statistical collection and analysis by the Bureau of Justice Statistics be preserved. The legislation currently requires BJS not only to collect but also to analyze data and produce reports on that analysis in a very short timeframe. We recognize the need for quick access to this information, but it must be balanced by providing BJS the opportunity to accurately and sufficiently analyze the data collected.
>
> Finally, the law authorizing BJS prohibits BJS from gathering data for any use other than statistical or research purposes. By requiring BJS to identify facilities "where the incidence of prison rape is significantly avoidable," the legislation calls for BJS to make judgments about what level of prison rape is "significantly avoidable". This responsibility goes beyond BJS's authorized statistical role.

BJS Data Collections and Reports in Support of the Act

In response to the enactment of PREA, BJS organized a series of data collection efforts, summarized in Bureau of Justice Statistics (2004c), that have been characterized as "a quantum leap in methodology and our knowledge about the problem" of prison rape (Dumond, 2006). The main efforts in the PREA-related data collections are an annual administrative-records-based inventory dubbed the Survey of Sexual Violence (SSV) and a recurring National Inmate Survey program. For the SSV, BJS contracted with the U.S.

---

[9]This grammatical insertion uses the wording of Henke's prepared statement, printed in the hearing record after the spoken remarks (U.S. House of Representatives, Committee on the Judiciary, 2003:16).

Census Bureau's Governments Division to collect records-based counts of reported incidents from federal and state prisons and a sample of local jails and private correctional facilities. Self-report personal interviewing contracts for the National Inmate Surveys were established with three separate contractors, corresponding to the specific populations and facility types envisioned by the act: RTI International (adult prisons and jails), Westat (juvenile facilities), and the National Opinion Research Center (soon-to-be released and former prisoners). In all of these self-report options, BJS settled on the use of audio computer-assisted self-interviewing (ACASI) as the best means to obtain personally sensitive information such as that called for in the inmate survey of sexual victimization. Under ACASI methods, respondents complete a questionnaire on a computer, following instructions played through earphones from the computer; in this way, respondents do not have to directly divulge embarrassing or sensitive information directly to another person, facilitating a more accurate response. Particularly for the adult prison populations, backup strategies for collection were also developed, including forms for administration to inmates considered too dangerous to interact with survey staff.

Beck and Hughes (2005) issued the first report on SSV data on victimization incidents reported to correctional facilities, corresponding to data collected in 2004. New reports on the SSV for 2005 and 2006 have since been issued. At this writing, two reports from National Inmate Surveys have been released. A December 2007 report (Beck and Harrison, 2007) described the results from interviewing at a sample of 146 state and federal prisons, a June 2008 report covered interview results at a sample of 282 local jails (Beck and Harrison, 2008), and a July 2008 report summarized results from interviews at juvenile correctional facilities (Beck et al., 2008).

Cognizant of BJS's legal reporting requirements, both the prison and jail releases from the National Inmate Surveys identified the names of institutions with high rates of offending; however, both have explicitly described an inability to identify the three highest-rate and two lowest-rate facilities as prescribed by the law. Table 5-1 reproduces the key table from Beck and Harrison (2007) on federal and state prisons, identifying 10 high-rate facilities. Noting the standard errors calculated for the estimates, the report carefully explains that, "statistically, the NIS is unable to identify the facility with the highest prevalence rate" or "provide an exact ranking for all facilities as required" under PREA "as a consequence of sampling error" (Beck and Harrison, 2007:3). The report is accompanied by spreadsheets tabulating facility-specific estimates and standard errors of reported sexual victimization for the full sample; the entries are presented alphabetically by state rather than the strict ranking suggested in the text of the act.

In the body of the report, BJS chose to tabulate the top 10 results. In a ranked list by weighted percentage of sexual victimization incidents, the

Table 5-1 Prison Facilities with Highest and Lowest Prevalence of Sexual Victimization, National Inmate Survey, 2007

Facility Name	Number of Respondents[b]	Response Rate (%)	Percent of Inmates Reporting Sexual Victimization[a]	
			Weighted Percent[c]	Standard Error[d]
U.S. total	23,398	72	4.5	0.3
10 highest				
Estelle Unit, TX	197	84	15.7	2.6
Clements Unit, TX	142	59	13.9	2.9
Tecumseh State Corr. Inst., NE	85	39	13.4	4.0
Charlotte Corr. Inst., FL	163	73	12.1	2.7
Great Meadow Corr. Fac., NY	144	62	11.3	2.7
Rockville Corr. Fac., IN[e]	169	79	10.8	2.4
Valley State Prison for Women, CA[e]	181	78	10.3	2.3
Allred Unit, TX	186	71	9.9	2.2
Mountain View Unit, TX[e]	154	80	9.5	1.9
Coffield Unit, TX	194	76	9.3	2.1
6 lowest[f]				
Ironwood State Prison, CA	141	60	0.0	—
Penitentiary of New Mexico, NM	83	38	0.0	—
Gates Corr. Ctr., NC	52	74	0.0	—
Bennettsville-Camp, BOP	77	69	0.0	—
Big Spring Corr. Inst., BOP	155	66	0.0	—
Schuylkill Fed. Corr. Inst., BOP	174	70	0.0	—

[a] Percent of inmates reporting one or more incidents of sexual victimization involving another inmate or facility staff in past 12 months or since admission to the facility, if shorter.

[b] Number of respondents selected for the National Inmate Survey on sexual victimization.

[c] Weights were applied so that inmates who responded accurately reflected the entire population of each facility on selected characteristics, including age, gender, race, time served, and sentence length.

[d] Standard errors may be used to construct confidence intervals around the weighted survey estimates. For example, the 95% confidence interval around the total percent is 4.5% plus or minus 1.96 times 0.3% (or 3.9% to 5.1%).

[e] Female facility.

[f] Facilities in which no incidents of sexual victimization were reported by inmates.

NOTES: —, Not applicable. BOP, Bureau of Prisons.

SOURCE: Reproduced from Beck and Harrison (2007:Table 1).

11th-ranked facility (the Hays State Prison in Georgia) is the first whose difference from the highest-ranked facility (the Estelle Unit in Texas) is statistically significant ($\alpha = 0.05$). Hence, the top 10 results constitute a group whose overall sexual violence victimization rates are high relative to others, even if they are not statistically distinguishable from each other. Following the same logic, Beck and Harrison (2008) tabulated results for 18 high-rate local jails; in compliance with PREA requirements, Beck and Harrison (2008) also list sampled jails that declined to participate and permit interviewing in the survey. BJS's report on the survey administration in juvenile facilities (Beck et al., 2008) differs from the other reports in the series in that it does not attempt any tabular listing of specific facilities or ranking of highest-offense facilities, instead reporting summary statistics from the sample as a whole. However, the report does identify those juvenile facilities that declined to participate as well as those that reported no victimization incidents.

In addition to our panel's concern about the use of BJS data for regulatory or administrative uses, we are also critical of the procedures used for this part of reporting pursuant to PREA. Specifically, we are concerned that the approach greatly understates the variability inherent in the data; see Box 5-2.

Developments Following the First PREA Report Releases

Following the release of Beck and Harrison (2007), the Review Panel on Prison Rape established by PREA in the Department of Justice (DOJ) held 7 days of hearings in March 2008, in Washington, DC, and Houston, Texas, to obtain testimony from each of the adult federal and state prisons identified in Table 5-1. The National Prison Rape Elimination Commission issued press releases on the occasion of the BJS report releases and the start of the Review Panel hearing. The Commission's June 25, 2008, release noted that:

> Even with margins of error, the study reveals that these facilities have extraordinarily high rates of sexual assault, highlighting the severity of this national problem. . . . We welcome BJS's stated willingness to adjust future surveys to gather additional information. We hope the agency will develop more questions about inmate reporting efforts, the response of officials and factors that may play into reporting, such as threats of retaliation.

Since the enactment of PREA, similar legislative calls for expanded data collection on inmate health conditions have been introduced in Congress but have not advanced beyond referral to committee. For instance, the proposed Justice for the Unprotected Against Sexually Transmitted Infections Among the Confined and Exposed (JUSTICE) Act introduced as H.R. 178 in January 2007 requires an annual survey of correctional facilities. In addition to

Box 5-2 Critique of the Reported Rankings in the Prison Rape Elimination Act Inmate Surveys

In its reports on the PREA inmate surveys in prisons and jails (Beck and Harrison, 2007, 2008), the Bureau of Justice Statistics (BJS) did what it could to convey the basic idea that the survey sample sizes are too small (and the underlying phenomenon being measured is sufficiently "rare") to preclude identification of high-rate facilities with the precision called for by the law. However, the panel observes that BJS's chosen approach is, itself, partly inaccurate in that it understates the variability inherent in the data.

As it stands, BJS has ranked the correctional facilities solely on the basis of sample-based estimates of prison rape rates. Each correctional facility among the top 10 is associated both with its name and its ranking. This is not a fully valid approach given that many of these rates have large standard errors. The standard errors reflect the uncertainty due to observing only a portion of the prison population. The estimated rates could have differed if a different sample of institutions were selected. This sampling uncertainty must be taken into consideration while developing such rankings. In other words, a facility's name is fixed but its ranking is affected by the sampling error in the estimated rates. There are simple procedures to account for such sampling uncertainty. For example, consider the 20 facilities with highest rates and their standard errors. How fair is to label only the top 10 as "bad" (let us call this top-10 group "Tier A") when the bottom 10 ("Tier B") could have easily been in Tier A purely by chance?

This question can be answered using a simple bootstrap procedure by simulating what could have been the estimated prison rape rates and their ranking for these 20 facilities purely by chance. We used 10,000 parametric bootstrap draws from the sampling distribution of the estimated overall prevalence rates for prison rape in Beck and Harrison (2007), and ranked them. We then computed the proportion of times a facility labeled in Tier A in BJS's report would have been placed in Tier B in the simulation, and vice versa. The following table gives these estimated probabilities.

Tier A	Estimated Prison Rape Rates (%)	Standard Error (%)	Probability of Being in Tier B (%)	Tier B	Estimated Prison Rape Rates (%)	Standard Error (%)	Probability of Being in Tier A (%)
1	15.7	2.6	0.2	11	9.1	1.9	49.1
2	13.9	2.9	3.6	12	8.7	2.2	46.8
3	13.4	4.0	12.1	13	8.5	2.4	39.1
4	12.1	2.7	12.3	14	8.2	2.0	32.4
5	11.3	2.7	20.3	15	8.1	2.0	30.1
6	10.8	2.4	23.7	16	8.0	2.2	29.6
7	10.3	2.3	30.4	17	8.0	2.1	29.6
8	9.9	2.2	37.2	18	7.9	2.1	28.2
9	9.5	1.9	42.4	19	7.9	1.9	25.0
10	9.3	2.1	47.2	20	7.9	1.7	24.5

NOTES: Bootstrap estimate of the error or misclassification of rates purely based on the point estimates and ignoring the standard error.

The misclassification rates are disturbingly high and expected given the large standard error, partly due to inadequate sample size. If the error rates of 5 percent or more are not acceptable, then only the first two facilities stand out in that they would have remained in the Tier A set with high probability had a different sample been obtained. For many facilities the decision to label them as Tier A or Tier B is quite arbitrary and made purely by chance. This amply illustrates the problems of using statistical data for regulatory purposes.

(continued)

Box 5-2 (continued)

Alternatively, all possible comparisons between Tier A set rates and Tier B set rates may be considered simultaneously and Bonferroni bounds may be used to conservatively determine significance levels for all possible pairwise comparisons. Such an approach has been used, in particular, in other applications where extreme ranks on some variable carry particular political sensitivity. For instance, the National Center for Education Statistics has developed such approaches for ranking states according to sample-based educational testing results, as described by Wainer (1996).

PREA-type queries on the incidence of sexual assault, the proposed data collection would require information on facility policy on testing for sexually transmitted diseases and data on test results that are sufficiently detailed to support disaggregation by disease type, race and ethnicity, age, and gender.

Assessment

Whatever the reasons for the change in reporting requirements, both the original and the final versions of the PREA bill violated the expected principles and practices of a federal statistical agency. BJS directly contributed to regulatory activities affecting individual data providers: explicitly singling out individual facilities to receive a summons to public hearings. Arguably, that direct summons to appear before the Review Panel is a somewhat lesser burden than a compulsory appearance before a congressional committee or the fuller National Prison Rape Elimination Commission. Nonetheless, the provision does explicitly direct the usage of data reported to BJS for nonstatistical purposes, a basic violation of the role of federal statistical data. The final language of the act exacerbated this violation by putting undue weight on point estimates of incidences of sexual assault—estimates of (ideally) a relatively low-probability phenomenon based on a small sample—without accounting for the inherent variability in the estimates.

BJS and its major constituencies and stakeholders—chief among them Congress and the administration—must be mindful of the extensive legal mandates placed on the agency and how they correspond to the resources provided to BJS; it is crucial that BJS not be assigned duties that violate fundamental principles for statistical agency conduct.

> **Finding 5.1:** Under the terms of the Prison Rape Elimination Act of 2003, BJS was required to release the identity of selected responding institutions (i.e., facilities with the highest and lowest rates of sexual violence against inmates) for later regulatory action as part of a statistical program.

Recommendation 5.1: Congress and the Department of Justice should not require, and BJS should not provide, individually identified data in support of regulatory functions that compromise the independence of BJS or require BJS to violate any of the principles of a federal statistical agency.

To be sure, criticism of the reporting requirements of PREA should not be mistaken for criticism of the study of prison rape; the problem of sexual violence in correctional facilities is a valid and important one for inquiry. BJS's work in developing the suite of inmate prison rape surveys also had the benefit of pushing the agency to make major methodological improvements, relative to other federal surveys, in the use of techniques such as ACASI. However, it is also important to note that implementing PREA involves major opportunity costs to BJS, over and above the concerns over the regulatory flavor of the work. The separate and highly sensitive nature of PREA interviewing makes it infeasible for BJS to conduct its standard inmate interviewing programs at the same time. Further, it is still too early to assess whether the PREA interviewing has any chilling effect on response to BJS's conventional corrections data series; although one possible reason for a dampening effect on response to regular corrections series might be resentment at being "singled out" for inclusion in the PREA sample, another is simply the time and resource burden of brokering BJS access to inmates on a more frequent basis. To its credit, BJS has taken steps to convey PREA's requirements to individual facilities and elicited comments and feedback from facilities and administrators, including participation in relevant professional association meetings and conduct of stakeholder workshops.

The PREA data providers have the risk of public display of their estimates in an active attempt at regulatory intervention. Within a short period of time, BJS will ask the same facilities for data for another purpose. The potential impact of BJS's participation in regulatory actions is that the data providers will no longer believe that other data requested will not also be used in such a manner. Again, once data providers lose trust that cooperation with BJS will not lead to individual harmful actions on them, the agency faces large problems.

5–A.2   Strong Position of Independence

A federal statistical agency must have a strong position of independence within the government. To be credible and unhindered in its mission to provide objective, useful, high-quality information, a statistical agency must not only be distinct from those parts of a department that carry out law enforcement and policy-making activities but also have a widely acknowledged position of independence. It must be able to execute its mission without being subject to pressures to advance a political agenda. It must be impartial and avoid even the appearance that its collection, analysis, and reporting processes might be manipulated for political purposes or that individually identifiable data might be turned over for administrative, regulatory, or law enforcement purposes. (National Research Council, 2009:6)

The establishment and maintenance of an independent, objective, and credible voice is a central principle for statistical agency operations. To maintain that objectivity and credibility, a statistical agency is obliged to keep apart from the policy-making sphere of the executive branch; its products inform the development of policy, but they must not themselves be policy statements. Maintaining this arm's-length distance from policy development is particularly difficult for statistical agencies that are administratively housed with program agencies of the executive branch, whose purpose is the furtherance of specific objectives.

In recent years, administrative layering of statistical agencies has become a subtle, but increasingly common, threat to the position of independence of federal statistical agencies. Agencies are diminished in their perceived importance, their claim to budgetary resources, and their attention from departmental policy makers through placement further down in a department's organizational hierarchy. In 2002, the National Center for Education Statistics was redesignated by P.L. 107-279 as a unit of a new Institute of Education Sciences. In 2004, the National Center for Health Statistics— already administratively removed from the main Department of Health and Human Services by administrative placement in the Atlanta-based Centers for Disease Control and Prevention (CDC)—was placed under the further administrative layer of a "coordinating center," as part of a broader CDC reorganization. Also in 2004, the Bureau of Transportation Statistics was converted by P.L. 108-426 to become a unit under the new Research and Innovative Technology Administration, and its director was changed from a presidential appointee with Senate confirmation to a career appointee designated by the Secretary of Transportation. Of the current members of the Interagency Council on Statistical Policy (see Box 1-3), only the Bureau of Labor Statistics and the Energy Information Administration have direct reporting authority to their respective cabinet secretary or department head. BJS, through its administrative placement in OJP, is not the most heavily layered of statistical agencies, but it ranks among them.

In the panel's judgment, the principle of a strong position of independence of a statistical agency was seriously violated in BJS's recent past by the circumstances surrounding the release of data from the 2002 PPCS. This particular "flashpoint" in BJS's recent history centered on the wording of a press release to accompany the data release. In the balance of this section, we describe the path toward this breach in principle and the corrective measures that have since been taken.

OJP and Press Release Policy

In a 1991 U.S. House subcommittee hearing on criminal justice statistics, Acting BJS Director Joseph Bessette was asked about policy on the press releases accompanying new BJS data releases and, specifically, the role of the Justice Department in clearing those releases. Bessette answered that:

> There has never been a case in my time [at BJS (5 years, at that point)], and people there tell me never before then as well, of the Department interfering in any way with the reports, with the accuracy, with the nature of the numbers, anything of that sort. So, in that respect, we have been functioning as a kind of semiautonomous statistical agency quite well. However, the BJS press releases—I use the term "BJS press releases," but, actually, they are Department of Justice press releases officially, and they have always gone to the Department for clearance. We draft them in BJS [and they] go up the chain of command for clearance, and that has been the case right along. So, in that respect, the policy hasn't changed.

Pressed further, he noted that "last year, for the first time, the Attorney General was quoted in a BJS press release commenting on the numbers and recommending public policy. That happened that one time; that has not happened since" (U.S. House of Representatives, Committee on the Judiciary, 1991:216).[10]

Over the next decade, the protocol for issuing BJS press releases evolved into the flow pattern illustrated in Figure 5-2. BJS staff would typically take the lead in developing the press release, in cooperation with the OJP Office of Communications. In all, the typical approval process required signoff from five noncareer appointees in DOJ and OJP (including the presidentially appointed BJS director). Figure 5-3 illustrates the general formatting of the standard notice and page posted to BJS's website upon the release of a new product, and Figure 5-4 shows the formatting of a formal press release, for one recent BJS product for which OJP and DOJ elected to issue a press release.

The 2002 Police-Public Contact Survey

In August 2005, the *New York Times*, followed by other media outlets, reported on a string of events over the previous 4 months that culminated in the removal of BJS Director Lawrence Greenfeld (Lichtblau, 2005a; Eggen, 2005; Sniffen, 2005). The removal was precipitated by disputes within the Justice Department over the statement of findings from the

---

[10]In response, the questioner—then-Rep. Charles Schumer—commented before moving on to the next line of questioning: "I think it is a good idea to keep the two separate. The Attorney General should comment on policy but not in the statistical press releases" (U.S. House of Representatives, Committee on the Judiciary, 1991:216).

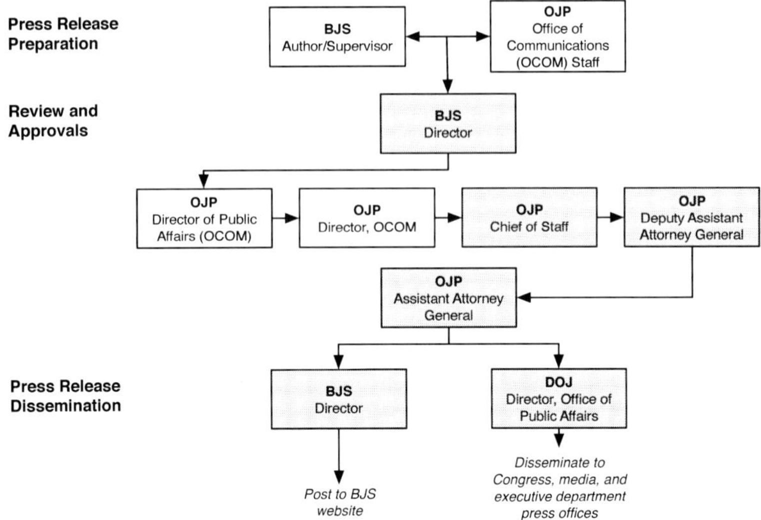

Figure 5-2 Review, approval, and dissemination process for BJS survey press releases, 2007

NOTE: BJS, Bureau of Justice Statistics. DOJ, Department of Justice. OJP, Office of Justice Programs. White boxes indicate career employees; grey boxes denote noncareer appointees.

SOURCE: Adapted from U.S. Government Accountability Office (2007:Fig. 3), based on information from BJS and OJP.

PPCS supplement to the NCVS. As described in Section 3–C.4, the PPCS was first fielded on a pilot basis in 1996, followed by full-scale implementation in 1999, 2002, and 2005. The events of 2005 concerned the release of information from the 2002 administration of the supplement (Durose et al., 2005). As indicated in the abstract of the report on the BJS website (http://www.ojp.usdoj.gov/bjs/abstract/cpp02.htm),

Highlights [from the 2002 PPCS] include the following:

- About 25% of the 45.3 million persons with a face-to-face contact indicated the reason for the contact was to report a crime or other problem.
- In 2002 about 1.3 million residents age 16 or older—2.9% of the 45.3 million persons with contact—were arrested by police.
- The likelihood of being stopped by police in 2002 did not differ significantly between white (8.7%), black (9.1%), and Hispanic (8.6%) drivers.
- During the traffic stop, police were more likely to carry out some type of search on a black (10.2%) or Hispanic (11.4%) than a white (3.5%).

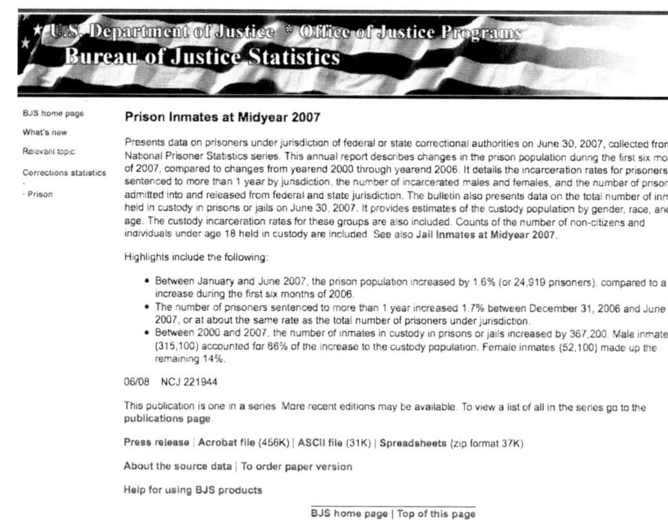

Figure 5-3   Example summary and links to report and data on Bureau of Justice Statistics website

Department of Justice

Office of Justice Programs

ADVANCE FOR RELEASE AT 12:01 A.M. EDT   BUREAU OF JUSTICE STATISTICS
FRIDAY, JUNE 6, 2008   Contact: Sheila Jerusalem: 202-616-3227
WWW.OJP.USDOJ.GOV/BJS   After hours: 202-598-3570

SLOWER GROWTH IN THE NATION'S PRISON AND JAIL POPULATIONS

WASHINGTON – The growth in the number of prisoners under state or federal jurisdiction slowed during the first six months of 2007, the Justice Department's Bureau of Justice Statistics (BJS) reported today. The number of prisoners rose 1.6 percent, which was lower than the 2.0 percent growth during the same period in 2006. In absolute numbers, prisoners under the legal jurisdiction of state or federal correctional authorities — some of whom were housed in local jails — increased by 24,919 prisoners to reach 1,595,037 prisoners.

On June 29, 2007, there were 780,581 inmates in local jails, which are correctional facilities operated by counties or municipal authorities. Growth slowed in the nation's jail population, from 2.5 percent in 2006 to 1.9 percent in 2007. This was the smallest annual rate of growth in the jail population since 2001 and the second smallest since 1981.

Despite slowing of growth in the number of jail inmates, local jails handled an estimated 13 million admissions during 2007. The volume of admissions was about 17 times larger than the number of inmates held in local jails on a given day.

Figure 5-4   Excerpt from example Office of Justice Programs press release accompanying new Bureau of Justice Statistics data release

After BJS staff developed a press release, the draft release was forwarded to OJP and then-Assistant Attorney General Tracy Henke. It was the inclusion of the last highlighted point—the finding of disparate levels of search (and related findings on use of force) by race and ethnicity—that led to a dispute. As Lichtblau (2005a) recounts,

> The planned announcement noted that the rate at which whites, blacks and Hispanics were stopped was "about the same," and that finding was left intact by Ms. Henke's office, according to a copy of the draft obtained by *The New York Times*.
>
> But the references in the draft to higher rates of searches and use of force for blacks and Hispanics were crossed out by hand, with a notation in the margin that read, "Do we need this?" A note affixed to the edited draft, which the officials said was written by Ms. Henke, read "Make the changes," and it was signed "Tracy." That led to a fierce dispute after Mr. Greenfeld refused to delete the references, officials said. ... Mr. Greenfeld refused to delete the racial references, arguing to his supervisors that the omissions would make the public announcement incomplete and misleading.

The report was publicly released—posted to the agency's website and disseminated through usual means—but without any accompanying news release or publicity. This decision "all but assured [that] the report would get lost amid the avalanche of studies issued by the government"; indeed, "a computer search of news articles [in August 2005] found no mentions of the study" (Lichtblau, 2005a). However, the study—and dispute over the press release—garnered considerable press attention after the *New York Times* story on the circumstances surrounding Greenfeld's dismissal as BJS director.

In the wake of these incidents, the U.S. Government Accountability Office (GAO) initiated a review of the conduct of the various administrations of the PPCS and the release of those data. Responding to a draft report, Assistant Attorney General Regina Schofield asserted that some of GAO's findings of interference were erroneous because they were "predicated on GAO's assumption that a press release is a statistical product." However, she continued (quoted in U.S. Government Accountability Office, 2007:48):

> We respectfully disagree with GAO's assumption. A press release simply is not a statistical product and thus should not be treated as a statistical product at all—let alone one that is somehow covered by the [CNSTAT guidelines in *Principles and Practices for a Federal Statistical Agency*.] A press release, rather, is a public relations announcement issued to encourage media coverage. The mere presence of statistics in a press release does not transform a press release into a statistical product.

Combining this argument with the legally nonbinding nature of the *Principles and Practices*, Schofield concluded (quoted in U.S. Government Accountability Office, 2007:50):

> By statute, 42 U.S.C. § 3732(b), the Director of BJS "shall be responsible for the integrity of data and statistics." In the exercise of such authority, he may elect to follow the NRC guidelines, but he is not and cannot be legally bound to do so, in the absence of some supervening statute. [Thus,] even if the [CNSTAT] written guidelines did apply to press releases (and they do not), the Director would and does decline, in the exercise of his statutory authority to apply them to BJS press releases.

In response, the GAO stood by its assumption, in large part for the simple reason that "the Police-Public Contact Survey press release was made up almost entirely of survey statistics, indicating to us that it was a statistical product" and that "the content of the press release was a more important determinant than the label attached to it" (U.S. Government Accountability Office, 2007:23–24). The GAO observed that "the role that certain noncareer appointees outside BJS have the ability to play, pursuant to Department of Justice policy, in the product issuance process" means that "BJS was not in a position to fully follow all guidelines related to agency independence," thus creating the potential for "future actual or perceived political interference" in BJS product releases (U.S. Government Accountability Office, 2007:14).[11]

Later, but too late to affect the DOJ actions, the U.S. Office of Management and Budget (OMB) issued formal guidance in early 2008 to clarify the gray-area dispute as to whether a press release constitutes a statistical product. In the March 7, 2008, *Federal Register*, OMB published Statistical Policy Directive 4 on the release and dissemination of products from the federal statistical agencies. Defining a "statistical press release" as one of the product types covered by the directive, OMB "encouraged" agencies to issue press releases to accompany the issuance of new data and reports. The directive does not speak directly to the issue of administrative review of the content of such press releases, advising only that:

> to maintain a clear distinction between statistical data and policy interpretations of such data, the statistical press release must be produced and issued by the statistical agency and must provide a policy-neutral description of the data; it must not include policy pronouncements.

---

[11] Reacting, most likely, to the GAO report, U.S. House appropriators issued an even stronger statement in its explanatory statement accompanying the fiscal year 2008 Commerce, Justice, and Science appropriations bill (H.Rept. 110-240):

> *Ensuring objective BJS studies*—The Committee directs that any statistical studies undertaken by the Bureau of Justice Statistics, as well as press releases describing the results of these studies, shall be publicly released by the Bureau without alteration or clearance by persons outside of the Bureau.

However, this provision was not repeated in the explanatory statement for the consolidated appropriations act that eventually funded BJS and DOJ.

The issuance of this guidance appears to have improved the release process for BJS products in recent months, even though the guidance emphasizes the need to "coordinate with public affairs officials from the parent organization" in those "cases in which the statistical unit currently relies on the parent agency for the public affairs function."

Aftermath

In 2007, when data from the 2005 administration of the supplement were made available, the report (Durose et al., 2007) was accompanied by a press release (http://www.ojp.usdoj.gov/bjs/pub/press/cpp05pr.htm). Entitled "Police Stop White, Black, and Hispanic Drivers at Similar Rates According to Department of Justice Report," the release observed:

> The 2002 and 2005 surveys found that white, blacks and Hispanics were stopped at similar rates. . . . In both 2002 and 2005 police searched about 5 percent of stopped drivers. . . . While the survey found that black and Hispanic drivers were more likely than whites to be searched, such racial disparities do not necessarily demonstrate that police treat people differently based on race or other demographic characteristics. This study did not take into account other factors that might explain these disparities.

This press release—like others, issued in recent years and even subsequent to the March 2008 OMB guidance—was issued on OJP letterhead.

Immediately following the dispute over the 2002 PPCS press release, BJS Director Greenfeld resigned. As the narrative description by Lichtblau (2005a) continues:

> Amid the debate over the traffic stop study, Mr. Greenfeld was called to the office of Robert D. McCallum Jr., then the third-ranking Justice Department official, and questioned about his handling of the matter, people involved in the episode said. Some weeks later, he was called to the White House, where personnel officials told him he was being replaced as director and was urged to resign, six months before he was scheduled to retire with full pension benefits, the officials said.
>
> After Mr. Greenfeld invoked his right as a former senior executive to move to a lesser position, the administration agreed to allow him to seek another job, and he is likely to be detailed to the Bureau of Prisons, the officials said.

After the appearance of the *Times* article, numerous newspapers ran editorials critical of Greenfeld's departure (see, e.g., *Hartford Courant*, 2005; *Houston Chronicle*, 2005; Joiner, 2005; Love, 2005; *Miami Herald*, 2005; *Tennessee Tribune*, 2005). Although some members of Congress called for Greenfeld to be reinstated (Lichtblau, 2005b), no such reversal was made.[12]

---

[12]At the time of his dismissal, Greenfeld authored a farewell letter to members of the Justice Research and Statistics Association (JRSA) that noted a positive aspect of the flare-up over

The wave of publicity concerning these events reinforced the perception that BJS's position of independence had been threatened.

Assessment

One immediate recommendation that is appropriate in light of the PPCS incident is to express formal concurrence with the OMB guidance that eventually followed. The press release associated with a new statistical series or the latest release of data from a continuing series is, properly, a statistical product. Taking care always to be policy-neutral, the press release is the agency's first chance (and sometimes the only and best chance) for a statistical agency to highlight its findings from the data, any methodological concerns that the new data may raise, and to promote accurate reporting and publicity of new results. Accordingly, press releases should share the same protections from interference as other BJS reports and releases.

> **Finding 5.2:** The appearance of political interference in release of statistical information undermines public trust in that information and in the entire agency.

> *Recommendation 5.2:* **The Department of Justice review of any BJS statistical product and related communications should not require changes to the content, the release schedule, or the mode of dissemination planned by BJS.**

The promulgation of the OMB guidance solves, or at least ameliorates, the immediate cause of this most glaring violation to BJS's position of independence, but a larger problem remains. Independence is an ever-present tension that exists when a statistical agency is administratively nested in a program agency, as BJS is within OJP. The OMB guidance is a useful safeguard but, by its nature and the nature of the decentralized statistical system, it is necessarily somewhat passive and advisory. That is, its successful implementation in BJS's case hinges on the compliance and goodwill of the leadership of BJS, OJP, and the broader DOJ to ensure that boundaries are not blurred. Though it is very welcome, the guidance makes no specific reference to the circumstances that befell BJS concerning the 2002 PPCS release and, accordingly, falls short of a forceful statement by the statistical system

---

press release language. "There is a good reason that more than 20,000 people a day turn to BJS for information on crime and the administration of justice; there is a good reason that no Congressional bill on crime and justice ever ignores our data on a subject and that we are repeatedly asked to gather even more data; there is a good reason that hundreds of thousands of newspaper and electronic media citations and numerous court decisions refer to BJS findings; there is a good reason that Office of Management and Budget regards our activities as the 'most effective' in all of the Department of Justice; and finally, there is a good reason that so many have expressed such concern about a few lines in a BJS press release, evidence of the importance of what we say" (Greenfeld, 2005).

that OJP's intervention in the PPCS press release violated the basic practices of an official statistical system.

The panel concludes that the current organizational arrangement under which BJS is administratively housed in a program agency (OJP) and the fact that its director serves at the pleasure of the president is a continuing and pressing threat to BJS's position of independence as a provider of objective statistical information. It is critically important that, whatever organizational structures or reporting requirements may apply, BJS function independently and be allowed to function independently. We also recognize that there exists no organizational arrangement that—on its own—can completely shield a statistical agency from threats to its independence and guarantee freedom from political or structural interference (or the appearance thereof). However, in our assessment, the continuing threat to BJS's independence is sufficiently dire—and the past violations sufficiently severe—as to warrant what we believe to be the strongest possible corrective actions and deterrents to incursions on independent functioning: moving BJS out of OJP and fixing the term of service of the BJS director.

BJS and the Office of Justice Programs BJS's functions are unique in its parent branch, OJP, with respect to both mission and technical requirements. Since grantmaking overwhelmingly drives the OJP organization and service-delivery infrastructure, OJP is ill-suited to address the needs of BJS to produce data and statistical reports and provides minimal support for carrying out these functions (although it does, with contributions from BJS and other OJP bureaus, operate the National Criminal Justice Reference Service for dissemination of BJS results). BJS's administrative placement within OJP is doubly a hindrance on BJS's effective function as the principal data-gathering unit within the Justice Department: first, by putting it into competition for funds and resources with popular grantmaking functions that provide assistance to state and local law enforcement and, second, by diminishing BJS's position within the Department. Other Justice Department divisions perform fairly major statistical and data collection functions—among them, the Civil Rights Division, the Justice Management Division, the Executive Office for U.S. Attorneys, and the Federal Bureau of Investigation (FBI). These units utilize statistical analysis for performance measurement, examination of voting issues, review of discrimination concerns, and so forth—major issues in which BJS's ability to offer advice or coordination is impaired by BJS's positioning within the department.

Although the historical reason for BJS being positioned within OJP is fairly clear, inheriting as both entities do from the Law Enforcement Assistance Administration (LEAA), the administrative positioning raises technical and practical concerns. The basic purpose of OJP is to promote certain activities, strategies, or interventions related to crime, primarily through financial assistance to state and local authorities. Statistics should serve as an independent way of assessing those practices by measuring whether crime problems are worsening or improving; that statistical activities are under the direction and funding of OJP creates the appearance and, at times, the reality of conflict and questionable integrity.

Moreover, BJS's placement within OJP forces it to compete for resources with grant monies that are popular with and coveted by local authorities and congressional representatives alike. In terms of budget, the Justice Department tends to view BJS as a small line entry in an overall OJP appropriation. The general process is such that OJP is budgeted or appropriated at a certain funding level and largely makes the internal distribution among component agencies; it is the assistant attorney general for OJP, and not the director of BJS, who is permitted to testify before congressional appropriations committees. Put into head-to-head competition with grant programs to "put cops on the street" or fund crime assistance programs, sustaining the growing costs of BJS statistical programs become a lower-order concern. Worse, in recent years, OJP has taken steps to make explicit BJS's subservience within a larger OJP appropriation, undercutting BJS's presence as even a simple line item in annual spending bills. In the 2003 House appropriations subcommittee report for the fiscal year 2004 Commerce, Justice, and Science spending bill, appropriators took note of a change in the budget request it received from the Justice Department (H.Rept. 108-221, p. 36):

> The fiscal year 2004 budget request proposed merging all programs administered by the Office of Justice Programs (OJP) under the Justice Assistance heading. The Committee recommendation retains the account structure used in previous years and funds State and local law enforcement programs under seven appropriation accounts.

House-Senate conferees on the final consolidated appropriations bill for fiscal year 2004 also noted that they "do not adopt the Administration's proposal to consolidate all [OJP] activities" under the single "Justice Assistance" heading (H.Rept. 108-401, p. 533). Similar attempts to consolidate accounts were noted by House appropriators in the fiscal year 2005 submission (H.Rept. 108-576, pp. 33–34; H.Rept. 108-792, p. 738). The attempt to consolidate OJP funding into a single pool has continued in each subsequent year, including submissions for fiscal year 2009 (e.g., Senate appropriators commented that "the Committee again rejects the Department's proposed merger of all OJP programs under this heading and instead has maintained the [previous] account structure;" S.Rept 110-397, p. 64).

The problem of BJS funding as it is currently situated within OJP is analogous to problems encountered in other governmental programs where new initiatives often receive greater attention than the existing responsibilities—fixing potholes often takes a back seat to more glamorous new construction projects. In the case of BJS, after passage of the multibillion-dollar Violent Crime Control and Law Enforcement Act of 1994, OJP funding expanded greatly and external grants flowed freely, yet BJS received no enhancements to its appropriated funding and, indeed, had difficulty even securing additional funding to cover cost-of-living adjustment increases payable to the Census Bureau for data collection. On one hand, statistical data collection activities should be seen as long-term activities requiring predictable funding so that they may be carried out on recurring schedules. On the other hand, the grant programs of the larger OJP have impermanence in both mission and appropriations; BJS's base function is jeopardized from being tied to an administrative parent whose resources can rise or fall dramatically and whose local-assistance grants are more popular funding targets than continuing statistical activities.

Statistical analysis and research have also been strikingly undervalued by OJP, as evidenced by attempts to "outsource" most BJS staff positions and functions. In August 2002, OJP was said to have issued a directive stating that jobs within BJS would be turned over for competitive bid to the private sector (Butterfield, 2002). Under the terms of the Federal Activities Inventory Reform (FAIR) Act, positions within a government agency must be characterized as either "commercial" or "inherently governmental"; in late 2002, OMB was in the process of revising its Circular A-76 to more directly require that those positions classified as "commercial" be opened to competitive bid with private-sector companies. As described by the Consortium of Social Science Associations (2002:5) in February 2003, the FAIR Act inventory developed by OJP classified 51 out of 57 positions as "commercial" and thus designated for outsourcing. Several statistician positions within BJS were classified in the inventory as being "grants monitoring and evaluation"; 20 of 23 jobs labeled "statistical analysis" and 18 of 20 "grants monitoring" positions were labeled commercial. This classification drew protest from several social science organizations including the American Society of Criminology, whose executive board passed a resolution in November 2002 arguing that "the compilation, analysis, interpretation, reporting, monitoring, and management of crime and justice statistics . . . are inherently governmental functions" (http://www.asc41.com/boardmin.annual022.htm).

This outsourcing effort was blocked by congressional appropriators: in explaining the fiscal year 2003 omnibus appropriations bill, House-Senate conferees insisted that the appropriations committees "must be assured that effectiveness is improved and savings are attained" through the OJP outsourcing plan before proceeding with changes (H.Rept. 108-10, p. 635), a provision repeated by House appropriators the following year (H.Rept. 108-221, p. 40). In the fiscal year 2006 appropriations round, House-Senate conferees specifically directed that "any action taken by OJP relating to [OMB's] Circular A-76 shall be subject to" a general provision requiring advance notice and special justification to Congress for program changes that, among other conditions, would reduce the personnel of an agency by 10 percent or more (H.Rept. 109-272, pp. 46, 86). However, implementation of outsourcing is still possible, and would still be damaging to BJS. The Justice Department's most recent publicly posted FAIR Act inventory listed commercial and inherently governmental activities for 2007 (http://www.usdoj.gov/jmd/pe/preface.htm); this roster lists 32 of 57 BJS positions (and 20 of 33 "statistical analysis" positions) as commercial, with the reason for classification as commercial listed as "pending an agency approved restructuring decision (e.g., closure, realignment)." In our assessment, the collection and analysis of statistical data by federal statistical agencies is an essential government function; that OJP has not more fully realized this point suggests a continued incompatibility of functions between BJS and its administrative parent.

Exacerbating this mismatch in functions between OJP as a program agency and BJS as a statistical agency, two threads of legislative text that have developed since the late 1990s have suggested attempts to tether BJS closer to OJP objectives and diminish BJS's functional independence. Both of these threads have involved wording changes that may appear short and subtle but have great meaning, and both require some detailed attention to legislative history to be fully understood.

The first of these threads began in 1997 when House appropriators expressed concern that "the current structure of administration of grants within [OJP] produces a fragmented and possibly duplicative approach to disseminating information to State and local agencies on law enforcement programs and developing coordinated law enforcement strategies." Noting a 213 percent growth in overall OJP grant program funding since 1995, the appropriators directed the assistant attorney general (AAG) for OJP to prepare a report recommending actions "that will ensure coordination and reduce the possibility of duplication and overlap among the various OJP divisions" (H.Rept. 105-207, pp. 43–44). This language was preserved in the House-Senate conference on the fiscal year 1998 spending bill (H.Rept. 105-405) that became law. The AAG issued this requested report in January 1998; on the basis of the report, House and Senate appropriations conferees inserted a provision into the fiscal year 1999 omnibus spending act asserting an oversight role for the AAG in finalizing grants (*Congressional Record*, October 19, 1998, p. H11310). Specifically, the final act read (P.L. 105-277; 112 Stat. 2681-67; compressing a first clause that gives the AAG grantmaking authority):

> Notwithstanding any other provision of law, during fiscal year 1999,
> the Assistant Attorney General for the Office of Justice Programs of the
> Department of Justice [shall] have final authority over all grants, coop-
> erative agreements, and contracts made, or entered into, for the Office
> of Justice Programs and the component organizations of that Office.

Though it left intact language from BJS's creation in 1979 giving the BJS director "final authority for all grants, cooperative agreements, and contracts awarded by the Bureau" (93 Stat. 1176; 42 USC § 3732(b)), this provision made OJP's "final authority" for grants primary to BJS's "final authority"—albeit only for fiscal year 1999.

BJS briefly won exemption from this provision when new appropriations language changed the effective date from fiscal year 1999 to 2000 but added a caveat that the AAG's final authority did not apply to grants made under certain sections of law (113 Stat. 1501A-20), including the section asserting BJS's "final authority" for its own grants. In 2000, appropriations language made no further changes to the text but indicated that it "shall apply here-after" (114 Stat. 2762A-68), which led to the language being codified as 42 USC § 3715. However, section 614 of the 2001 USA PATRIOT Act (P.L. 107-56; 115 Stat. 370) made two critical changes:

- By adding three words, the revised law gave the AAG "final authority over all *functions, including any* grants" (emphasis added), a much wider sweep of authority over BJS and other OJP-component offices.

- The revised law amended "component organizations of that Office" to read "component organizations of that Office (including, notwith-standing any contrary provision of law (unless the same should ex-pressly refer to this section), any organization that administers any program established in title 1 of Public Law 90-351)"—a rather con-voluted way of making explicit that OJP's "final authority" supersedes BJS's (which still exists, albeit as an "other provision of law").

One year later in September 2002, this perceived takeover of BJS authority was exacerbated by one final small but sweeping change included in reau-thorization language for the Department of Justice. Reference to the AAG was stricken and the text amended to read that, "during any fiscal year, the Attorney General" shall have final authority—asserting strong Justice De-partment control over BJS and other OJP offices (P.L. 107-273; 116 Stat. 1778).

The second legal thread deals with a clause in the enumerated powers of the AAG. The Justice System Improvement Act of 1979 that created BJS also created OJP's predecessor, the Office of Justice Assistance, Research, and Statistics (OJARS), but did so in an interesting way: defining the LEAA, BJS, and the National Institute of Justice (NIJ) up front in sections A–C but only specifying OJARS in a catch-all Part H on "Administrative Provisions."

Specifically, section 802(b) of the Act (93 Stat. 1201) directed that (emphasis added):

> The *Office of Justice Assistance, Research, and Statistics shall directly* provide staff support to, and *coordinate the activities of*, the National Institute of Justice, *the Bureau of Justice Statistics*, and the Law Enforcement Assistance Administration.

The Justice Assistance Act of 1984 substantially rewrote and reorganized the existing law, creating OJP in its current form and pointedly giving it primacy by defining it in Part A (where the LEAA was previously defined). The 1984 act also made explicit that "the Director [of BJS] shall report to the Attorney General through the [AAG]" (98 Stat. 2079). In place of the above-quoted 1979 language, the 1984 act specified duties of the AAG including (98 Stat. 2078; emphasis added):

> The *Assistant Attorney General shall* . . . (5) *provide staff support to coordinate the activities of the Office and the* Bureau of Justice Assistance, the National Institute of Justice, the *Bureau of Justice Statistics*, and the Office of Juvenile Justice and Delinquency Prevention. . . .

The Homeland Security Act of 2002 (P.L. 107-296; 116 Stat. 2162) made a small but telling change to point (5), simply inserting the words "coordinate and" at the beginning to give the phrase its current form (42 USC § 3712(a), emphasis added):

> The *Assistant Attorney General shall* . . . (5) *coordinate* and provide staff support to coordinate *the activities of the Office [and the] Bureau of Justice Statistics*. . . .

In isolation, these legislative changes might appear to be relatively innocuous. In terms of strict legislative text, the 2002 Homeland Security Act's provision did nothing but restore a "coordination" function held by OJARS at its (and BJS's) founding in 1979—at which point it was arguably a worse situation for BJS, given OJARS's more weakly defined position. However, in context and in combination, the changes convey an intent by OJP to take a more heavy-handed role in BJS activities. A press account at the height of this legislative activity in 2002 noted a statement by then-AAG Deborah Daniels, suggesting that stronger OJP control over BJS and NIJ was desirable in order to ensure that DOJ "speaks with one voice" on crime and justice issues (Butterfield, 2002:33). This rationale is antithetical to the position of independence that statistical and research agencies must have in order to be most effective; statistical agencies must have the latitude to release findings that run counter to the policy of their parent departments, if those findings are borne out by the data. Consequently, taken together, OJP's legislative assertion of "final authority" over BJS functions and its intent to "coordinate" BJS activities constitute dangerous infringements of BJS's proper function.

Conceptually, the current organizational structure under which BJS is housed within OJP along with other research and subject-matter bureaus does have certain advantages. If heavy-handed "coordination" gave way to real synergy—full collaboration between BJS and sister bureaus such as the Office of Juvenile Justice and Delinquency Prevention or the Office for Victims of Crime—BJS data and analysis could meaningfully inform OJP policy development. Likewise, in such a true synergistic environment, the AAG for OJP could provide strong and visible advocacy for BJS concerns. However, we believe that such an effective and beneficial implementation of the status quo organizational arrangement hinges critically on the priorities and temperaments of the AAG and other top officials in the Justice Department and the strength of the BJS director to function independently. In our assessment, the inherent conflicts between the priorities of a program office such as OJP and a statistical agency such as BJS—and the too-fine line between synergistic work by OJP offices and attempts to make those offices "speak with one voice"—makes the status quo untenable in the long run. On the basis of these arguments, we conclude that BJS's administrative placement in OJP is detrimental:

> **Finding 5.3:** The placement of BJS within the Office of Justice Programs has harmed the agency's ability to innovate in data collections and expand the efficiency of achieving its statistical mission. It suffers from a zero-sum game in competition with programs of direct financial benefit to states and localities.

In the panel's assessment, a BJS that is better established as an independent structure within the DOJ infrastructure would have an enhanced ability to support and sustain statistical programs. We also expect that a higher-placed BJS—ideally as a direct report to the attorney general or the deputy attorney general—would have a powerful effect on the timeliness of information released by BJS, because it would be called upon to provide more contemporaneous information to the highest levels in the department. Such an administrative move would make clear the permanence of data-gathering functions and the need to use the resulting information in policy development and review; it would also provide a clear separation from competing interests who wish to advocate for certain programs or initiatives. In terms of data collection, a more-prominent and higher-profile BJS would also be helpful in dealing with balky or resistant data suppliers. To be sure, administrative attachment of BJS to the office of the attorney general runs the risk of politicization—far from the intended effect. However, in our judgment, such a high-level attachment would afford BJS the most prominence and stature and, hence, be the strongest corrective remedy for past breaches of BJS's independence. Accordingly, we recommend:

> *Recommendation 5.3:* **BJS should be administratively moved out of the Office of Justice Programs, reporting to the attorney general or deputy attorney general.**

It follows that this administrative change involves removing the legislative language asserting a strong OJP oversight role over BJS functions.

To this general recommendation, we add two corollaries:

- In foregoing ties to OJP, it is important for BJS to retain the capacity for letting contracts. In particular, it is vital that BJS retain full ability for administering grants such as those that maintain the state Statistical Analysis Center (SAC) network and that support development of and improvement to source criminal justice databases, as described in Chapter 4.

- The problems faced by BJS in its administrative nesting within a program agency are similar to those faced by some other OJP units, notably NIJ: a research agency embedded within a program agency. In November 2008, John Jay College of Criminal Justice president and former NIJ Director Jeremy Travis issued an open letter to the membership of the American Society of Criminology urging the creation of an Office of Justice Research within DOJ. This new office would include BJS and NIJ, elevating NIJ's Office of Science and Technology to become the National Institute of Justice Technology; all three agencies would report to an assistant attorney general for justice research, appointed by the president with Senate confirmation. Relevant excerpts from this letter are shown in Box 5-3.

  Determining the administrative placement of NIJ is beyond this panel's scope; a parallel National Research Council panel is currently evaluating NIJ's research program, and NIJ's structure is more the province of that panel. However, we note that an approach by which both BJS and NIJ report to an assistant attorney general for research is certainly consistent with our own recommendation; our guidance in this report is intended to speak to a choice between BJS remaining in OJP versus moving out of OJP, and the Travis proposal would also achieve the result we think is best for BJS. A separate office including both a research agency and a statistical agency would also be uniquely poised to develop research programs in justice-related issues that have received relatively little rigorous empirical treatment, such as the extent to which forensic evidence (e.g., fingerprints or firearm-related toolmarks) are introduced in judicial proceedings (and the effectiveness of that evidence) or the perceived fairness of court verdicts.

Term of Appointment of BJS Director   To provide an added measure of insularity, the panel further concludes that BJS would benefit from the designation of the BJS directorship as a fixed-term appointment by the president, with the advice and consent of the Senate.

Box 5-3   Excerpts from Travis (2008) Open Letter on an Office of Justice Research

I propose that the Congress create, with support from the new Administration, a new office in the Department of Justice, called the Office of Justice Research, to be headed by an Assistant Attorney General for Justice Research. This office would be separate from the Office of Justice Programs, which would continue to administer the funding programs that support reform efforts by state and local law enforcement and criminal justice agencies. . . .

The argument for creation of the new Office of Justice Research, separate from the Office of Justice Programs, is very straightforward: if the research, statistics, and scientific development functions of the federal government are located within an office that is primarily responsible for the administration of assistance programs, three risks are created. First, the scientific integrity of the research functions is vulnerable to compromise. Second, the research and development function will never be given the priority treatment that is needed to meet the enormous crime challenges facing the country. Third, the research agenda on crime and justice will more likely reflect short-term programmatic needs rather than the long-term need to develop a better understanding of the phenomenon of crime in America and the best ways to prevent and respond to crime. . . .

[As part of this new office,] the **Bureau of Justice Statistics** would continue all of the functions currently carried out by BJS. [But] the current constellation of data collections systems on crime and justice are fragmented and incomplete. To remedy this situation—and to provide the nation the capability to track crime trends in a timely manner—the mandate of BJS should be expanded significantly. First, BJS should be authorized to work closely with the Federal Bureau of Investigation to improve the timeliness and completeness of the Uniform Crime Reports. Similarly, responsibility for the ADAM program [(see Section 2–C.4)] should be transferred from ONDCP (it was originally housed at NIJ), and responsibility for the statistical series on juvenile justice should be transferred from the Office of Juvenile Justice and Delinquency Prevention (a component of OJP). But the new BJS would be more than a manager of existing statistical series. It should also develop new initiatives to track crime trends, drawing on capabilities of police departments that now post crime trends close to real time. It would develop new protocols for tracking critical crime issues, such as the level of illegal drug selling activity, public confidence in the criminal justice system, the operations of the federal law enforcement agencies, etc. This expanded portfolio would clearly require additional funding, but there are compelling arguments for creating a robust national capacity to improve our understanding of crime trends. . . .

If we were designing a federal research and development capacity on crime and justice today, we would probably not propose the current structure that houses NIJ and BJS within the Office of Justice Programs, three levels below the Attorney General, with a focus on state and local criminal justice. Rather, we would create a scientific branch of government that operates under scientific principles reporting directly to the Attorney General. We would recognize that crime is now a transnational phenomenon and we need to understand human trafficking, drug smuggling, immigration trends and terrorism. We would examine the many systems of justice—civil justice, immigration courts, the federal justice system, in addition to state and local justice systems. We would develop a modern capacity to understand local crime conditions using high-tech surveys. We would develop creative ways to measure

(continued)

Box 5-3 (continued)

non-traditional crimes, such as identity theft, corporate and white collar crime, and transnational crime. We would design a research and development program that would harness the power of technology so the agencies that enforce the law can benefit from the scientific and technological revolution. This ambitious agenda clearly requires additional resources. But it also requires a new structure within the Department of Justice, a structure that guarantees both scientific integrity and policy relevance.

SOURCE: Excerpted from Travis (2008:1, 4, 5); emphasis in the original.

Finding 5.4: Under current law, the director of the Bureau of Justice Statistics serves at the pleasure of the president; the director is nominated to an unspecified term by the president, with the advice and consent of the Senate (42 USC § 3732(b)).

It is worth noting that fixed-term appointments are relatively rare in the federal statistical system. Currently, only two of the nation's principal statistical agencies—the Bureau of Labor Statistics and the National Center for Education Statistics—have heads who are appointed and confirmed to fixed terms of 4 and 6 years, respectively (29 USC § 3 and 20 USC § 9517(b)).[13] The heads of BJS, the Census Bureau, and the Energy Information Administration are appointees (with Senate confirmation) who serve at the pleasure of the president; the other nine heads of Interagency Council on Statistical Policy member organizations are career employees and departmental appointments. Bills to create a termed appointment for the director of the Census Bureau have been introduced, but not enacted, in recent Congresses—most recently, one that would fix the term at 5 years (at the same time that it would remove the Census Bureau from the Department of Commerce and establish it as an independent executive agency).[14]

The range of models for the term of appointment of a BJS director can be expressed simply:

- Presidential appointment with Senate confirmation, at pleasure (the status quo);

- Presidential appointment with Senate confirmation, fixed term; and

- Career employee, appointed by the president, cabinet secretary, or other official.

---

[13] Ironically, the same legislation that positioned the National Center for Education Statistics under a new administrative layer—the Institute of Education Sciences—also extended the length of the fixed term for the commissioner of education statistics. Prior to 2003, commissioners served a 4-year term rather than a 6-year term.

[14] See H.R. 7069, introduced by Rep. Carolyn Maloney (D-N.Y.) on September 25, 2008, in the 110th Congress.

In the right environment—with a strong and well-defined position of independence and the latitude for innovation—the career employee directorship is an attractive option that has the added advantage of ensuring that a director is well versed in the agency's existing work and subject-matter domain. Indeed, among BJS's fellow statistical agencies, career employee appointments such as the directorship of the Bureau of Economic Analysis rank among the most effective leadership models. However, as we described in arguing for an administrative move out of OJP, BJS does not enjoy such an environment. We view a presidential appointment with Senate confirmation as a necessity for the BJS directorship, carrying with it the stature to interact effectively with the appointees at the top ranks of the Justice Department.

The events of 2005 demonstrated that BJS can be and has been harmed by the current arrangement by which the BJS director serves strictly at the pleasure of the administration. The circumstances of Director Greenfeld's dismissal—in the immediate aftermath of refusing to alter a press release to address political concerns—fostered the appearance of formal and structural interference in BJS's operations. In our assessment, a fixed-term appointment for the BJS directorship would be the best and strongest palliative measure to put some distance between BJS and its political superiors in the Justice Department (whether BJS remains in OJP or not). The model of the directorship of the Bureau of *Labor* Statistics is the one that we find most compelling for BJS: In our judgment, it makes sense for the federal officer directly tasked with reporting key indicators of social justice in America to have stature, political insularity, and term of service commensurate with the federal officer directly responsible for reporting key economic indicators such as unemployment and job growth.[15] The director of BJS must have the capability to objectively report both good news and bad news—to provide information on crime and justice in the United States, even when the findings are politically inconvenient or unappealing. We believe that a presidential appointment with confirmation provides the appropriate stature for such a position, and that the specification of a fixed term of service prevents the kinds of attempted interference that has harmed BJS in recent years.

Accordingly, we recommend:

> *Recommendation 5.4:* **Congress and the administration should make the BJS director a fixed-term presidential appointee with the advice and consent of the Senate. To insulate the BJS director from political interference, the term of service should be no less than 4 years.**

---

[15]Though the jobs are obviously much different in scope, it is worth noting that the other principal federal officer tasked with reporting statistics on crime in the United States—the director of the FBI, reporting results from the Uniform Crime Reporting program—holds the relative insularity of a 10-year fixed-term appointment, nonrenewable, with Senate confirmation (P.L. 90-351 § 1101).

It would make sense for the term to be about 6 years because that would take the director to a new administration or to a second term of a incumbent administration.

5–A.3   Relevance to Policy Issues

A federal statistical agency must be in a position to provide objective information that is relevant to issues of public policy. A statistical agency must be knowledgeable about the issues and requirements of public policy and federal programs and able to provide objective information that is relevant to policy and program needs. . . . In establishing priorities for statistical programs for this purpose, a statistical agency must work closely with the users of such information in the executive branch, Congress, and interested nongovernmental groups. (National Research Council, 2009:4)

This principle has implications, both for the parent department of a statistical agency and for the actions of the agency itself. The parent department must take the agency seriously. Statistical units, when best used by their parent agency, are the window into the performance of their agency in addressing key issues facing the society. When intelligently used, the statistical agency can measure the prevalence and importance of different issues tasked to the department. When intelligently used, they can be the management dashboard to guide allocation of budget to different activities. When intelligently used, they can assemble information about likely trends of future phenomena within the mission of the department.

However, achievement of such a role is not merely dependent on outreach by the leadership of the parent department. Rarely are the government officials appointed to departmental leadership aware of the utility of statistical information to guide the work of the department. The director and senior staff of the statistical agency have an obligation to be outwardly-focused, to become expert in the program mission of the agency. Only with such substantive expertise can the department's statistical agency produce optimally relevant statistical information to the policy makers of the department. Statistical agencies are part of the management information system for policy making in program departments. Senior statistical staff must have the skills, time resources, and mandate to develop relationships with the policy-making units to provide information relevant (not necessarily supporting, but relevant) to the policy makers' tasks.

In the judgment of the panel, BJS's ability to carry out this role of providing policy-relevant data is impaired by its relatively low profile within the agency. At one of its plenary meetings, the panel met with senior DOJ officials and discussed past and current roles of BJS within DOJ policy-making activities; from those discussions, it was apparent that BJS was not viewed as a relevant player in many of the key initiatives of DOJ. Indeed, there did not seem to be high awareness of the range of BJS activities or the ways in which data could be brought to bear in broader DOJ activities. In the panel's view, BJS has not been perceived as an important asset in assembling relevant information for key policy initiatives; fault for this is undoubtedly shared by BJS (for limited "promotion" of its work within the department) and by higher officials in DOJ.

There are two potential, relevant solutions, the first of which looks at BJS activities within DOJ. The panel believes that the BJS director should be a very visible and active promoter of the value of objective statistical information for use in policy decisions within DOJ. Every budget initiative of DOJ is a potential opportunity for enriched statistical information about the status of the justice system. The BJS director and his or her senior staff should increase their outreach to sister DOJ units.

> **Recommendation 5.5:** The BJS director needs to reach out to other agencies within DOJ, forming partnerships to propose initiatives for information collection that are relevant to policy needs.

> **Recommendation 5.6:** The Department of Justice should build provisions for BJS collection of data and statistical information into its program initiatives aimed at crime reduction. These are not intended as program evaluation funds, but rather as funds for the basic monitoring and assessment of the phenomena targeted by the initiative.

Although this recommendation is a necessary step to achieve more relevance to DOJ, the panel believes that it may not be sufficient. Effective outreach by BJS depends on willingness to receive such outreach and respect for BJS expertise. The visibility of BJS within DOJ and in the legislature appears to be quite low. On budget initiatives the BJS director rarely meets directly with legislative staff; the BJS budget is reviewed as part of the OJP budget, and so those discussions are held at the OJP level. Hence, our previous recommendation to administratively move BJS out of OJP—giving the BJS director the authority (and the duty) to interact directly with congressional appropriators and overseers—would also contribute greatly to BJS's ability to provide policy-relevant data.

In Section 5–B.8 below, we discuss the need for an effective research program as another means of bolstering the relevance of BJS and its data products.

5–A.4   Credibility Among Data Users

A federal statistical agency must have credibility with those who use its data and information. . . . To have credibility, an agency must be free—

> and must be perceived to be free—of political interference and policy
> advocacy. Also important for credibility is for an agency to follow such
> practices as wide dissemination of data on an equal basis to all users,
> openness about the data provided, and a commitment to quality and
> professional practice. (National Research Council, 2009:5)

Credibility is a reputational attribute of a statistical agency. It is frequently argued that the credibility of the statistical products is partly derived from sound statistical properties (high precision and low statistical bias) and from perceptions that the source of the information has no point of view or ideological lens on the information (National Research Council, 2005b:5). Thus credibility is enhanced with sound professional practice and widespread recognition of this professionalism. It is also enhanced by demonstration of independence from influence from policy viewpoints.

Panel members and staff were active observers in a workshop of users of BJS data, conducted by the Council of Professional Associations for Federal Statistics (COPAFS) with BJS sponsorship, in February 2008. Attendees at the workshop included members of BJS's state SAC network, academic researchers, representatives of police chiefs, representatives of state courts, and others, along with BJS staff and officials. There was general high praise for BJS, some calls for increased timeliness of BJS data (for enhanced law enforcement management purposes), and finer granularity of estimates for local uses. For some panel members in the audience of the workshop, some of the law enforcement community were asking for almost real-time event data—a goal that is difficult for any statistical agency to achieve. Despite these types of critiques of BJS, panel after panel at the workshop expressed great belief that the BJS data series were credible, valued, and relevant to their work.

> **Finding 5.5:** BJS enjoys high credibility but often is critiqued for
> missing fine-grained data by geography or time.

5–B PRACTICES OF A FEDERAL STATISTICAL AGENCY

5–B.1 Clearly Defined and Well-Accepted Mission

> An agency's mission should include responsibility for all elements of
> its programs for providing statistical information—determining sources
> of data, measurement methods, efficient methods of data collection and
> processing, and appropriate methods of analysis—and ensuring the public availability not only of the data, but also of documentation of the
> methods used to obtain the data and their quality. (National Research
> Council, 2009:7)

That BJS's mission and basic functions are clearly defined is virtually indisputable. We have frequently referred to Box 1-2, BJS's extensive list of authorized activities under its enabling legislation, which is testament to the detail in BJS's defining mission. Whether they are clearly accepted is quite another matter. As we discussed in Section 5–A.3, the panel was disappointed by the apparent lack of understanding of BJS's role and its potential when it met with higher-level Justice Department officials. Although expressions of support were plentiful, an understanding of the importance of high-quality data for shaping policy was generally lacking.

BJS's recent history in the appropriations process is also, potentially, evidence that its range of existing data collections—and the cost of data collection, generally—is not well understood in important places. In summer 2006, the appropriations committees in both houses of Congress processed BJS's budget request of about $60 million. While the House sought to keep BJS funding at about fiscal year 2006 levels ($36 million, compared to final 2006 allocation of $34.6 million; H.Rept. 109-520), the Senate's mark came in considerably lower at $20 million (S.Rept. 109-280). (No final appropriations bill for DOJ was passed for fiscal year 2007; like many other federal agencies, it was funded through a series of continuing resolutions at fiscal year 2006 levels, with some exceptions). A brief explanatory note in the Senate committee's report acknowledged BJS's role in collecting the NCVS and other data programs but did not explain the reason for the reduction. The problem was exacerbated in the fiscal year 2008 appropriations process: House appropriators provided $45 million for BJS (H.Rept. 110-240) but the Senate appropriators, with no explanatory statement whatsoever, included only $10 million for BJS: a funding level that would have terminated the NCVS, if not much of BJS's activities. Inquiries by the Consortium of Social Science Associations yielded the explanation from Senate subcommittee staff that the $10 million figure was a "misprint" that would be corrected and replaced by "full funding" later in the process (Consortium of Social Science Associations, 2007:3). It was not corrected in the version of the bill that finally passed the Senate; in the final consolidated appropriations bill that included DOJ, BJS funding came closer to the House mark than the Senate mark.[16]

As before, the panel concludes that a clear separation between BJS and OJP and placement of BJS elsewhere in the DOJ hierarchy would help clarify the mission of BJS and strengthen its profile as a principal statistical agency. Given congressional stalemate and the inability to pass most individual appropriations bills, the particular budget climate in recent fiscal years would be difficult for any organizational configuration of BJS within DOJ. Still, the story of the varying appropriations marks suggests that, in at least one important circle, knowledge of the basic cost of data collection and the value

---

[16]For fiscal year 2009, Senate appropriators recommended $40 million for BJS (S.Rept. 110-397).

(and cost) of BJS's flagship data collection was sufficiently weak as to put BJS's viability at stake. BJS's mission is not well served by having its interests solely represented and managed by OJP in the budget and planning arenas, precisely because BJS's own mission is not well articulated by OJP's general mission "to increase public safety and improve the fair administration of justice across America through innovative leadership and programs" (U.S. Department of Justice, Office of Justice Programs, 2006:3), principally through financial assistance.

5–B.2 Continual Development of More Useful Data

> Statistical agencies must continually look to improve their data systems to provide information that is accurate, timely, and relevant for changing public policy needs. They should also continually seek to improve the efficiency of their programs for collecting, analyzing, and disseminating statistical information. (National Research Council, 2009:7)

The February 2008 data users workshop, sponsored by BJS and conducted by COPAFS, was a good step for BJS in carrying out the practice of improving and modifying its data collections to be more useful and relevant. The session suggested both useful analyses and extracts that could be made from existing data series (e.g., tailoring analyses and sponsoring research on the NCVS; Heimer, 2008) and wholesale revisions to collection methodologies to improve timeliness or relevance (e.g., an NCVS-type survey of experiences in *civil* justice matters; Eisenberg, 2008). As we observed in Chapter 4, BJS's state SACs, and its coordination through JRSA, provide it with a mechanism for ready communication and interaction with state-level practitioners, all of which contribute to reevaluation of individual BJS programs and reports.

Although BJS has done well on this score, we encourage it to push further and develop the tools that other statistical agencies use to inform themselves of changing data needs of their user bases. Specifically:

1. As BJS staff indicated at the time, the February 2008 users workshop should be seen as a first step and not a one-time conversation. BJS could sponsor an annual users conference, perhaps drawing from a larger base of downstream users than JRSA's annual research conference. These user meetings could be similar to those routinely held by the National Center for Health Statistics, CDC (for the Behavioral Risk Factor Surveillance System), and the Census Bureau.

2. Through JRSA, BJS sponsors a journal (*Justice Research and Policy*), much as the Bureau of Transportation Statistics has done for its related fields. BJS's role in such a journal or statistical publication—and knowledge of strengths and weaknesses in BJS data—could be enhanced by encouraging BJS staff or grantees to seek publication in

the journal or developing "special users" on specific user constituency needs.

3. Consistent with item 21 in BJS's legally authorized duties (Box 1-2), BJS could convene meetings of official justice statisticians from other countries, charged with missions similar to that of BJS, to apprise itself of international comparability.

4. BJS could commission small "white papers" from key leaders in the justice systems about future data needs.

5. BJS should continue, and interact with, informal advisory mechanisms that have developed over the years, such as the Committee on Law and Justice Statistics of the American Statistical Association.

Historically, BJS has convened periodic expert workshops as a first step in scoping out new work. McEwen (1996) summarized the 1995 workshop on police use of force that contributed to the development of PPCS, and BJS partnered with SEARCH, the National Consortium for Justice Information and Statistics, on a series of workshops on law enforcement databases such as criminal history records and sex offender registries (Bureau of Justice Statistics, 1995, 1997b, 1998a). However, such workshops have become rarer events in light of funding resources. As suggested by the first point in our list above, we think that these workshops are an important mechanism that would have the added benefit of improving concerns about the timeliness of content in BJS data collections; they would provide for regular input and feedback on emerging problems and views. One possible topic on which such a stakeholder workshop could be beneficial is to review content in the correctional data series and the NCVS in order to ensure that definitions and concepts of "mental health" are consistent with current practitioner usage.

> ***Recommendation 5.7:*** **To effectively get input on contemporaneous topics of interest, BJS should regularly convene ad hoc stakeholder workshops to suggest areas of immediate data needs.**

However, we also believe that BJS would strongly benefit from a more formal means of obtaining user input: therefore, we recommend that BJS establish a standing technical advisory committee, appointed under the terms of the Federal Advisory Committee Act (5 USC App. 1). The legislation that created BJS, the Justice System Improvement Act of 1979, originally mandated a 21-member BJS Advisory Board, with members appointed to 3-year terms by the attorney general; this board was directed to review and make recommendations on BJS programs as well as to recommend candidates in the event of a vacancy in the BJS directorship (93 Stat. 1178–1179). However, this provision for an advisory board was removed in the 1984 reauthorization (see notes at 42 USC § 3734). Although BJS receives valuable advice through informal means, we conclude that there would be real value in having a standing advisory committee, including members with substantive expertise, operating staff within justice system institutions, statistical experts, and others who could articulate future needs. It is important that such an advisory board contain high-level policy makers and justice system practitioners as well as methodologists and statisticians so that detailed research-specific recommendations are paired with input on the timeliness and usefulness of the data in the field.[17]

The Census Bureau organizes several such advisory committees (including, for instance, groups specifically focused on input from diverse race and ethnicity groups and on advice from relevant professional associations); another model is the Board of Scientific Counselors of the National Center for Health Statistics. Both of these advisory structures in the statistical system provide written recommendations to their respective agencies and, in the case of the Board on Scientific Counselors, undertake program reviews of parts of the agency's portfolio; this kind of regular feedback would greatly benefit BJS operations.

> **Recommendation 5.8:** BJS should establish an Advisory Group under the Federal Advisory Committee Act to provide guidance to BJS on the addition of new data collection efforts and the modification of current ones in light of needs identified by the group. Membership in the group should include, at a minimum, leaders and practitioners from each of the major subject matters covered by BJS data, as well as those with statistical and other types of academic expertise in these subject matters. The members of the group should be selected by the BJS director and the group should provide the director with at least two reports each year that contain its recommendations.

This recommendation is consistent with, but more fully articulated than, Recommendation 5.1 in our interim report (National Research Council, 2008b).

A standing advisory committee could be designed with subgroups of topic specialties in mind so that, for instance, the committee is poised to render NCVS-specific methodological advice without having to convene separate committees for each major collection. By having both coverage and depth in topic areas, a standing advisory committee would be useful as a means for suggesting new directions for research. One specific example

---

[17]As reference, the original BJS Advisory Board specified in the Justice Systems Improvement Act was to have members including "representatives of States and units of local government, representatives of police, prosecutors, defense attorneys, courts, corrections, experts in the area of victim and witness assistance, and other components of the justice system at all levels of government, representatives of professional organizations, members of the academic, research, and statistics community, officials of neighborhood and community organizations, members of the business community, and the general public" (93 Stat. 1178).

where a formal advisory committee would be useful would be in revisiting content in the Law Enforcement Management and Administrative Statistics (LEMAS) survey, as part of implementing a core-supplement design. By its nature, LEMAS is an establishment survey that is targeted at a wide variety of individual law enforcement agencies. However, these agencies may differ in their usage and basic definition of terms; for example, depending on the prevailing definition of "community-oriented policing," *all* departments might consider themselves to follow that practice whereas others (possibly confounding the term with specific grant/funding streams) may think that they do not. Regular review of the basic language used in the data collection is important to avoid the perception that questions are overly blunt or are confusing.

In developing its outreach to its user base, it is important that BJS not neglect the needs and interests of a critical user constituency: members of Congress and their staffs. Steps to assess the issues of interest to the House and Senate Judiciary committees would be useful to build awareness of and interest in BJS products, promote a clearer understanding of what is and is not possible in statistical data collections (as did not seem to occur in developing the PREA reporting requirements), and gain critical support for new and continuing data collections.

> **Recommendation 5.9:** **DOJ should take steps to ensure that congressional staff are aware of BJS data that could be used in developing legislation; DOJ and BJS should learn from congressional staff how their data are needed to inform/support legislation so that they can improve the utility of their current data and so that they can develop new data sets that could enhance policy development.**

5–B.3 Openness About Sources and Limitations of Data

A statistical agency should be open about its data and their strengths and limitations, taking as much care to understand and explain how its statistics may fall short of accuracy as it does to produce accurate data in the first place. Data releases from a statistical program should be accompanied by a full description of the purpose of the program; the methods and assumptions used for data collection, processing, and reporting; what is known and not known about the quality and relevance of the data; sufficient information for estimating variability in the data; appropriate methods for analysis that take account of variability and other sources of error; and the results of research on the methods and data. (National Research Council, 2009:8)

In general, the panel believes that the BJS staff is fully open regarding the strengths and weaknesses of its data series. Its house style for the preparation of the report emphasizes that even short reports contain a fairly detailed section on methodology; these sections generally do a good job at presenting synopses of the design of data collections. The recent episodes concerning the 2006 and 2007 releases of data from the NCVS—culminating in the conclusion that 2006 data constituted a "break in series" (see Section 3–A.3)—is illustrative in this regard. Recognizing the presence of a problem, BJS staff sought external opinions and worked closely with the Census Bureau to try to understand what had occurred. The declaration of a "break in series" was not an easy one to make, but BJS's descriptions of the circumstances in its reports (and the documentation accompanying the archived data file) are certainly candid about the limitations of the data.

However, the "break in series" incident also illustrates a point that we make later in this chapter concerning the technical skill mix of the BJS staff. In such an incident, it would be useful for BJS to have more in-house staff with advanced technical skills, to more completely understand how design changes and sample size reductions combine to produce discrepant effects. BJS shares with other federal statistical agencies a fundamental problem that it has insufficient numbers of technical staff whose primary job is to focus on *evaluation* of the quality of data collected by and for BJS. Because of this absence, the outside user of BJS data has no set of working papers, methodological briefs, or quality profiles that may be consulted to inform themselves of the characteristics of particular data sets or the potential strengths and weaknesses for their specific uses of the data.

The lack of routine evaluation and quality assessments of BJS data is problematic because of the wide variety of sources from which BJS data series are drawn; BJS's correctional data provide a useful example. Much of the correctional data are collected from agencies and institutions that rely on varied local systems of record-keeping. Heterogeneity in record-keeping standards produces heterogeneity in responses to administrative surveys. For some data collections, such as the National Corrections Reporting Program (NCRP), states may have varying definitions of the race, ethnicity, and schooling of admitted and released prisoners. Detailed instructions for classification and measurement would improve the quality of corrections data reporting.

> *Recommendation 5.10:* **To improve the utility and accuracy of the National Corrections Reporting Program (NCRP), BJS should work with correctional agencies to develop their own internal records to promote consistent data collections and expand coverage beyond the 41 states covered in the most recent NCRP.**

It follows that the same kind of evaluation of the raw data provided by state and local authorities, coupled with work to promote consistent reporting, would also benefit BJS's other correctional, law enforcement, and adjudication data series.

5–B.4   Wide Dissemination of Data

A statistical agency should strive for the widest possible dissemination of the data it compiles. . . . Elements of an effective dissemination program [include] a variety of avenues for data dissemination [including, but] not limited to, an agency's Internet website, government depository libraries, conference exhibits and programs, newsletters and journals, e-mail address lists, and the media for regular communication of major findings. (National Research Council, 2009:9)

BJS deserves great credit for its data dissemination efforts, several of which are described in Box 1-1. It makes good use of public use data set archiving through the National Archive of Criminal Justice Data (NACJD); its own website and the OJP-sponsored National Criminal Justice Reference Service provide ready access to an extensive backfile of reports; its website entries for individual reports generally provide the reports in text or print formats and typically include either plain text or spreadsheet tables corresponding to key data tables. As noted in Chapter 4, the state SAC network also provides a means for the dissemination of BJS data and products (and SAC analyses thereof) to local audiences. All of these steps have been a great service to the user community and represent shrewd use of partnerships with outside groups with specific expertise that in-house BJS staff could not not do in isolation. The coupling of the public data archive with the regular instructional workshops conducted by the Inter-university Consortium for Political and Social Research is a very valuable service, opening BJS resources to new researchers.

Timeliness of Data Release

Although we laud BJS for its work in data dissemination, this principle does suggest three areas where some further comment is necessary, the first of which concerns the timeliness of data release. Once a report is prepared and new data are ready for release, BJS is very good at executing the release; the problem is that the lag times between data collection and the time of report and data release can be considerable, sometimes taking several years, which hurts the freshness and timeliness of the new results. This is, of course, a fundamental problem that applies to statistical agencies other than BJS: timely release of data is essential for those data to be useful in policy formulation and in research, yet the process of collecting high-quality data, ensuring that quality, and protecting the confidentiality of response takes time and is not one that can readily be rushed without overburdening respondents.

Finding 5.6: A recurring criticism of BJS data products is that their quality is highly valued but that they are not sufficiently timely to meet user needs. All statistical agencies are attempting to grapple with new data collection designs that offer more timely estimates.

Delays in the release of data arise—and can be particularly pronounced—in those circumstances where BJS is dependent on other agencies, especially the Census Bureau. By this, we do not impugn the Census Bureau but merely note that it has its own privacy protection protocols and data quality procedures that, combined with BJS's own review, can add substantially to processing time. In some instances where the Census Bureau has been the data collection agent, release of data can be obstructed because of post hoc determinations that a particular release format would threaten confidentiality. This has been the case with collection and coding of the industry and occupation data from the Workplace Risk Supplement, for which the Census Bureau has opposed release because the cell sizes for certain occupations are too small. Negotiations with the Census Bureau have continued for years, to the extent that these data collected in 2002 have not yet (late 2008/early 2009) been released.

Another case of the Census Bureau restricting or impeding the timely availability of data is the removing of the area-identified NCVS from Census Analysis Centers. These data were available in analysis centers around the nation for a number of years but were subsequently withdrawn amidst concerns about confidentiality and documentation. Similar issues have barred the release of a special area-identified data file from the NCVS. Such a file is critical to studying the prospects for local-area estimation from the NCVS, and the file was once made available through BJS's data archive, but it has now been offline and unavailable for about 4 years. Delays of this extent suggest that something is broken in the relationship between BJS and the Census Bureau, which is obstructing the timely release of these data.

In cases where other agencies provide the funding of a data collection, such as supplements to the NCVS, the release of the data can be delayed because both BJS and this other agency must issue a "first" release and because there can be ambiguity with regard to which agencies have "control" over the data. All of these factors delay release of the data and should be scrutinized to see if there could be joint "first" releases or other streamlining of this process. Agreements on supplements or other joint ventures with other agencies could include time limits on the release of data and clearer lines of authority for release.

Maximizing the use of BJS data requires that it be released in a timely and equitable fashion and in formats that facilitate its use, while protecting the confidentiality of the data and furthering the goals of the agency. These objectives are often conflicting, and balancing them is no simple matter. It would benefit BJS to track the processing that occurs after data collection is complete and document the times of data collection, report preparation, report release, and data archival to study which components of processing are most time-consuming (and which may be made more efficient).

More generally, BJS should work to confront the challenge of timely data release in creative ways. One mechanism to consider is issuance of preliminary estimates—labeled as such and clearly noted as being subject to future revision—that could be issued on a quick basis and separate from a fuller and more detailed report that would contain final estimates. Another (and more elaborate) idea that is worthy of consideration is adoption of continuous data collection designs. These designs spread diffuse sample; information is collected from a smaller number of respondents at any given time but collectors are in the field on as continual a time basis as possible. These designs have the advantage of avoiding the startup costs of reinventing survey design machinery and sample from scratch every time a new round of data is collected, but their continuous streams of data can also be combined and pooled to produce more timely estimates. With the Census Bureau's introduction of the American Community Survey (ACS), the U.S. public will become accustomed to interpreting the period estimates that span several time periods (e.g., 3-year or 5-year averages); opportunities to present BJS data in similar structures should be considered.

> *Recommendation 5.11:* **BJS should evaluate each of its data programs to inquire whether more timely estimates might be obtained by (a) making discrete data collections into more continuous operations and (b) issuing preliminary estimates, to be followed by final estimates.**

Equitable Release of Data

A second area of discussion about data dissemination is the equitable release of data, meaning that all of the public should generally have access to data releases at the same time, in formats that are conducive to use and interpretation. There may be instances in which some individuals outside of BJS should have access to some data before general release because it furthers the goals of the agency, such as evaluating and maintaining data quality. In those cases, priority access should be available and granted. Joint publications where BJS staff collaborate with persons outside the agency may also be an acceptable form of early release. Except in circumstances such as these, statistical agencies such as BJS should strive to ensure that individuals' access to data files are on an equal footing.

The formats in which BJS are data are released should facilitate the use of these data. Here, format includes the medium by which the data are made available as well as the content of the releases. BJS, like all statistical agencies, has different formats for different user communities. Written reports and electronic versions of written reports are available for readers who do not wish to manipulate the data. Spreadsheet versions of key tables and some Web-based tools for simple online analysis are provided for users who want to manipulate but may lack the sophistication to do so in a complex way. The full data sets are available for the most sophisticated users who are interested in manipulating the microdata substantially. It would be helpful for BJS to be more direct in spelling out the logic and connectedness of their product lines and formats. It would be useful for the website for a LEMAS report to indicate that users can go to a separate part of the BJS website to access online anlaysis options if they cannot find the particular rate or cross-tabulation in the hard-copy and electronic reports; this clue is not immediately obvious. If the online analytical capabilities cannot answer their question, then consumers should be explicitly referred to the NACJD where they can download data sets. This search logic may be obvious to some but not to others who visit the BJS website, and it is not clear that the formats and product lines currently available have a coherent and integrated dissemination plan or strategy.

Increasing sophistication of the public with regard to electronic access to information may warrant a reevaluation of the mix of media used to disseminate BJS data. BJS has already taken steps to reduce the number of reports produced in hard copy by emphasizing online distribution as Portable Document Format (PDF) files; a next step would be to consider ways to reduce paper format even more and to make better use of hyperlink facilities in the PDF files to point users to related reports. Some BJS publications such as the Fireram Inquiry Statistics summarizing background checks for handgun purchases have moved to a release format where the "report" release consists almost entirely of data tables, with minimal prose. Finding additional avenues for this format would have the dual benefit of potentially providing more timely release (as discussed above) and freeing staff to spend less time on standard report writing and more on innovation and evaluation; that said, careful prose summaries are also very important, and we do not want to be construed as saying that the standard written reports should be abandoned.

These suggestions for improving the dissemination of BJS data will put more strain on an overworked staff. Some of the format changes may free up some resources, if they reduce the amount of time required in the editorial process. In the short run, the agency may consider making greater use of the NACJD to develop some of the format changes mentioned in the foregoing paragraphs.

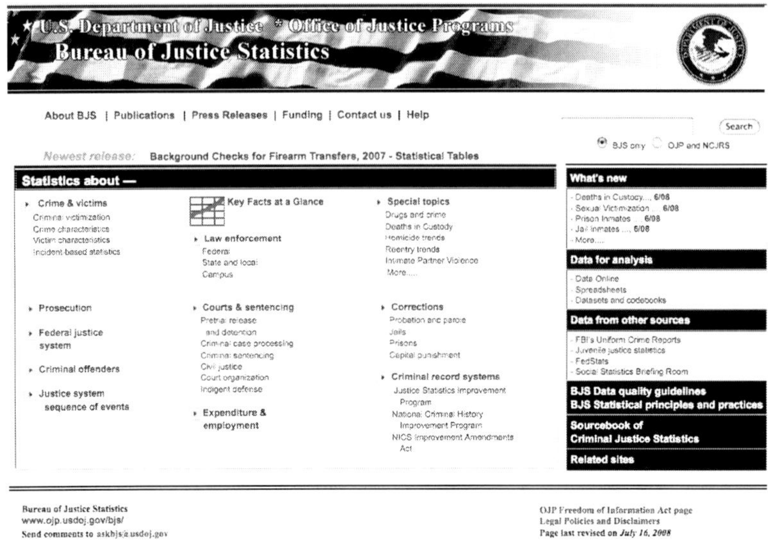

Figure 5-5   Bureau of Justice Statistics home page, July 2008

NOTE: URL for home page is http://www.ojp.usdoj.gov/bjs/; this version accessed July 21, 2008.

BJS Web Presence

A third, and final, discussion topic under the general heading of data dissemination concerns an essential tool for such dissemination: BJS's presence on the World Wide Web, the current front page of which is illustrated in Figure 5-5. As suggested by our comments earlier in this section, BJS recognizes the importance of its Web presence to the spread of its information. Former BJS Director Jeffrey Sedgwick (2008:2) commented that:

> Over the past year, we have continued to develop a new website that will more effectively connect our users with the information they need. The website restructures the way our information is presented, giving users a more intuitive way to retrieve the data they need. Future development will include enhancing our ability to generate custom data tables and other interactive products online.

No website design is perfect in the eyes of every user. It is unclear how useful specific design suggestions would be, though we have indicated preferences for some additional topic pages throughout this report (e.g., summarizing what data are and are not available concerning white-collar crime; Section 2–C.1). However, one point that we do want to raise as BJS revamps its Web presence is to suggest emphasis on data sourcing and external collaboration. It is worthwhile to frame this discussion by stating a conclusion that is consistent with the principles expected of a federal statistical agency:

> **Finding 5.7:** The credibility of BJS's products is a function of its quality review procedures.

It follows that the BJS "brand"—explicitly being labeled as a BJS product—carries weight and is a meaningful distinction. Hence, there is a need to take care in what gets designated and explicitly linked to as a BJS product. BJS's collaborative projects, such as the Federal Justice Statistics program with the Urban Institute and the Court Statistics Project with the National Center for State Courts, are prone to ambiguity and confusion on this score: BJS's website is sometimes abrupt in linking users to the Urban Institute-hosted Web hub for the federal system statistics. Likewise, the Court Statistics reports sometimes carry a BJS logo but BJS's sponsorship role (and use of some of the data) is not immediately apparent. To be clear, we do not argue that the reports and portals on non-BJS Web servers are bad in any sense or that the BJS "brand" is being misused by these external placements. Quite to the contrary, the hope is for both BJS and its data collection partners to receive appropriate credit for good work.

Accordingly, we conclude and recommend as follows:

> **Finding 5.8:** Several BJS data series are collected and maintained by external organizations linked to the BJS website (e.g., Federal Justice System statistics). It is not clear why some data and reports reside on external websites, rather than on the BJS website. It is unclear whether such data and reports achieve the quality standards used by BJS. It is not apparent why some websites are permitted to use the BJS label (http://fjsrc.urban.org).

> *Recommendation 5.12:* **BJS should articulate why some data collections are housed on external websites and describe the process by which links to external websites are allowed. BJS should articulate and justify the use of its insignia on external websites.**

We also endorse BJS's efforts to develop the capability for users to perform custom tabulations and data summaries directly through the BJS website, as envisioned by former Director Sedgwick's comments above. By doing so, BJS would establish a more full-fledged Web presence rather than serving principally as a document repository. The current model under which (generally) some set tabulations are available as spreadsheets but more advanced data users are directed to download raw data files through the NACJD may actually be said to minimize BJS's presence somewhat: the precise information is available but not (directly) from BJS. Some of the larger federal statistical agencies—notably the Bureau of Labor Statistics and the Census Bureau (the latter through its "American FactFinder" interface)—have made considerable efforts in permitting website users to tabulate (and even to plot on a map) their own queries of interest. Clearly, the same level of interactive features cannot be expected without commensurate resources, but developing means by which steady streams of researchers, reporters, students, or congressional staff could readily obtain BJS information directly from the BJS site would ultimately be beneficial.

5–B.5   Cooperation with Data Users

> [A statistical agency should] seek advice on data concepts, statistical methods, and data products from data users as well as from other professional and technical subject-matter and methodological experts, using a variety of formal and informal means of communication that are appropriate to the types of input sought. (National Research Council, 2009:9–10)

We have described BJS's existing programs for outreach to and feedback from user groups and key constituencies in Section 5–B.2, in the context of the continual search to provide more useful data. Hence, our comments in this section are brief: BJS deserves credit for implementing a variety of outreach venues and the discussion at the February 2008 users workshop provided ample testimony that there is widespread appreciation of BJS among the user base. BJS's performance is certainly within the norms of other principal statistical agencies and we suggest that it could be improved still further through the recommendations we offer in the earlier section.

5–B.6   Fair Treatment of Data Providers

> [Fair treatment practices include] policies and procedures to maintain the confidentiality of data, whether collected directly or obtained from administrative record sources, [and to] inform data providers of the purposes of data collection and the anticipated uses of the information. . . . [They also include] respecting the privacy of respondents by minimizing the contribution of time and effort asked of them, consistent with the purposes of the data collection activity. (National Research Council, 2009:10)

Fair treatment practice is fairly synonymous with the principle of establishing a relationship of mutual respect and trust with data providers, described in detail in Section 5–A.1. The same general messages apply: BJS is generally very diligent and fair in its relationship with both establishment (state agency or individual facility) and person respondents. However, in our assessment, the PREA reporting requirements to which BJS is currently subject constitute a direct violation of this practice. The relationship of trust within which BJS collects information from its data providers is threatened by PREA because this data collection directly threatens and sanctions the data providers in ways that others do not. When there is direct harm from PREA participation perceived by a data provider, the other BJS data collections are threatened. Fair treatment of data providers is one of the foundations of trust. Violating this practice can have consequences that take decades to undo.

5–B.7 Commitment to Quality and Professional Standards of Practice

A statistical agency should:

- use modern statistical theory and sound statistical practice in all technical work.

- develop strong staff expertise in the disciplines relevant to its mission, in the theory and practice of statistics, and in data collection, processing, analysis, and dissemination techniques.

- develop an understanding of the validity and accuracy of its data and convey the resulting measures of quality to users in ways that are comprehensible to nonexperts. . . . (National Research Council, 2009:11)

As indicated at several points in this chapter, in our judgment, BJS has high standards for quality that are generally well understood. For this, BJS deserves considerable credit but, having expressed the point already, we do not reiterate at length here.

In the area of using modern statistical techniques and data collection practices, we worry that BJS is somewhat out of touch with current developments in statistical data collection. For instance, as described in Box 5-2, the PREA reporting requirements put BJS in a position where the inherent variability in estimates is such that it could not identify the highest- and lowest-ranked facilities as specified by the act (flawed and inappropriate though that requirement is). Instead, BJS chose to list a group of high-incidence facilities that, in some sense, are indistinguishable from each other. Yet this approach still has the effect of suggesting a level of precision that the estimates simply do not support; though we recognize that BJS faced difficult choices in issuing its PREA reports and that it was undoubtedly correct not to try to match the exact letter of the requirements in the law, the release would have benefited from very rigorous review of other approaches for presenting high-sensitivity data and attention to issues of multiple comparisons.

Similarly, in our interim report (National Research Council, 2008b:119) we expressed concern about the lack of mathematical statistics and survey practitioner expertise on the BJS staff; in its recent problems with the NCVS and the possible "break in series," BJS was possibly too dependent on the Census Bureau's (unfortunately post hoc) analyses of the effects of design changes and sample size reductions on the final NCVS estimates. Subsequent to our interim report, BJS has created a "senior leader" position among its top management with the idea of bolstering its survey management expertise. This is a very positive development, yet we still suggest that the absence of a chief mathematical statistician is troubling because such a post (as well as a chief survey methodologist) tends to lead the agency's attention to continual statistical improvements over time. (We return to the issue of staff expertise in Section 5–B.8.)

One way to judge professionalism is to look at methodological contributions made by an agency's staff with the intent of making it easier for users to correctly use and interpret data. One major contribution in this regard was BJS's sponsorship of development of a "crosswalk" data set by NACJD staff between the FBI's Originating Agency Identifier (ORI) codes and more standard geographic constructs such as cities and counties (Bureau of Justice Statistics, 2004d). The service populations of law enforcement agencies with ORI codes do not necessarily correspond neatly with official geographies and, in many cases, may overlap each other. The crosswalk file approximates the service populations to facilitate some direct comparisons between the FBI's UCR data and other data sources. Other user-oriented methodology contributions include the summary by Langan and Levin (1999) of differences in state prisoner counts when prison records (NCRP) or court records (National Judicial Reporting Program) are used and a series of clear, approachable pieces on the conceptual differences between the UCR and the NCVS (U.S. Department of Justice, Bureau of Justice Statistics, 2004).

The panel also requested of BJS a summary of professional activities of the staff, in an effort to evaluate whether the staff was connected with networks that would alert them to new developments in statistical design, data collection, and estimation. BJS staff are frequent participants in interagency working groups of staff from the range of federal statistical agencies. Several of these activities are topic working groups of the Federal Committee on Statistical Methodology, itself an interagency working group coordinated by OMB. Other interagency groups to which BJS contributes members are the Federal Interagency Forum on Aging Related Statistics, the Interagency Forum on Child and Family Statistics, and the Interagency Subcommittee on Disability Statistics. As a stakeholder and sponsor of the Census Bureau's demographic surveys program, it also participates in several interagency working groups organized by the Census Bureau, specifically those on the ACS and Sample Survey Redesign (updating sample and addresses for demographic surveys based on new census results). On the international level, BJS staff have also participated in relevant statistical programs of the United Nations Economic Commission on Europe (UNECE) and the Organisation for Economic Co-operation and Development, including specific UNECE task forces on victimization surveys and statistical dissemination and communication. However, the bulk of its staff professional and working group activities are internal to DOJ, ranging from membership on NIJ committees on drugs and crime and evaluation of justice on American Indian reserva-

---

Box 5-4   Review Process for an Information Collection by a Federal Agency

The U.S. Office of Management and Budget (OMB) is responsible for reviewing and approving any information collection activity—not only surveys for statistical purposes, but any form or application—that will be administered to 10 or more respondents (44 USC § 3502(3)(A)(1)). This "clearance" process can be time-consuming, because it must include two postings in the *Federal Register* for public comment as well as time for OMB's Office of Information and Regulatory Affairs to render its decision.

Agencies develop and submit an Information Collection Review (ICR) request or, more colloquially, a "clearance package," to OMB. In this process, surveys and any information collection making use of statistical methodology (for editing, imputation, or sample selection) are held to a higher standard. All ICRs must include a Part A, giving a detailed justification for the collection, indicating how and for what purpose the data will be used (or, if the ICR is reauthorizing an existing collection, how the data have been used); Part A also includes cost and time burden estimates. Statistical collections must also include a Part B report, which must include details on the sampling strategy for the collection and procedures for handling nonresponse, as well as descriptions of any tests to be conducted prior to full fielding of a collection. Names and contact information of any person consulted on the design of the collection are also required in Part B.

OMB maintains a publicly accessible database of pending and completed ICRs, including links to agency-submitted supporting statements, at http://www.reginfo.gov.

---

tions and tribal lands to membership on the Bureau of Prisons' institutional review board. Collectively, these efforts suggest attempts to build ties and outreach to other units in DOJ—and hence increase BJS's relevance to DOJ, which we encouraged and recommended above. However, the range of these activities is largely insular to the Justice Department and the executive branch; this bolsters the importance of the outreach efforts, including an advisory panel, suggested above.

Though there is much to commend in BJS's professional standards of practice, there is one area where BJS often displays, publicly, a marked weakness: the preparation of supporting statements for its information collections. As described in Box 5-4, all federal agency requests to collect information from 10 or more respondents must be cleared with OMB, in compliance with the Paperwork Reduction Act. For collections involving statistical methodology, the bar for approval is set higher; the Information Collection Review (ICR) packages submitted to OMB must include a "Part B" return providing details on sample construction, procedures for collecting and processing information, and pretests of survey instruments. Public versions of the ICRs are browseable online at http://www.reginfo.gov by searching for data collections listed under OJP.

On one hand, preparation of ICR supporting statements could be seen as no more and no less than clearing a bureaucratic hurdle. On the other, however, a reading of many of BJS's submissions over the past few years suggests a surprising and disappointing lack of specificity, as well as less-than-compelling arguments for the necessity and utility of the studies. Questions on the justification for the information collection are usually answered along strictly legal lines, citing BJS's general mandate to "collect and analyze statistical information, concerning the operations of the criminal justice system at the Federal, State, and local levels" (see Box 1-2) and usually including a copy of that section of the U.S. Code as an attachment. Rarely does the justification section indicate how the collection fits with, supplants, or is superior to existing data series, and information on the uses to which the data will be put is sparing. As strong as the methodology sections of BJS's final reports are, its technical specifications in the information collection requests—language that ought to be, effectively, the first draft of the technical documentation for a new data set—are strikingly weak. Examples of these ICR packages, and deficiencies in their support documentation, are described in Box 5-5.

Though they are, functionally, a bureaucratic step, the ICRs that BJS develops to obtain clearance from OMB are also a first opportunity to carefully explain the rationale for data collections from the substantive and technical viewpoints. They are also first drafts of the technical documentation for new data series and templates for actual data collection efforts. On these dimensions, neither new nor continuing BJS data collections are helped by having weak and deficient supporting statements made for them in a public (if not widely viewed) forum.

5–B.8   Active Research Program

> A statistical agency should have a research program that is integral to its activities. Because smaller agencies may not be able to afford as extensive a research program as larger agencies, agencies should share research results and methods. Agencies can also augment their staff resources for research by obtaining the services of experts not on the agency's staff through consulting or other arrangements as appropriate. (National Research Council, 2009:11)

Some of the estimates produced from BJS data have acquired a status as national benchmarks that should be preserved, and the agency's products are known for their quality standards and objectivity. To be sure, maintenance of series continuity is, properly, a high priority; this is because estimates of change, and especially change over a relatively long period of time, are among the most important pieces of information that these long-term data resources can provide.

At the same time, it is important for statistical agencies to ensure that their product lines are current both substantively and methodologically. As

---

Box 5-5 Problems in Bureau of Justice Statistics Information Collection Requests

BJS's Information Collection Request (ICR) package for the proposed Census of Law Enforcement Aviation Units (ICR 200708-1121-002) is a useful example. The abstract mentions that collection is a "part of the BJS Law Enforcement Management and Administrative Statistics program," and the statement on the necessity of the collection references 2003 LEMAS data:

> It is estimated that about 250 law enforcement aviation units are in operation among State and local agencies in the United States. These units operate an estimated 1,000 aircraft, including about 600 helicopters and 450 fixed-wing aircraft. The 2007 Census of Law Enforcement Aviation Units will be a census of all agencies, sampled in the 2004 LEMAS survey, which reported having either a fixed-wing aircraft or helicopter. It will be the most comprehensive study conducted in this area to date. The data collection will include detailed items on the functions, personnel, equipment, record keeping, expenditures, and safety records of these units.

The basic cited need for the collection is homeland security-tinged—"it is important to know the location and nature of available assets that could be mobilized in the event of large-scale regional or National emergencies"—with the add-on mention that "this information is also critical to law enforcement policy development, planning, and budgeting at all levels of government." The description is muddled as to whether the data are intended to draw some inference about characteristics of agencies that maintain aviation units (e.g., through the detailed items on equipment and safety records) or as a convenient directory of relevant agencies (for mobilization purposes). The statement is further unclear about how the collection fits with the broader LEMAS program, whether the information is sufficiently important that it should be collected on a regular basis, and whether there is any auxiliary information to evaluate the accuracy of the 2003 estimate that about 250 agencies have such units.

Most disappointing in this ICR, however, is the Part B return on statistical methods. Save for BJS contact information, what is supposed to be a fairly detailed technical specification of data collection techniques and planned methodologies runs about half a page, as follows:

> **Universe and Respondent Selection:** This data collection will be a census of law enforcement aviation units from among agencies with 100 or more officers. No sampling is involved with this collection.
>
> **Procedures for Collecting Information:** The census will be conducted initially by mailout. The address mailing list will be updated prior to mailout in order to maintain a current list of the respondents. Personal telephone interviews will be conducted for non-respondents.
>
> **Methods to Maximize Response:** We will do everything possible to maximize response, including telephone facsimile transmission, telephone interviews, and on-site assistance. Response rates for prior BJS law enforcement surveys and censuses have typically been 95% and above.
>
> **Testing of Procedures:** The census instrument has been pretested in three selected jurisdictions by individuals that will be receiving the final census instrument. Comments received as a result of that testing have been incorporated into the census instrument accompanying this ICR.

(continued)

Box 5-5   (continued)

The grounds for criticism of this extremely scant statement are numerous:

- The proposed collection shares with its fellow special-agency censuses a lack of clarity over whether the collection is intended as a "survey" or a "census." Throughout the rest of the document, and in the title of the collection, "survey" had been used; in the Universe and Respondent Selection section, "census" suddenly becomes the preferred choice.

- Regardless of the "survey" or "census" label, the primary source of contact information is the existing LEMAS survey; even if the aviation unit study is meant as a census, the method of construction of its frame/address list (LEMAS) should be described in more detail. Part A suggests that the LEMAS listings would be supplemented by listings from the Airborne Law Enforcement Association and International Association of Chiefs of Police; coverage properties for either of those lists is missing, as is any hint of how many additional units might be added through reference to those lists. The restriction to agencies with 100 or more officers is not previously mentioned, or described further.

- The statement is absent of any notion of whether and how the contact strategy differs from that of the main LEMAS collection or, indeed, who will carry out the collection. Likewise, any formal connection to the basic LEMAS survey (e.g., whether the results of the aviation-specific study might be used to revise questions on the main survey) is unspecified.

- The reference to providing "on-site assistance" is vague—does it refer to follow-up by a field interviewer?

- The reference to response rates in previous law enforcement surveys is interesting but unpersuasive; a better point of comparison might be similarly scoped attempts to canvass special units within departments rather than the main LEMAS survey.

- The final section, on testing of procedures, is particularly uninformative. How were the pilot jurisdictions chosen? Were there any difficulties encountered in the questionnaire, such as terminology usage? Did specific comments from the pilot respondents lead to changes in the contact strategy?

The Law Enforcement Aviation Unit ICR is an example of particularly weak justification and technical specification statements, but reading of other BJS-prepared ICRs show similar deficiencies. BJS's request for clearance of the 2007 Survey of Law Enforcement Gang Units (ICR 200705-1121-001) shared some gross features of the aviation unit ICR, again using the "survey" nomenclature but describing the effort as a "nationwide census of all law enforcement gang units operating within police agencies of 100 or more officers." The supporting statement for the gang unit study does not explain whether any other data sources besides previous LEMAS returns are to be used to build the frame of dedicated gang units, leaving it unclear whether the collection is indeed a census (a canvass of all known gang units) or a survey (probability sample). In another example, the section on testing of procedures in the ICR for the 2007/2008 National Survey of Prosecutors says that "the survey instrument was previously pretested with 310 jurisdictions during the 2005 data collection whereby BJS received a 99% response rate " (ICR 200704-1121-004). However, other portions of the statement make clear that the newer 2007/2008 version was purposely designed as a complete census of prosecutor offices, meaning that questions were revised and the number of questions was scaled back. Since this makes the newer survey different in scope and character than the 2005 version, the 2005 response rate—though impressive—fails to answer the question of experience in pretesting the questionnaire.

is true of other statistical agencies facing tight resources, BJS has been forced into an overriding focus on basic production of a set of data series and standard reports, at the expense of research, development, and innovation. As we discussed in Section 3–F.2, the performance measures in BJS's strategic plan are largely ones of volume and throughput—counts of file access on the NACJD, number of reports and supporting material accessible on the BJS website, number of data collections performed or updated each year—that lack a forward-looking focus on improvements in methodology and options for improving content.

A statistical agency should be among the most intensive and creative users of its own data, both to formally evaluate the quality and properties of its data series but also to understand the findings from those data and shape future refinements. BJS's "Special Reports" series have, in the past, gone into depth on topics not routinely studied by the agency's standard reports or have taken unique looks at BJS data, such as age effects in intimate partner violence (Rennison, 2001), the interaction between alcohol and criminal behavior (Greenfeld, 1998; Greenfeld and Henneberg, 2001), and the prevalence of ever having served time in prison among the U.S. population (Bonczar and Beck, 1997; Bonczar, 2003). They have also provided some opportunity for BJS analysts to make use of multiple BJS data sets or combine BJS data with non-BJS data sets in interesting ways:

- To study educational attainment in the correctional population, Harlow (2003) studied data from BJS's prisoner and jail inmate surveys, its 1995 Survey of Adults on Probation, the Current Population Survey of the Bureau of Labor Statistics, and the 1992 National Adult Literacy Survey of the National Center for Education Statistics.

- Zawitz and Strom (2000) combined data from the NCVS and multiple data series from the National Center for Health Statistics to describe both lethal and nonlethal violent crime incidents involving firearms.

- Greenfeld (1997) combined information from the UCR, the NCVS, and BJS's corrections and adjudications to summarize the state of quantitative information on sex offenses including rape and sexual assault.

Moreover, in fairness, BJS deserves credit for several innovative tacks that it has taken. Although full use of electronic questionnaires took considerable time, BJS and the NCVS were (through its work with the Census Bureau) relatively early adopters of computer-assisted methods in major federal household surveys. And, though we have argued at length that the reporting requirements are inappropriate, BJS's work on data collections in support of PREA led the agency to make great strides in the use of ACASI and other techniques for interviewing on sensitive topics. BJS has also demonstrated itself to be effective and innovative in developing data collection instruments to confront very tough methodological problems: identity theft, hate crimes, police-public contact, and crimes against the developmentally disabled.

But innovative in-house data analyses by BJS have slowed in recent years as the focus on production has increased and resources have tightened; major methodological innovations such as the use of ACASI were possible because PREA carried with it substantial funding. BJS's need to update longstanding products and keep activities in place, for basic organizational survival, has too frequently trumped innovative research and intensive exploration of new and emerging topic areas. Indeed, the principal means for identifying "emerging data needs" cited in BJS's strategic plan is not examination of the criminological literature or frequent interaction with criminal justice practitioner communities, but rather "emerging data needs as expressed through Attorney General priorities and Congressional mandates" (Bureau of Justice Statistics, 2005a:32).[18] In our assessment, the lack of a research program (and the capacity for a research program) puts BJS and its data products at risk of growing stagnant and becoming less relevant.

> **Finding 5.9:** The active investigation of new ways of measuring and understanding crime and criminal justice issues is a critical responsibility of BJS. The agency has lacked the resources needed to fully meet this responsibility and, for some issues, has fallen behind in developing such innovations.

> **Finding 5.10:** BJS has lacked the resources to sufficiently produce new topical reports with data it currently gathers. It also lacks the resources and staff to routinely conduct methodological analyses of changes in the quality of its existing data series and to fully document those issues. Instead, the BJS production portfolio primarily is limited to a routine set of annual, biannual, and periodic reports and for some topics, the posting of updated data points in online spreadsheets.

In our interim report, we made specific recommendations to stimulate research directly related to the NCVS, specifically calling for BJS to initiate studies of changes in survey reference period, improvements to sample efficiency, effects of mixed-mode data collection, and studies of nonresponse bias (National Research Council, 2008b:Recs. 4.2, 4.7, 4.8, 4.9). In response, BJS quickly issued requests for proposals for external researchers to conduct such studies, and has also signaled its intent to conduct a survey design competition to evaluate broad redesign options (Rec. 5.8 in National Research Council, 2008b). This is a laudable reaction that is a step toward laying out more concrete options for and future activities related to

---

[18]"In addition," the plan notes shortly thereafter, "BJS staff meet regularly with Federal, State, and local officials to identify emerging data needs or desirable modifications to existing collection and reporting programs" (Bureau of Justice Statistics, 2005a:32).

the NCVS, BJS's largest data program, but a fuller research program is critical to future-oriented option development for BJS's non-NCVS programs. It is also critical to avoid implementation problems such as those experienced in the 2006 administration of the NCVS. As we noted in our interim report, "design changes made (or forced) in the name of fiscal expediency, without grounding in testing and evaluation, are highly inadvisable" (National Research Council, 2008b:83). To this end, a short recommendation that we offered in our interim report (National Research Council, 2008b:Rec. 4.1) is worth formally restating here:

> ***Recommendation 5.13:*** **BJS should carefully study changes in the NCVS survey design before implementing them.**

It follows that this guidance can be applied to changes to other BJS data collections, and that such evaluative studies are not possible without the resources necessary to make innovative research a priority for the agency.

Congress and the administration cannot reasonably expect BJS to shoulder daunting data collection requests without the agency engaging in ongoing research, development, and evaluation. Going forward, a key priority should be detailed error analysis of the NCVS to get a sense of how big a problem survey nonobservation may be in specific socioeconomic subgroups, as the basis for understanding where improvements may most properly be made. On a related matter, BJS research activities should also be directed at improving outreach and data collection coverage of groups that are traditionally hard to reach by survey methods; such groups include new immigrant groups and persons and households where English is not the primary spoken language, young minorities in urban centers, and the homeless.

> ***Recommendation 5.14:*** **BJS should study the measurement of emerging or hard-to-reach groups and should develop more appropriate approaches to sampling and measurement of these populations.**

In the following, we suggest a few selected areas for a BJS research program. These should not necessarily be interpreted as the only or as the most pressing research priorities, but we believe they are all important directions.

In terms of methodological innovations, BJS should consider greater use of model-based estimation. In our interim report, we recommended investigation of such modeling for the generation of subnational estimates from the NCVS (National Research Council, 2008b:Rec. 4.5); improving the spatial and, perhaps, temporal resolution of estimates from the NCVS remains the highest priority in this regard, but the methodology could be brought to bear in other areas. The development of small-area estimates is particularly pressing because the agency is often criticized for not being able to speak to subnational areas. Modeling can also refer to the use of multivariate analyses to control for factors that mask real changes in the phenomenon of interest. Just as many economic indicators are adjusted for inflation or seasonal fluctuation, it would make sense to adjust crime rates for factors that mask important variation. Age-adjusting crime rates, for example, would help separate the effects of macro-level social changes (over which one has little control) from more troubling and actionable changes in the incidence of crime. The same can be said of incarceration rates: adjusting admission rates for the volume of crime would provide a perspective on the use of incarceration not available in simple population-based rates. Modeled data should surely be used when we know that right or left censoring of data makes data incomplete and inaccurate. For years BJS published estimates of time served in prison using exiting cohorts when they knew that this seriously underestimated the time served. This is a case where model-based estimates would most certainly have been more accurate than data-based estimates.

However, greater use of model-based estimates must be done with caution, for several reasons. One is the challenge of interpretation: Modeling may not be understood by many consumers of BJS data. This may be largely a presentational problem that can be solved by presenting the estimates simply and then providing the detailed description of modeling elsewhere. The use of double-decrement life tables by Bonczar and Beck (1997) (later updated by Bonczar, 2003) is a good illustration of how modeling could be presented in BJS reports. Another challenge is that models are always based on assumptions, assumptions that can be more or less accurate or robust (and there can be wide disagreement over what is accurate or robust). Hence, situations where the choice of assumptions may be interpreted as reflecting political or other bias should be avoided.

A more basic methodological development, but still complex research effort, would be for BJS to invest in the creation and revision of basic classifications and typologies for crime and criminal justice matters. Its role in coordinating information from a variety of justice-related agencies and promoting standards through National Criminal History Improvement Program–type grants for improvement of source databases gives BJS unique advantages in taking on such an effort. The classification of "index crimes" used in the UCR has changed little in 80 years and remains the nation's major crime classification; its implications for what crimes are most serious are central to the definitions used in the NCVS and other BJS collections. Yet the interest in crime and the amount of information available on crime has changed greatly over those 80 years, and the basic classification of crime should be revisited to keep pace with these changes.

BJS should also invest some effort in getting denominators for risk rates that are more reflective of the at-risk population. Major cities, for example, are disadvantaged in the annual crime rankings of jurisdictions based on UCR data because their rates are based upon their residential population—a base that excludes the commuters, shoppers, and culture seekers who contribute to the numerators of the rates. Likewise, incarceration rates based on the entire population are technically correct but may be otherwise misleading, because very young and very old populations are not at risk. The generation of risk rates should not be restricted to the data generated by BJS but should use other data as long as the quality and periodicity of those data are acceptable. To report estimates from BJS's inmate surveys as proportions of the prison population misses a great opportunity to understand much better how the nation uses its prison resources; incarceration rates reflecting the general household population (as in Bonczar, 2003) may be uniquely informative.

5–B.9   Strong Internal and External Evaluation Program

> Statistical agencies that fully follow [this set of prescribed practices] will likely be in a good position to make continuous assessments of and improvements in the relevance and quality of their data collection systems. . . . Regular, well-designed program evaluations, with adequate budget support, are key to ensuring that data collection programs do not deteriorate. (National Research Council, 2009:47, 48)

The practice of instituting a strong internal and external evaluation program is a new addition to the fourth edition of *Principles and Practices of a Federal Statistical Agency*. It is similar to the practice of an ongoing research program (Section 5–B.8) but has slightly different connotations, emphasizing not only the continuous quality assessment of individual data collection programs but periodic examination of the quality and relevance of an agency's entire data collection portfolio. It is very much to BJS's credit with respect to following this practice that it has periodically sought the advice of external users and methodologists on specific methodological problems, that it engaged in the intensive rounds of testing and evaluation that led to the redesigned NCVS in the early 1990s, that it regularly receives feedback on data quality from its state SAC network and JRSA, and that it actively sought and encouraged this panel's review of the full BJS portfolio.

Like other small statistical agencies, BJS's ability to mount large-scale evaluation efforts is limited by available resources. Still, attention to internal and external evaluation is critical. Indeed, some of the guidance we offer in this report—for instance, on emphasizing the flows from step to step in the justice system within existing BJS data sets and facilitating linkage between current data sets (Section 3–F.1)—depends critically on careful evaluation of the strengths and limitations of current data collections and structures as a first step.

One general direction for improvement by statistical agencies, including BJS, is greater attention to known data quality issues and comparisons with other data resources as part of the general documentation of data sets. BJS reports are generally careful to include a concise methodology section, and the public-use data files that are accessible at the NACJD typically include additional detail in their codebooks. Still, as a general practice, BJS should work to find ways to improve the documentation on its major data holdings that is directly accessible from BJS. This could include developing and making available technical reports based on specific user experiences and providing direct links to Census Bureau (and other BJS-contracted data collection agents) technical reports on developing specific survey instruments.

As part of an evaluation program, it would also be useful for BJS to move beyond individual series examinations and approach critiques of the relative quality of multiple sources. This work should be done in partnership with other statistical agencies or data users, such as we describe below in Section 5–B.11 for comparing BJS's prison and jail censuses with the data quality and resolution provided by the Census Bureau's ACS. Other examples for multiple-source evaluation include:

- Examination of differences between homicide rates computed from the UCR data and those from the cause-of-death data coded in the vital statistics that are compiled by the National Center for Health Statistics;

- Reconciliation of the number of gunshot victims known to the police (or measured in emergency room admissions data) with the number of self-reported gunshot victims in the NCVS (see, e.g. Cook, 1985); and

- Examination of the reasons why serious-violence victimization rates from the NCVS and School Crime Supplement differ from those derived from CDC's Youth Risk Behavior Surveillance System.

5–B.10 Professional Advancement of Staff

To develop and maintain a high-caliber staff, a statistical agency must recruit and retain qualified people with the relevant skills for its efficient and effective operation, including analysts in fields relevant to its mission (e.g., demographers, economists), statistical methodologists who specialize in data collection and analysis, and other specialized staff (e.g., computer specialists). (National Research Council, 2009:12)

At the panel's request, BJS supplied biographical information for its staff members as of fall 2008. A total of 32 of the 53 staff members hold positions with labels connoting direct statistical work (statistician, senior statistician, or branch chief); 12 have doctoral degrees (with an additional five listed as being Ph.D. candidates) and nearly all list master's degrees. However, none holds a doctoral or master's degree in statistics, although two statisticians have completed master's degrees in the Joint Program in Survey Methodology of the University of Maryland, the University of Michigan, and Westat.

Indeed, the only formal statistics degree on the full BJS staff is a bachelor's degree, held by a specialist on the support staff. As is not surprising, advanced degrees in criminology (or criminal justice) or sociology abound, though other fields such as social psychology, social welfare, and public affairs are also included. The statistician ranks in BJS also include one holder of a law degree.

Our review of the staff biographies—and of BJS's publications, throughout this report—suggests a very capable and dedicated staff, with a median length of service of about 8 years and including several career staff members of 20 years or more. Our intent is not to impugn the good work of the BJS staff. However, in Section 5–B.7 and our interim report, we commented on the need for more highly skilled technical leaders within BJS; we think this is necessary to put BJS on a better footing in dealing with its external data collection agents, to cultivate a climate of research and innovation, and to safeguard the continued credibility and quality of BJS data. Going further, we suggest that BJS would benefit from additional staff expertise in mathematical and survey statistics; computer science and database management are also notable deficiencies in staff expertise, given the agency's role in executing grants to improve criminal justice databases and the importance of record linkage for conducting longitudinal studies of flows in the justice system.

> *Recommendation 5.15:* **BJS must improve the technical skills of its staff, including mathematical statisticians, computer scientists, survey methodologists, and criminologists.**

At the same time, the panel notes that the recruitment problem for technical staff to all statistical agencies is a large one. The agencies in the federal statistical system that seem to do better on this score are those who are actively supporting advanced degrees among their junior staff—that is, making human capital investments in bachelor's-level staff and assisting their graduate studies to yield more technically astute staff in 2–4 years. In addition, agencies have sponsored dissertation fellowships on their own data, using the contact with the Ph.D. candidate to recruit talented staff.

5–B.11 Coordination and Cooperation with Other Statistical Agencies

> Although agencies differ in their subject-matter focus, there is overlap in their missions and a common interest in serving the public need for credible, high-quality statistics gathered as efficiently and fairly as possible.
>
> When possible and appropriate, federal statistical agencies should cooperate not only with each other, but also with state and local statistical agencies in the provision of data for subnational areas. (National Research Council, 2009:13)

There are some valuable and mutually productive partnerships between BJS and other statistical agencies. These include relatively long-term arrangements such as the National Center for Education Statistics' sponsorship of the School Crime Supplement as well as one-time collaborations, such as a joint report by BJS and CDC staff on findings from the NCVS on injuries sustained in the course of violent crime victimizations (Simon et al., 2001). BJS has also enjoyed some collaborative work with the National Center for Health Statistics, including use of vital statistics data collected from state public health departments and registrars. BJS has also, on occasion, worked with agencies that are not principal statistical agencies but that do conduct statistical work; for instance, BJS sponsored the Consumer Product Safety Commission to add a Survey of Injured Victims of Violence as a module to the commission's National Electronic Injury Surveillance System—a sample of hospitals that provide their emergency department records for coding and analysis (Rand, 1997).

Of course, BJS's most intensive relationship with another statistical agency is with the Census Bureau. Although there are some cooperative aspects of the partnership between the two agencies, the panel believes that there are some fundamental strains in the relationship. One is that, as noted in the preceding section, BJS has lacked the strong statistical expertise to fully engage with the Census Bureau staff on design (and redesign) issues, and so its role in modifying the NCVS to fit within budgetary constraints has largely been one of deciding which Census Bureau–developed cost-saving options are least objectionable. Another element of strain is discussed in our interim report (National Research Council, 2008b:Sec. 5–D): the failure of the Census Bureau to provide transparency in its costs and charges for data collection to BJS (or its other federal agency sponsors), which makes assessments of the trade-offs between survey costs and errors impossible.

Agencies that contract out much of their work—and BJS is one of the extreme cases within the statistical system in that regard—can easily evolve into ones where contract management is the dominant focus. While more (and more sophisticated) technical staff will not solve the BJS budget problems, they can make BJS a stronger partner to the other statistical agencies with which it works.

On substantive grounds, an important area in which a healthy BJS–Census Bureau relationship and collaboration would beneficial is in reconciling BJS's corrections data series with the Census Bureau's measures of the correctional institution population. The American correctional apparatus has grown enormously since the mid-1970s; there are now on the order of 2.3 million persons in prison or jail, and the incarceration rate has grown fourfold since 1980. Another 800,000 people are on parole, and 4.2 million are on probation. Virtually all the growth in incarceration since 1980 has been among those with less than a high school education. In this context, the

BJS data collections are a valuable supplement to the large Census Bureau household surveys which are drawn exclusively (or nearly so) from the non-institutional household population. BJS collections on the population under correctional supervision are not just an important part of an accounting of the criminal justice system, but an increasingly important part of the nation's accounting for the population as a whole. Those groups overrepresented in prison populations—minority men under age 40 with little schooling—are also significantly undercounted in household surveys and other data collections from the general population.

The Census Bureau's ACS now contains the detailed social, demographic, and economic questions that were traditionally asked of a sample of the population through the "long form" questionnaire of the decennial census. When the ACS entered full-scale collection earlier this decade, it also included coverage of the group quarters (nonhousehold) population, including prisoners. The first 3-year-average estimates from the ACS for areas with populations of 20,000–65,000 only became available in 2008, and the first 5-year-average estimates for all geographic areas (including those under 20,000 population) are only slated for release in 2010. Hence, the properties of these estimates—much less their accuracy for segments of the relatively small group quarters population—are only beginning to be studied and understood. Going forward, an important question will be how the most accurate picture of the prison and jail population can be derived, balancing the ACS estimates with the annual count (and basic demographic) information from BJS's prison and jail censuses and the detailed information available from BJS's inmate surveys.

5–C SUMMARY

The panel believes that BJS and DOJ should conduct continual examination of BJS's fulfillment of the principles and practices of a federal statistical agency. Our panel's review found that the perceived independence of the agency was severely shaken by recent events. We found that the trust of data providers is threatened by BJS directly assisting regulatory activities. We also found that a renewed emphasis on increasing the technical and research skills of BJS's staff is needed.

– 6 –

Strategic Goals for the Bureau of Justice Statistics

6–A PRIORITY SETTING AND CONSTRAINED RESOURCES

A POSSIBLE CRITICISM of this report—reviewing the detailed recommendations made throughout the text—is that it tends to suggest *adding* to the Bureau of Justice Statistics' (BJS's) inventory of collections rather than *subtracting*. The recommendations are generally geared to improvements within BJS's various existing data collections—for instance, ensuring a high-quality independent measure of crime in the National Crime Victimization Survey (NCVS), emphasizing conceptual frameworks in BJS's adjudication and law enforcement collections, and expanding its corrections series to study prisoner reentry into society. (There are exceptions, such as Recommendation 3.10's suggestion to discontinue special-focus law enforcement agency surveys if they cannot be fitted into a regular, structured Law Enforcement Management and Administrative Statistics [LEMAS] supplement plan.)

The panel raised its own questions regarding this: How should these improvements be paid for? Should some existing series be cut to make ways for new ones? And, given the disproportionate share of BJS's current resources consumed by the NCVS, should some smaller collections be stopped to free up additional resources for the NCVS or should some NCVS resources be steered to other areas?

The reason why we rejected explicitly suggesting that some data series be cut in order to pay for others is certainly not that the current collections are perfect. The improvements we suggest in the recommendations are testimony to that. Rather, we cannot accept the underlying premise and inherent assumption that BJS can achieve its legislated goals by cutting programs. In our assessment, we think it can be stated as a fact: BJS has been given more responsibilities than can be achieved with current resources. The resources provided to BJS to carry out its work are not commensurate with the breadth—and importance—of the responsibilities assigned to the agency by its authorizing legislation.

Because of this, the agency has for some years walked a tight line of small cuts of sample or measurement, short delays of publications, and temporary hiring freezes—each of these tolerable in itself, but cumulating over the years such that core functions have broken down. On a routine basis, decisions must be made about addressing certain responsibilities and not addressing others; trade-offs must be made in the periodicity and completeness of data collections. Maintenance and continuation of existing data collections must also be balanced with the need to comply with directives from Congress and the Department of Justice, complicating resource allocation decisions and the setting of priorities. Such decisions are hardly unique to BJS—at some point, all organizations must make such trade-offs—but BJS's mismatch between resources and responsibility makes the decisions particularly difficult.

Thus, in setting priorities, BJS directors have perforce had a short time horizon—responding to a certain set of demands even though those decisions may have negative long-term consequences for individual data collections and the health of the agency. Certainly, in the midst of year-to-year juggling of data series in order to keep production moving, longer-term investments in research and innovation become difficult or impossible to make. The most striking example of the consequences of this extremely tough climate is the current state of the NCVS: What was once, clearly, the best victimization survey in the world is now unable to satisfy its basic function of providing annual estimates of level and change in common-law crime. This decay happened gradually as BJS administrators were attempting to respond to immediate exigencies, aggravated by an overly broad mandate. Each single cut in sample size, or other cost-cutting measure, was justifiable given then-present alternatives. Cumulatively—as demonstrated most vividly by the declared "break in series" with the 2006 NCVS data—they lead to the conclusion that "the [current] NCVS is not achieving and cannot achieve BJS's legislatively mandated goals" (National Research Council, 2008b:Finding 3.1).

Statistical infrastructure (data collections and records series) shares with physical infrastructure (such as roads, bridges, sewers, and cable) the fundamental problem that it lacks glamor and can be difficult to champion as a government spending priority. Yet it is undeniably important to the public knowledge and the public good; BJS's purview includes topics important to general welfare, spanning the measurement of interpersonal violence, the function and magnitude of law enforcement and corrections, and the operations of the judicial branch of government. The problem is not that there are parts of BJS's legal mandate that are unimportant or unworthy, but that its current resources do not permit it to cover its mandates as effectively as possible. Given this finding, we are loath to construct a list of series to terminate or reduce out of concern that the agency fail even more visibly to fulfill the charge given to it in legislation.

Another reason for not considering a specific ranking of data collections by importance, or some metric, in order to suggest possible cuts is consistency with the approach we took in our interim report on options for the NCVS. In that report, we presented an array of possible options and described how specific design choices corresponded to particular goals for the survey. However, we made a point of "not suggest[ing] one single path as the ideal for a redesigned NCVS" (National Research Council, 2008b:4). Different people and decision makers do not necessarily put the exact same weights on the goals and objectives of a program such as the NCVS, and we did not want presume "that our preferred set of NCVS goals is correct to the exclusion of all others" (National Research Council, 2008b:4). The same logic applies to considering other collections in the BJS portfolio: deeming one collection more worthy than another involves complicated value judgments, not science. Specific constituencies for BJS data that are not represented on the panel or by groups who have spoken before the panel could make eloquent and compelling cases for their particular favored set of statistics; again, we do not wish to suggest that any weighting we could suggest is somehow paramount.

Ideally, in evaluating and weighing individual data collection programs, it would be possible to know the causal impact of particular BJS statistics—for instance, if BJS were to collect data and report estimates on X rather than Y, certain policy actions would result (or have different probabilities of resulting). Such causal analysis is very difficult and, aside from some studies on federal fund allocation based on government estimates (see, e.g., Spencer, 1980), very little work has been done to attempt even a partial understanding of the causal impact of government statistics. Although such analyses would be a worthwhile undertaking for BJS and the federal statistical system as a whole, it would involve such an extensive research agenda as to be infeasible for a small agency such as BJS. That said, there is merit in BJS devoting resources to understanding and documenting the ways in which its estimates are actually used in making policy decisions at the federal, state, and local levels, and using that information to assess how well its statistical portfolio matches the needs of policy analysts.

> **Finding 6.1:** Relatively little is understood about how BJS data are used in developing policy and could be used in improving policy development.

While hoping that tight fiscal constraints may be alleviated somewhat in coming years, we cannot assume infinite resources either. We recognize that our charge obliges us to "assist BJS to refine its priorities and goals, as embodied in its strategic plan, both in the short and longer terms." Hence, in this chapter, we consider basic strategic goals for BJS, formalizing some notions inherent in our detailed recommendations and deriving from the broader themes we noted in Chapter 1.

6–B   CURRENT BJS STRATEGIC GOALS

To begin, it is useful to review BJS's current plans and goals. The agency's 2005–2008 strategic plan (Bureau of Justice Statistics, 2005a) articulates three goals for the agency:

1. To produce national statistics on crime and the administration of justice that facilitate measurement over time and across geographic areas;

2. To improve record-keeping by state and local governments and to improve the ability of states and localities to produce statistics on crime and the administration of justice;

3. To ensure public access to statistics and criminal justice data.

These goals mainly serve to reinforce the two parts of the agency's formal mission statement—service as a statistical agency and support of state and local governments—with an added goal of public access and dissemination. As stated above, the first goal adds emphasis on "measurement over time," though it is unclear whether this refers to simply continuing collections (thus building longer time series of data) or longitudinal measurement within the criminal justice system.

Although they are not written into the agency's current strategic plan, other strategic priorities are evidenced by changes in BJS's organizational structure (see Figure 1-3) and commentary at the panel's public meetings. In particular, the organization was revised to create three distinct units—law enforcement, courts and adjudication, and "recidivism, reentry, and special projects" (Sedgwick, 2008:2). These changes provide resources separate from BJS's existing correctional statistics unit to study prisoner reentry issues; they are also consistent with comments made by BJS staff in remarks to the panel about the agency's interest in broadening its collections in the area of law enforcement.

Box 6-1 Summary of Recommendations and Commentary on Strategic Goal 1: A Strong Position of Independence

Move BJS Out of the Office of Justice Programs

> **Recommendation 5.3: BJS should be administratively moved out of the Office of Justice Programs, reporting to the attorney general or deputy attorney general.**

Make the BJS Directorship a Fixed-Term Appointment

> **Recommendation 5.4: Congress and the administration should make the BJS director a fixed-term presidential appointee with the advice and consent of the Senate. To insulate the BJS director from political interference, the term of service should be no less than 4 years.**

Strengthen BJS's Independent Function as a Statistical Agency

> **Recommendation 5.2: The Department of Justice review of any BJS statistical product and related communications should not require changes to the content, the release schedule, or the mode of dissemination planned by BJS.**

> **Recommendation 5.1: Congress and the Department of Justice should not require, and BJS should not provide, individually identified data in support of regulatory functions that compromise the independence of BJS or require BJS to violate any of the principles of a federal statistical agency.**

Position BJS as a Statistical Resource to the Department of Justice

> **Recommendation 5.5: The BJS director needs to reach out to other agencies within DOJ, forming partnerships to propose initiatives for information collection that are relevant to policy needs.**

> **Recommendation 5.6: The Department of Justice should build provisions for BJS collection of data and statistical information into its program initiatives aimed at crime reduction. These are not intended as program evaluation funds, but rather as funds for the basic monitoring and assessment of the phenomena targeted by the initiative.**

(continued)

6–C   SUGGESTED BJS STRATEGIC GOALS

In the following section, we propose four strategic goals for the Bureau of Justice Statistics. Each section includes a text box (Boxes 6-1, 6-2, 6-3, and 6-4) illustrating how our recommendations and commentary throughout this report support or relate to Goals 1–4, respectively.

6–C.1   A Strong Position of Independence

Goal 1: To establish and maintain a strong position of independence as a statistical agency; to serve as an independent and

---

Box 6-1   (continued)

Improve Ties Between BJS and the Federal Bureau of Investigation

As discussed in Section 4–C.4, a fully-functional National Incident-Based Reporting System (NIBRS) would be a valuable research tool. Accordingly, BJS should facilitate more universal implementation of NIBRS and offer technical support on issues such as small-agency bias, in order to promote better understanding of instances in which NIBRS data may best be used.

Build the National Crime Victimization Survey's (NCVS's) Standing as a Critical Social Indicator

Recommendation 6.1: BJS must ensure that the nation has quality annual estimates of levels and changes in criminal victimization.

Recommendation 6.2: Congress and the administration should ensure that BJS has a budget that is adequate to field a survey that satisfies the goal in Recommendation 6.1.

Recommendation 6.4: Additional resources made available for the NCVS should be used not only to increase the reliability of annual estimates but also to supplement the survey in ways that increase our understanding of criminal victimization.

Explore Means for Dedicated Funding of Victimization Studies

Recommendation 6.3: More information about the needs of victims is essential to the compensation and assistance goals of the Office of Victims of Crime. Congress should allow additional funding for the collection and improvement of victimization data to be obtained from funds obtained through the Victims of Crime Act.

Continue and Strengthen State Justice Statistics Program and Statistical Analysis Center Partnerships

Recommendation 4.1: Through its Statistical Analysis Center and State Justice Statistics programs, BJS should continue to develop its ties with the states, and more fully exploit the potential for using states as partners in data collections.

Recommendation 4.2: Developments toward longitudinal and small-area measurement systems should involve state partners who are active in data collection and knowledgeable about state justice systems.

---

objective source of statistical information on crime and the administration of justice.

In light of the discussion in Chapter 5 on BJS's performance relative to the expected principles and practices of a federal statistical agency, it should not be surprising that building (and rebuilding) BJS's position of independence ranks high in our suggested directions for the agency. In recent years, BJS has endured strong challenges to its position of independence, as documented in Section 5–A.2. Arguably, BJS's most pressing priority in the coming years is restorative: rebuilding BJS's position of independence as a statistical agency. We suggest an administrative shift of BJS out of the Office of Justice Programs (OJP) and designation of a termed appointment for the BJS director as corrective measures. By this goal statement, we suggest that preserving a role of independence also be BJS's first criterion for undertaking new data collections or revising existing ones. We recommend (Recommendation 5.1) that BJS should not provide individually identified data in support of regulatory functions that compromise BJS's role; this goal statement formalizes and extends this clause to include acceptance of functions that involve collecting and analyzing data for policy-furthering, tactical, and operational purposes. Research on policy-specific evaluation issues should not be done by BJS but by other units within the Justice Department that are formally charged with this responsibility.

In contrast to the "statistical arm of the U.S. Department of Justice" verbiage that BJS has sometimes used to describe itself, it is useful to consider as a model the mission statement of the Bureau of Labor Statistics (BLS). In past years, BLS used similar "statistical arm" language. However, at least since 2001, BLS has taken the following as its mission statement (http://stats.bls.gov/bls/blsmissn.htm):[1]

> The Bureau of Labor Statistics (BLS) is the principal fact-finding agency for the Federal Government in the broad field of labor economics and statistics. The BLS is an independent national statistical agency that collects, processes, analyzes, and disseminates essential statistical data to the American public, the U.S. Congress, other Federal agencies, State and local governments, business, and labor. The BLS also serves as a statistical resource to the Department of Labor.

BJS would be well served to adopt similar language in its self-description, emphasizing that it is first and foremost a statistical agency, and that it is a statistical resource to the Justice Department but not an "arm" to further any policy objective.

6–C.2 Building Statistical Systems and Conceptual Frameworks

Goal 2: To build, maintain, and utilize statistical systems that describe the extent and characteristics of crime in our nation and the status and response of the justice system.

More than a basic statement of topic, the language of this goal is intended to reframe BJS's grantmaking programs such as the National Criminal History Improvement Program. As discussed in Chapter 4, a major problem with the current administration of such programs is that they are strictly a money-transfer operation with no statistical product given back to

---

[1]The full statement includes the second paragraph: "BLS data must satisfy a number of criteria, including relevance to current social and economic issues, timeliness in reflecting today's rapidly changing economic conditions, accuracy and consistently high statistical quality, and impartiality in both subject matter and presentation."

Box 6-2 Summary of Recommendations and Commentary on Strategic Goal 2: Building Statistical Systems and Conceptual Frameworks

Articulate Interrelationships of Data Collections in Strategic Plan

Recommendation 3.7: To be useful, a BJS strategic plan must articulate a blueprint of interrelated data collection and product activities, including both current and potentially new data products. This blueprint would be used to evaluate new opportunities.

Use Criminal History Record Databases for Research Studies

Recommendation 4.3: BJS should actively utilize the NCHIP program to improve criminal history records necessary for longitudinal studies of crime.

Implement Core-and-Supplement Frameworks for the National Crime Victimization Survey (NCVS) and Law Enforcement Management and Administrative Statistics (LEMAS)

Recommendation 3.8: (Rec 4.3 from National Research Council, 2008b) BJS should make supplements a regular feature of the NCVS. Procedures should be developed for soliciting ideas for supplements from outside BJS and for evaluating these supplements for inclusion in the survey.

Recommendation 3.9: To maximize both utility and timeliness of information, the LEMAS survey should be conducted as core-supplement design in the context of a continuous data collection.

Recommendation 3.10: To improve the utility of censuses of law enforcement agencies, BJS should develop an integrated conceptual plan for their periodicity, publish a 5-year schedule of their publication, and integrate their measurement into the LEMAS as supplements.

Develop Broader Vision of Law Enforcement Than Management and Administration

As discussed in Section 3–F.2, BJS's current data holdings in law enforcement are overly limited to administrative data (e.g., number of sworn officers or presence of certain policies or technologies). BJS should perform analyses linking these data to crime data (thus coming closer to informing assessment of the *effectiveness* of police policies).

Recommendation 3.11: The NCVS (and its supplements) should be more effectively used as a tool for studying law enforcement, both in terms of the types of crime that are reported (and not reported) to police and the action that results from the reporting of a crime (e.g., the Police-Public Contact Survey).

Pursue a Records-Based Law Enforcement Data Series

Recommendation 4.4: To improve the timeliness of crime statistics, BJS should explore the development of a crime reporting system based on a probability sample of police administrative records. The goals of such a system would be national representativeness, high response, high data quality, timeliness and flexibility in terms of crime classification and analysis, and national statistics for the monitoring of crime trends.

(continued)

Box 6-2   (continued)

Explore Role of BJS's Corrections Data Expertise Within the Federal Statistical System

As discussed in Section 5–B.11, BJS and the Census Bureau should reconcile the relative roles of BJS's prison and jail censuses and inmate surveys and the group quarters portion of the Census Bureau's American Community Survey to determine which series (or combination) provides the most accurate picture of the correctional population.

Stabilize and Improve Sampling in Adjudication Series

Recommendation 3.12: As court records become more accessible through computerized case management systems, BJS should implement more rigorous methods of probability sampling in its adjudication series.

Recommendation 3.13: To inform future revisions to its adjudication portfolio and to more efficiently acquire and work with court data in the future (including longitudinal analysis), BJS should develop a research program to build representative samples of courts and to assess strategies for collection of case records.

Develop Better Understanding of Prosecutor Declinations and Other Adjudication Decisions

As discussed in Section 3–F.3, prosecutors' declinations to prosecute certain cases remain a major gap in understanding in the justice system flow model. Ideally, fuller data will arise as participation in State Court Processing Statistics–type collection becomes more common and standard; as a first check, BJS should add some basic questions on case processing to the National Prosecutors Survey, which currently has a purely administrative focus.

Outline Research Agenda for Civil Justice Collections

As discussed in Sections 2–C.2 and 2–D, civil proceedings are a major part of the overall justice system—and a major gap in BJS's data portfolio, relying essentially on one collection. But an expanded role for BJS in covering civil justice issues raises serious definitional issues, and developing a measurement system capable of taking an accurate reading of civil disputes settled privately is a very difficult prospect. BJS should clearly articulate the strengths and weaknesses of its current work in civil justice and identify the highest priority for expanded data collection, should resources become available.

---

the taxpayer. Other than generating rough summary statistics of firearm-purchase background checks, BJS has not been able to utilize the criminal history record data that its grant monies help to develop at the state and local levels. This failure has occurred, despite a formal set-aside of funds in all of those grants for BJS evaluation purposes. However, as also described in Chapter 4, recent developments have finally made it possible for BJS to access the compiled history record data for research purposes, opening very exciting avenues for study.

To be clear, we consider methodological research to be an essential part of building and maintaining statistical systems. Although attention to innovation and methodological development is difficult in a climate of constrained resources, it is essential to the effective improvement or expansion of statistical systems. A higher priority on methodological research is not a panacea; it might not have prevented the necessity of the cost-cutting measures that produced the recent "break in series" in the NCVS but it could have provided a fuller assessment of possible risks before those changes were finalized.

By use of the term "statistical systems," this goal statement would give lesser priority to one-time data collections that have no enduring benefit to ongoing series or planned data collections. This is not meant to stifle new one-time collections but to ensure that additions to the portfolio are done in a broader context. For example, it is not meant to say that first-time efforts like a census of law enforcement aviation units should not be undertaken, but rather that their role should be made clear (as part of the larger LEMAS survey and as a means of building and improving the frame for that broader survey) and that its utility in describing part of the justice system should be explicated.

We also consider "statistical systems" to include the partnerships between BJS and state and local authorities—partnerships that, properly, go beyond "improve[ment of] record-keeping" and dissemination of BJS products.

6–C.3   Improving Coverage of the Justice System

Goal 3: To provide comprehensive statistical coverage of all parts of the criminal justice system, including the longitudinal flow of persons and cases through the system and their return to the community.

The specific extents of what is called "the criminal justice system" can be debated. In general, the criminal justice system includes the criminal act, consequences of the act, and responses to the act by public agencies and other parties. We think the 1967 funnel model of the system remains useful and have taken it as a point of reference. Accepting this model as a premise, this goal might be restated as saying that data collections rise and fall in importance relative to their proximity to the funnel: the further a specific phenomenon is from the activities described by the funnel model, the less central it is to the mission of BJS.

By its nature, the funnel model has strong implications for setting priorities that merit further discussion. First, it suggests that BJS should focus on activities that affect the most people and affect them most extensively. By definition, this means the incidence of crime and victimization, which the funnel graphically describes as the most prevalent phenomenon in the crim-

Box 6-3 Summary of Recommendations and Commentary on Strategic Goal 3: Improving Coverage of the Justice System

Emphasize Longitudinal Flows and Structures

Recommendation 3.4: BJS should develop an approach to measure the experiences of individuals through the criminal justice system on a prospective, longitudinal basis, beginning as early as practicable in the process (arrest) and ending with their eventual exit (ranging from early dismissal of charge through completion of sentence).

Recommendation 3.5: BJS should develop an approach to measure the victimization experiences of individuals on a prospective, longitudinal basis, beginning from a focal victimization and following the victim forward in time measuring subsequent victimizations and possible consequences of victimization. The NCVS may be used to recruit respondents to a panel survey of crime victims.

Recommendation 3.1: BJS's goal in providing statistics from basic administrative data on corrections should be the development of a yearly count of correctional populations capable of disaggregation and cross-tabulation by state, offense categories, and by demographic groups (age, race, gender, education).

Recommendation 3.2: BJS should produce yearly transition rates between steps in the corrections process capable of disaggregation and cross-tabulation by state, offense categories, and demographic groups.

Recommendation 3.6: BJS should develop a panel survey of people under correctional supervision to understand how individuals move between institutional and community settings, and to understand the social contexts of correctional supervision.

Facilitate Linkage in Existing Data Sets

Recommendation 3.3: BJS should explore the possibilities of increasing the utility of their correctional data collections by facilitating the linkage of records across the data series. For example, the ability to link records from the Recidivism Studies or from NCRP to the Census of Adult Correctional Facilities (CACF) would increase the ability to understand how correctional facilities contribute to recidivism.

Build Capacity for Studying New and Emerging Types of Crime

Recommendation 2.1: Consistent with its legal mandate to collect, analyze, and disseminate statistical information on all aspects of the justice system, BJS should (a) document and organize the available statistics on forms of crime not covered by the NCVS, the FBI's UCR and NIBRS data systems, and other major data series maintained by other statistical agencies, (b) pursue research on what new statistics could be feasibly and usefully developed, and (c) propose such new data collections as the research suggests to be both feasible and useful. BJS should strive to function as a clearinghouse of justice-related statistical information, including reference to data not directly collected by BJS.

(continued)

Box 6-3 (continued)

Build Capacity for Covering Hard-to-Reach Populations

> ***Recommendation 5.14:*** **BJS should study the measurement of emerging or hard-to-reach groups and should develop more appropriate approaches to sampling and measurement of these populations.**

Continue Coordination With Office of Juvenile Justice and Delinquency Prevention on Juvenile Justice Data

> ***Recommendation 2.2:*** **In line with its original charge and to better document and understand the contribution of juveniles to street crime and violence, the victimization of youth, and the consequences for youth and society of their victimization and offending, BJS should develop juvenile victimization, crime, and justice statistical series suitable for describing the patterns of offending and victimization of youth, longitudinal progression of youth through the juvenile and criminal justice systems, and reentry into the community and criminal system. Taking on this responsibility would require additional resources.**

Leverage Second Chance Act Responsibilities to Build Active Program on Recidivism and Reentry

> ***Recommendation 3.14:*** **BJS should mount a feasibility study of the flow of individuals between correctional supervision and community settings. Repeated interviews of samples of about-to-be-released prisoners that track their successes and failures in reintegrating with the community would enhance understanding of this critical policy issue.**

---

inal justice system. Following this logic, and all other things being equal, describing crime, victimization, and the immediate consequences should be the principal focus of BJS. Second, however, the nature of a funnel also directs attention to the late stages of the process. As offenders make their way through the funnel, their numbers decrease but the consequences of justice system decisions become increasingly consequential in terms of impacts on lives and public budgets. At the far end of the criminal justice system, in the correctional area, the effects of criminal justice decisions are so extensive that obtaining adequate data on these populations takes on an importance well beyond their numbers. This principle for establishing the importance of any given activity need not correspond to the allocation of resources to that activity because some data can be collected more efficiently than others. Nonetheless, it should guide decisions as to where to put the intellectual energy of the agency if not in a one-to-one correspondence to its fiscal resources. A logical decision process may be to separately value activities by their proximity to the funnel, then assess their relative cost; the final portfolio is a function of both importance and cost.

Two of these goals may be combined and interpreted as providing an answer to one of the questions raised at the outset of this section: the primacy of the NCVS in BJS's resource allocations. In terms of BJS's function as an independent statistical agency (Goal 1), the NCVS can play a major role. The survey's role as an independent counterpoint to the count of crimes officially reported to police and its capacity to assess public opinions of and interactions with justice officials make a vigorous NCVS a cornerstone of BJS's reputation for objectivity and data quality. The NCVS's role in measuring the input to, and largest part of, the justice system funnel of Goal 3 further argues for NCVS-intensive funding.

This interpretation is valid and we accept it. In our interim report, we considered a number of alternative structures for conducting the survey and the ways in which specific design features satisfy desired measurement goals. In doing so, one fundamental option that had to be considered "is *not* to conduct the survey at all"—to decide that the collection is, for whatever reasons, not worth its considerable expense. As we noted in that report (National Research Council, 2008b:79), and reaffirm here:

> We reject that option. To take that option violates the legislative responsibilities of the Bureau of Justice Statistics. Furthermore, the panel thinks that BJS is the appropriate locus of responsibility for victimization measurement. As a federal statistical agency, it alone has the mandate for independent, objective, statistical measurement, with the transparency that can establish public trust in the information.

In our assessment, the NCVS is sufficiently core to BJS's legally mandated duties and it's basic function as a statistical agency that it is difficult to imagine an effective BJS without a strong and continuing NCVS.

For reasons we describe in more detail in our first report (National Research Council, 2008b), the costs of personal survey interviewing are increasing and response rates to surveys in general are declining (even though some federal surveys such as the NCVS continue to maintain very high response rates). The NCVS requires large sample sizes in order to derive information on relatively low-incidence phenomena, and so these broader forces on survey research render impossible the truly ideal situation: survey data on victimization with the same quality as the existing NCVS but gathered at a small fraction of the cost. Congress and the administration need to be aware that a high-quality victimization survey is an investment and that accuracy comes with a price.

Accordingly, we reiterate two recommendations from our interim report (National Research Council, 2008b:Recs. 3.1 and 3.2)—one on the overriding goal of the NCVS to serve as a continuous, quality indicator of criminal victimization in the United States and another on the need to maintain the NCVS with proper resources.

> *Recommendation 6.1:* **BJS must ensure that the nation has quality annual estimates of levels and changes in criminal victimization.**

> *Recommendation 6.2:* **Congress and the administration should ensure that BJS has a budget that is adequate to field a survey that satisfies the goal in Recommendation 6.1.**

We see the development of a core-supplement structure for the NCVS as one means of remedying the chronic funding difficulties of the NCVS. By seeking commitments from federal agency partners for regular and recurring topic supplements to the NCVS, with some portion of the payment for supplements also contributing to the core costs of the survey, BJS has opportunities to more aggressively position NCVS as a flexible and effective data collection vehicle. Another partial solution to the chronic funding difficulties of the NCVS is to consider alternative funding streams, one possibility of which is the Crime Victims Fund created by the Victims of Crime Act of 1984 (42 USC § 10601). The fund consists of monies paid as various federal fines, penalty assessments, and the proceeds from federal asset forfeitures (as well as gifts from private entities), and grants for victim compensation and support services are made by OJP's Office of Victims of Crime from the fund. In 1999, the law enabling the fund was changed to authorize, in particular, U.S. Attorneys Offices and the Federal Bureau of Investigation (FBI) to "such sums as may be necessary" in order "to improve services for the benefit of crime victims in the Federal criminal justice system, and for a Victim Notification System." Although creating similar access for BJS would require language in a legislative reauthorization, it stands to reason that some provision for a modest—but, importantly, stable and recurring—allocation from the Crime Victim Fund to defray NCVS expenses would be consonant with the intent of the fund.

> *Recommendation 6.3:* **More information about the needs of victims is essential to the compensation and assistance goals of the Office of Victims of Crime. Congress should allow additional funding for the collection and improvement of victimization data to be obtained from funds obtained through the Victims of Crime Act.**

However, we also reiterate—as we described in Section 3–F—that more should also be expected from that investment in terms of timeliness of analysis, content of topic supplements, and methodological improvement, and we offer a follow-up recommendation:

> *Recommendation 6.4:* **Additional resources made available for the NCVS should be used not only to increase the reliability of annual estimates but also to supplement the survey in ways that increase our understanding of criminal victimization.**

Although the NCVS may, fairly, enjoy some primacy in resource allocations, the other parts of the justice system funnel are also important. Accordingly, BJS should also give priority to filling in the funnel, that is, ensuring that data exist on every important decision made throughout the process. Some decisions—notably the declination decisions made by prosecutors and the revocation decisions made by parole and probation officers—are not well described by BJS data systems at this time. Increasing the coverage of decision points in the system may not happen overnight and the quality of these data could be designed to improve over time, but BJS should have a plan and should gather the information necessary for planning data collections or other means of describing all important decisions in the criminal justice system. The panel has judged that the most cost-efficient method to study the key transition points in the funnel requires new longitudinal measurements.

This simple statement of priorities also relegates to a position of lesser importance matters that are not criminal in their nature. We reiterate the point made in Section 2–D that civil justice issues are very important, representing a major share of the judicial proceedings in the nation's courts. The definitional problems inherent in scoping out major new data collection in civil justice, as well as the problems of accurately measuring private, out-of-court settlements, are such that expanded data collection in the civil justice area is only practicable with substantial resources and commitment from Congress and the Justice Department. More work in civil justice is certainly worth doing if such resources become available, but a strict interpretation of priorities makes it a lower priority for the agency in the immediate future.

6–C.4 Facilitating Access and Improving Dissemination and Outreach

Goal 4: To ensure access to statistics and data on crime and justice by the American public, the U.S. Congress, the U.S. Department of Justice and other executive agencies, and state and local government agencies.

This goal is a direct carryover from BJS's self-identified strategic goals, and we recognize BJS for including the notion of outreach in its original listing; we have adapted language from BLS's mission statement to develop our wording.

BJS has made laudable strides in promoting access to its data and reports. Its website provides ready and relatively easy access to an extensive backfile of BJS reports, summary tabulations, and data collection instruments; so, too, does the OJP–sponsored National Criminal Justice Reference Service which also includes content from the National Institute of Justice and other agencies. The summary tabulations included with report releases are accessible to more casual data users while high-end users are well served by BJS's

Box 6-4   Summary of Recommendations and Commentary on Strategic Goal 4: Facilitating Access and Improving Dissemination and Outreach

Create and Utilize a Professional Advisory Committee

Recommendation 5.8: BJS should establish an Advisory Group under the Federal Advisory Committee Act to provide guidance to BJS on the addition of new data collection efforts and the modification of current ones in light of needs identified by the group. Membership in the group should include, at a minimum, leaders and practitioners from each of the major subject matters covered by BJS data, as well as those with statistical and other types of academic expertise in these subject matters. The members of the group should be selected by the BJS director and the group should provide the director with at least two reports each year that contain its recommendations.

Institute Ad Hoc User and Stakeholder Workshops

Recommendation 5.7: To effectively get input on contemporaneous topics of interest, BJS should regularly convene ad hoc stakeholder workshops to suggest areas of immediate data needs.

Continue to Develop Mechanisms for More Accurate Data Compilation

Recommendation 5.10: To improve the utility and accuracy of the National Corrections Reporting Program (NCRP), BJS should work with correctional agencies to develop their own internal records to promote consistent data collections and expand coverage beyond the 41 states covered in the most recent NCRP.

Nurture an Active, Continuous Research Program

Recommendation 5.13: BJS should carefully study changes in the NCVS survey design before implementing them.

Improve Technical Skill Mix of BJS Staff

Recommendation 5.15: BJS must improve the technical skills of its staff, including mathematical statisticians, computer scientists, survey methodologists, and criminologists.

Build Ties With Congress

Recommendation 5.9: DOJ should take steps to ensure that congressional staff are aware of BJS data that could be used in developing legislation; DOJ and BJS should learn from congressional staff how their data are needed to inform/support legislation so that they can improve the utility of their current data and so that they can develop new data sets that could enhance policy development.

Consider Means to Improve Timeliness of BJS Estimates

Recommendation 5.11: BJS should evaluate each of its data programs to inquire whether more timely estimates might be obtained by (a) making discrete data collections into more continuous operations and (b) issuing preliminary estimates, to be followed by final estimates.

(continued)

Box 6-4 (continued)

Improve Equitable Release of BJS Data for Analytical Purposes

As discussed in Section 5–B.4, and in keeping with the expected practice of wide dissemination of data, BJS should continue to work to ensure that all of the public (including external researchers) should have access to data releases at the same time, in formats that are conducive to use and analysis.

Enhance BJS Web Presence

Recommendation 5.12: BJS should articulate why some data collections are housed on external websites and describe the process by which links to external websites are allowed. BJS should articulate and justify the use of its insignia on external websites.

---

microdata holdings at the National Archive of Criminal Justice Data. It has also, in recent years, taken steps to add to the feedback and expert knowledge contribution that it already obtains through its state Statistical Analysis Center networks; its sponsorship of a data user conference in February 2008 was an instructive exercise.

Going forward, the challenge is for BJS to increase its public profile and its relevance in policy debates while maintaining the high quality standards it has set for itself. In the panel's assessment, the creation of a formal advisory committee would benefit BJS as a sounding board for specific data collection revisions, a reviewer of new research priorities, and a source of intelligence on new and emerging topics in criminal justice. A formal advisory committee would usefully complement feedback from ad hoc user and stakeholder workshops on specific data needs. From the technical standpoint, an important means for BJS to increase its public relevance is to find ways to address long-standing concerns about the agency's data products: though they are generally held in high regard, they are frequently seen as lacking timeliness and, particularly for the NCVS, subnational geographic detail. In addition to work on statistical modeling to improve the temporal and spatial resolution of estimates (a search in which BJS could partner with federal statistical agencies, most of which face similar pressures from its user constituencies), BJS should also consider a release schedule involving preliminary estimates subject to final review and possible revision.

6–D SUMMARY

Congress has given the Bureau of Justice Statistics a comprehensive mandate to collect data and disseminate policy-relevant statistical information on a broad array of domains. However, BJS does not have the human and fiscal resources to fulfill this mandate. Recent history shows evidence of interference from within the Department of Justice in its attempts to report objectively the information it *does* collect.

As a principal federal statistical agency BJS must become an independent, objective collector and reporter to the nation about the justice system. To achieve that independence, in the judgment of the panel, BJS should be organizationally separated from units of the Justice Department that administer programs; its director should be a presidential, fixed-term appointment; it should not participate in regulatory activities; its products must not be censored; it should renew the NCVS; it should enrich its ties with state justice systems to the benefit of national statistical information.

BJS should shape its statistical series with the entire justice system in mind, from the moment of a criminal act through all stages of the administrative consequences. This means a new blueprint for its products, exploiting new developments in computerized record systems, more timely supplements to existing surveys on new developments, guiding conceptual and statistical principles for its adjudication series, and revived attention to court statistics. The panel urges BJS to develop more longitudinal measurement tools, following the course of a criminal event and the actors involved in the event over time. Only with such longitudinal observation can the country learn about the outcome of various policy interventions. This longitudinal measurement will probably involve the mix of administrative data and self-report data from sample surveys, and should apply to all stages of the justice system. Such a plan will probably involve active partnerships with the FBI, the Office of Juvenile Justice and Delinquency Prevention, and the courts. It will clearly entail studies of the movement in and out of incarceration and in and out of the community. The panel believes that such a longitudinal vision requires new funding models to achieve success.

One action step to achieve BJS's appropriate stature is to expand BJS's role as a clearinghouse for all justice-related statistical information. To do this, BJS must develop mechanisms to be informed about the key information needs of the country. BJS needs a permanent technical advisory committee, with rotating membership of key stakeholders and statistical experts; it needs ongoing user workshops; it needs human resources to continuously examine new ways to utilize records and innovative measurement designs for its statistical series. The agency needs to add more highly skilled statistical staff so that it can design its own statistical programs instead of relying on other agencies and contractors. In short, BJS should be able to provide the technical leadership for the country's justice statistics.

As it is currently equipped, BJS is not fulfilling its legal mandate. This is due neither to the current staff of BJS, who have proven capable and resilient over the 30 years since the agency's creation, nor to its existing set of data collections. Nor is it due to the legal mandate itself: the tasks and expectations placed on BJS are appropriately broad and sprawling, reflecting the complexity of crime and the importance of information on the administration of justice to the public good. Rather, the problem is that BJS lacks the position of independence and culture of innovation that are—with commensurate resources—necessary for the agency to more fully meet its legally defined expectations. BJS has accomplished a great deal in 30 years, but much work remains to be done to ensure the quality, credibility, and relevance of statistics on justice in the United States—to achieve the BJS the country deserves.

References

Addington, L. A. (2003). Students' fear after Columbine: Findings from a randomized experiment. *Journal of Quantitative Criminology 19*(4), 367–387.

Addington, L. A. (2007a). Hot vs. cold cases: Examining time to clearance for homicides using NIBRS data. *Justice Research and Policy 9*(2), 87–112.

Addington, L. A. (2007b). Using NIBRS to study methodological sources of divergence between the UCR and NCVS. See Lynch and Addington (2007), Chapter 8.

Addington, L. A. and C. Rennison (2008). Rape co-occurrence: Do additional crimes affect victim reporting and police clearance of rape? *Journal of Quantitative Criminology 24*(2), 205–226.

Barnett, C. (2000). *The Measurement of White-Collar Crime Using Uniform Crime Reporting (UCR) Data*. Unnumbered; NIBRS Publication Series. Clarksburg, WV: Federal Bureau of Investigation, U.S. Department of Justice.

Bauer, L. (2004, September). *Local Law Enforcement Block Grant Program, 1996–2004*. NCJ 203096. Washington, DC: U.S. Department of Justice, Bureau of Justice Statistics.

Baum, K. (2005, August). *Juvenile Victimization and Offending, 1993–2003*. NCJ 209468. Washington, DC: U.S. Department of Justice, Bureau of Justice Statistics.

Baum, K. (2007, November). *National Crime Victimization Survey: Identity Theft, 2005*. NCJ 219411. Washington, DC: U.S. Department of Justice, Bureau of Justice Statistics.

Baum, K., S. Catalano, M. R. Rand, and K. Rose (2009, January). *National Crime Victimization Survey: Stalking Victimization in the United States*. NCJ 24527. Washington, DC: U.S. Department of Justice, Bureau of Justice Statistics.

Baumer, E. P. (2002). Neighborhood disadvantage and police notification by victims of violence. *Criminology 40*(3), 579–616.

Baumer, E. P., R. B. Felson, and S. Messner (2003). Changes in police notification for rape, 1973–2000. *Criminology 41*(3), 841–872.

Baumer, E. P., J. Horney, R. B. Felson, and J. L. Lauritsen (2003). Neighborhood disadvantage and the nature of violence. *Criminology 41*(1), 39–71.

Beattie, R. H. (1959). Sources of statistics on crime and correction. *Journal of the American Statistical Association 54*(287), 582–592.

Beck, A. J., D. B. Adams, and P. Guerino (2008, July). *Prison Rape Elimination Act of 2003: Sexual Victimization Reported by Juvenile Correctional Authorities, 2005– 06*. NCJ 215337. Washington, DC: U.S. Department of Justice, Bureau of Justice Statistics.

Beck, A. J. and P. M. Harrison (2007, December). *Prison Rape Elimination Act of 2003: Sexual Victimization in State and Federal Prisons Reported by Inmates, 2007*. NCJ 219414. Individual pages marked as being revised in March and April, 2008. Washington, DC: U.S. Department of Justice, Bureau of Justice Statistics.

Beck, A. J. and P. M. Harrison (2008, June). *Prison Rape Elimination Act of 2003: Sexual Victimization in Local Jails Reported by Inmates, 2007*. NCJ 221946. Washington, DC: U.S. Department of Justice, Bureau of Justice Statistics.

Beck, A. J. and T. A. Hughes (2005, July). *Prison Rape Elimination Act of 2003: Sexual Violence Reported by Correctional Authorities, 2004*. NCJ 210333. Washington, DC: U.S. Department of Justice, Bureau of Justice Statistics.

Beck, A. J. and L. M. Maruschak (2001, July). *Mental Health Treatment in State Prisons, 2000*. NCJ 188215. Washington, DC: U.S. Department of Justice, Bureau of Justice Statistics.

Beck, A. J. and L. M. Maruschak (2004, April). *Hepatitis Testing and Treatment in State Prisons*. NCJ 199173C. Labeled "A Correction Reprint, October 2004: The original report contained data that a State subsequently changed." Washington, DC: U.S. Department of Justice, Bureau of Justice Statistics.

Biderman, A. D. and A. J. Reiss (1967, November). On exploring the "dark figure" of crime. *Annals of the American Academy of Political and Social Science 374*, 1–15.

Blumstein, A. and A. J. Beck (1999). Population growth in U.S. prisons, 1980–1996. *Crime and Justice 26*(17). Accessed by LexisNexis.

Blumstein, A. and A. J. Beck (2005). Reentry as a transient state between liberty and recommitment. In J. Travis and C. Visher (Eds.), *Prisoner Reentry and Crime in America*. New York: Cambridge University Press.

Bonczar, T. P. (2003, August). *Prevalence of Imprisonment in the U.S. Population, 1974–2001*. NCJ 197976. Washington, DC: U.S. Department of Justice, Bureau of Justice Statistics.

Bonczar, T. P. and A. J. Beck (1997). *Lifetime Likelihood of Going to State or Federal Prison*. NCJ 160092. Washington, DC: U.S. Department of Justice, Bureau of Justice Statistics.

Bridges, G. and R. D. Crutchfield (1988). Law, social standing, and racial disparities in imprisonment. *Social Forces 66*(3), 699–724.

Brien, P. M. (2005, July). *Improving Access to and Integrity of Criminal History Records*. NCJ 200581. Washington, DC: U.S. Department of Justice, Bureau of Justice Statistics.

Brown, J. M. and P. A. Langan (1999, July). *Felony Sentences in the United States, 1996*. NCJ 175045. Washington, DC: U.S. Department of Justice, Bureau of Justice Statistics.

Bureau of Justice Statistics (1995, January). *National Conference on Criminal History Records—Brady and Beyond: Proceedings of a BJS/SEARCH Group Conference*. NCJ 151263. Washington, DC: U.S. Department of Justice, Bureau of Justice Statistics.

Bureau of Justice Statistics (1996). *Murder Cases in 33 Large Urban Counties in the United States, 1988*. Computer file. ICPSR 9907. Washington, DC, and Ann Arbor, MI: U.S. Department of Justice, Bureau of Justice Statistics [producer] and Inter-university Consortium for Political and Social Research [distributor].

Bureau of Justice Statistics (1997a). *Bureau of Justice Statistics Style Guide*. Washington, DC: U.S. Department of Justice, Bureau of Justice Statistics.

Bureau of Justice Statistics (1997b, May). *National Conference on Juvenile Justice Records: Appropriate Criminal and Noncriminal Justice Uses—Proceedings of a BJS/SEARCH Conference*. NCJ 164269. Washington, DC: U.S. Department of Justice, Bureau of Justice Statistics.

Bureau of Justice Statistics (1997c). *Survey of Campus Law Enforcement Agencies, 1995: United States*. Computer file. Conducted by U.S. Department of Commerce, Bureau of the Census. ICPSR 06846-v1. Ann Arbor, MI: Inter-university Consortium for Political and Social Research [producer and distributor].

Bureau of Justice Statistics (1998a, April). *National Conference on Sex Offender Registries: Proceedings of a BJS/SEARCH Conference*. NCJ 168965. Washington, DC: U.S. Department of Justice, Bureau of Justice Statistics.

Bureau of Justice Statistics (1998b). *State and Federal Corrections Information Systems: An Inventory of Data Elements and an Assessment of Reporting Capabilities*. NCJ 170016. Joint project of Association of State Correctional Administrators, Corrections Program Office (Office of Justice Programs), Bureau of Justice Statistics, and National Institute of Justice. Washington, DC: U.S. Department of Justice, Bureau of Justice Statistics.

Bureau of Justice Statistics (2001, December). *Use and Management of Criminal History Record Information: A Comprehensive Report, 2001 Update.* NCJ 187670. Washington, DC: U.S. Department of Justice, Bureau of Justice Statistics.

Bureau of Justice Statistics (2002). *Recidivism of Prisoners Released in 1994: United States.* Computer file. U.S. Department of Justice, Bureau of Justice Statistics, and Regional Justice Information Services Commission [producers]. ICPSR 03355-v5. Ann Arbor, MI: Inter-university Consortium for Political and Social Research [distributor].

Bureau of Justice Statistics (2003a). *Census of State and Local Law Enforcement Agencies (CSLLEA), 2000: United States.* Computer file. Conducted by U.S. Department of Commerce, Bureau of the Census. ICPSR 03484-v3. Ann Arbor, MI: Inter-university Consortium for Political and Social Research [producer and distributor].

Bureau of Justice Statistics (2003b, November). *Compendium of State Security and Privacy Legislation: Overview 2002.* NCJ 200030. Washington, DC: U.S. Department of Justice, Bureau of Justice Statistics.

Bureau of Justice Statistics (2003c). *National Survey of DNA Crime Laboratories, 2001.* Computer file. Conducted by U.S. Department of Commerce, Bureau of the Census. ICPSR 03550. Ann Arbor, MI: Inter-university Consortium for Political and Social Research [distributor].

Bureau of Justice Statistics (2004a). *Census of State and Federal Adult Correctional Facilities, 2000.* Computer file. Conducted by U.S. Department of Commerce, Bureau of the Census. ICPSR 04021. Ann Arbor, MI: Inter-university Consortium for Political and Social Research [producer and distributor].

Bureau of Justice Statistics (2004b). *Civil Justice Survey of State Courts, 2001: United States.* Computer file. Conducted by National Center for State Courts. ICPSR 09357. Ann Arbor, MI: Inter-university Consortium for Political and Social Research [producer and distributor].

Bureau of Justice Statistics (2004c, June 30). *Data Collections for the Prison Rape Elimination Act of 2003.* Status Report. Washington, DC: U.S. Department of Justice, Bureau of Justice Statistics.

Bureau of Justice Statistics (2004d, September). *Law Enforcement Agency Identifiers Crosswalk [United States], 2000.* Computer file. ICPSR 04082. Ann Arbor, MI: Inter-university Consortium for Political and Social Research [producer and distributor].

Bureau of Justice Statistics (2005a, April). *Bureau of Justice Statistics: Strategic Plan FY 2005–2008.* Washington, DC: U.S. Department of Justice, Bureau of Justice Statistics.

Bureau of Justice Statistics (2005b). *Census of Publicly Funded Forensic Crime Laboratories, 2002*. Computer file. Conducted by the University of Illinois at Chicago. ICPSR 04287-v1. Ann Arbor, MI: Inter-university Consortium for Political and Social Research [distributor].

Bureau of Justice Statistics (2006a, December). *Compendium of Federal Justice Statistics, 2004*. NCJ 213476. Washington, DC: U.S. Department of Justice, Bureau of Justice Statistics.

Bureau of Justice Statistics (2006b, December). *Criminal Victimization in the United States, 2005—Statistical Tables*. NCJ 215244. Washington, DC: U.S. Department of Justice, Bureau of Justice Statistics.

Bureau of Justice Statistics (2006c). *Law Enforcement Management and Administrative Statistics (LEMAS): 2003 Sample Survey of Law Enforcement Agencies*. Computer file. Conducted by U.S. Department of Commerce, Bureau of the Census [producer]. ICPSR 04411-v1. Ann Arbor, MI: Inter-university Consortium for Political and Social Research [distributor].

Bureau of Justice Statistics (2006d). *Survey of Inmates in Local Jails, 2002: United States*. Computer file. Conducted by U.S. Department of Commerce, Bureau of the Census. ICPSR 04359-v2. Ann Arbor, MI: Inter-university Consortium for Political and Social Research [producer and distributor].

Bureau of Justice Statistics (2007a, August). 2005 Justice System Expenditure and Employment Extracts program. NCJ 219370. Published as comma-delimited spreadsheet and explanatory text files at http://www.ojp.usdoj.gov/bjs/pub/sheets/cjee05.zip.

Bureau of Justice Statistics (2007b). *Annual Survey of Jails: Jurisdiction-Level Data, 2006*. Computer file. Conducted by U.S. Department of Commerce, Bureau of the Census. ICPSR 20368-v1. Ann Arbor, MI: Inter-university Consortium for Political and Social Research [producer and distributor].

Bureau of Justice Statistics (2007c). *Capital Punishment in the United States, 1973–2005*. Computer file. Conducted by U.S. Department of Commerce, Bureau of the Census. ICPSR 20580-v1. Ann Arbor, MI: Inter-university Consortium for Political and Social Research [producer and distributor].

Bureau of Justice Statistics (2007d). *National Corrections Reporting Program, 2003: United States*. Computer file. Conducted by U.S. Department of Commerce, Bureau of the Census. ICPSR 20741-v1. Ann Arbor, MI: Inter-university Consortium for Political and Social Research [producer and distributor].

Bureau of Justice Statistics (2007e). *National Judicial Reporting Program, 2004: United States*. Computer file. Conducted by U.S. Department of Commerce, Bureau of the Census. ICPSR 20760-v1. Ann Arbor, MI: Inter-university Consortium for Political and Social Research [producer and distributor].

Bureau of Justice Statistics (2007f). *National Prosecutors Survey, 2005.* Computer file. Conducted by U.S. Department of Justice, Bureau of Justice Statistics. ICPSR 04600-v1. Ann Arbor, MI: Inter-university Consortium for Political and Social Research [producer and distributor].

Bureau of Justice Statistics (2007g). *State Court Processing Statistics, 1990–2004: Felony Defendants in Large Urban Counties.* Computer file. Conducted by Pretrial Justice Institute (formerly, the Pretrial Services Resource Center) [producer]. ICPSR 02038-v3. Ann Arbor, MI: Inter-university Consortium for Political and Social Research [distributor].

Bureau of Justice Statistics (2007h). *Survey of Inmates in State and Federal Correctional Facilities, 2004.* Computer file. ICPSR 04572-v1. Ann Arbor, MI: Inter-university Consortium for Political and Social Research [producer and distributor].

Bureau of Justice Statistics (2008a, July). *Background Checks for Firearm Transfers, 2007—Statistical Tables.* NCJ 223197. Electronic-only set of tables. Washington, DC: U.S. Department of Justice, Bureau of Justice Statistics.

Bureau of Justice Statistics (2008b). *Census of Publicly Funded Forensic Crime Laboratories, 2002 and 2005.* Computer file. Conducted by the University of Illinois at Chicago and Sam Houston State University. ICPSR 23120-v1. Ann Arbor, MI: Inter-university Consortium for Political and Social Research [distributor].

Bureau of Justice Statistics (2008c, August). *Criminal Victimization in the United States, 2006—Statistical Tables.* NCJ 223436. Washington, DC: U.S. Department of Justice, Bureau of Justice Statistics.

Bureau of Justice Statistics (2008d, August 4). *National Crime Victimization Survey, 2006 [Record-Type Files].* Computer file. ICPSR 22560-v1. Ann Arbor, MI: Inter-university Consortium for Political and Social Research [distributor].

Bureau of Justice Statistics (2008e). *National Crime Victimization Survey: School Crime Supplement, 2005.* Computer file. Conducted by U.S. Department of Commerce, Bureau of the Census. ICPSR 04429-v2. Ann Arbor, MI: Inter-university Consortium for Political and Social Research [producer and distributor].

Bureau of Justice Statistics (2008f). State Justice Statistics Program for Statistical Analysis Centers, 2008. Solicitation 2008-BJS-1747.

Burton, S. E., M. Finn, D. Livingston, K. Scully, W. D. Bales, and K. Padgett (2004). Applying a crime seriousness scale to measure changes in the severity of offenses by individuals arrested in Florida. *Justice Research and Policy* 6(1), 1–18.

Butterfield, F. (2002, September 22). Some experts fear political influence on crime data agencies. *New York Times*, 33.

Cahalan, M. W. (1986, December). *Historical Corrections Statistics in the United States, 1850–1984.* NCJ 102529. Washington, DC: U.S. Department of Justice, Bureau of Justice Statistics.

Cantor, D. and J. P. Lynch (2000). Self-report surveys as measures of crime and criminal victimization. See National Institute of Justice (2000), pp. 85–138.

Caulkins, J. P. (2000). Measurement and analysis of drug problems and drug control efforts. See National Institute of Justice (2000), pp. 391–449.

Caulkins, J. P., P. Ebener, and D. F. McCaffrey (1995). Describing DAWN's dominion. *Contemporary Drug Problems 22*, 547–567.

Chiricos, T. G. and M. A. Delone (1992). Labor surplus and punishment: A review and assessment of theory and evidence. *Social Problems 39*, 421–446.

Cohen, T. H. (2004, November). *Tort Trials and Verdicts in Large Counties, 2001.* NCJ 206240. Washington, DC: U.S. Department of Justice, Bureau of Justice Statistics.

Cohen, T. H. (2005a, January). *Contract Trials and Verdicts in Large Counties, 2001.* NCJ 207388. Washington, DC: U.S. Department of Justice, Bureau of Justice Statistics.

Cohen, T. H. (2005b, March). *Punitive Damage Awards in Large Counties, 2001.* NCJ 208445. Washington, DC: U.S. Department of Justice, Bureau of Justice Statistics.

Cohen, T. H. (2006). *Appeals from General Civil Trials in 46 Large Counties, 2001–2005.* NCJ 212979. Washington, DC: U.S. Department of Justice, Bureau of Justice Statistics.

Cohen, T. H. and K. A. Hughes (2007, March). *Medical Malpractice Insurance Claims in Seven States, 2000–2004.* NCJ 216339. Washington, DC: U.S. Department of Justice, Bureau of Justice Statistics.

Cohen, T. H. and B. A. Reaves (2006, February). *Felony Defendants in Large Urban Counties, 2002.* NCJ 210818. Washington, DC: U.S. Department of Justice, Bureau of Justice Statistics.

Cohen, T. H. and B. A. Reaves (2007, November). *State Court Processing Statistics, 1990–2004: Pretrial Release of Felony Defendants in State Courts.* NCJ 214994. Washington, DC: U.S. Department of Justice, Bureau of Justice Statistics.

Cohen, T. H. and S. K. Smith (2004). Research note: Examining civil trial litigation in state courts. *Justice Research and Policy 6*(2), 79–98.

Consortium of Social Science Associations (2002, February 10). Justice research and statistics agencies face competitive sourcing. *COSSA Washington Update 22*(3), 5.

Consortium of Social Science Associations (2007, July 24). House says NCVS 'a critical source;' huge 'senate' cut for BJS 'a misprint'. *COSSA Washington Update 26*(14), 3.

Cook, P. J. (1985). The case of the missing victims: Gunshot woundings in the National Crime Survey. *Journal of Quantitative Criminology 1*(1), 91–102.

Cook, P. J. (1991). The technology of personal violence. In M. Tonry (Ed.), *Crime and Justice: An Annual Review*, Volume 14, pp. 1–71. Chicago: University of Chicago Press.

Corlew, K. R. (2006). Congress attempts to shine a light on a dark problem: An in-depth look at the Prison Rape Elimination Act of 2003. *American Journal of Criminal Law 33*(157), Accessed by LexisNexis.

Court Statistics Project (2006). *State Court Caseload Statistics, 2006*. Williamsburg, VA: National Center for State Courts.

Cowan, C. D., L. R. Murphy, and J. Wiener (1984). Effects of supplemental questions on victimization estimates. See Lehnen and Skogan (1984), pp. 69–73.

Dantzler, M. L. (2008, September). NASS/NASDA orientation. Presentation to 2008 National Association of State Departments of Agriculture Annual Meeting.

Dawson, J. M. and B. Boland (1993, May). *Murder in Large Urban Counties, 1988*. NCJ 140614. Washington, DC: U.S. Department of Justice, Bureau of Justice Statistics.

Dawson, J. M. and P. A. Langan (1994, July). *Murder in Families*. NCJ 143498. Washington, DC: U.S. Department of Justice, Bureau of Justice Statistics.

DeFrances, C. J. (2001, September). *State-Funded Indigent Defense Services, 1999*. NCJ 188464. Revised October 1, 2001. Washington, DC: U.S. Department of Justice, Bureau of Justice Statistics.

DeFrances, C. J. and M. F. Litras (2000, November). *Indigent Defense Services in Large Counties, 1999*. NCJ 184932. Washington, DC: U.S. Department of Justice, Bureau of Justice Statistics.

DeFrances, C. J. and K. J. Strom (1997, March). *National Survey of Prosecutors, 1994: Juveniles Prosecuted in State Criminal Courts*. NCJ 164265. Washington, DC: U.S. Department of Justice, Bureau of Justice Statistics.

Dinkes, R., E. F. Cataldi, W. Lin-Kelly, and T. D. Snyder (2007, December). *Indicators of School Crime and Safety: 2007*. NCES 2008-021; NCJ 219553. Washington, DC: U.S. Department of Education and U.S. Department of Justice, Office of Justice Programs.

Dorsey, T. L., M. W. Zawitz, and P. Middleton (2004, April). *Drugs and Crime Facts*. Washington, DC: U.S. Department of Justice, Bureau of Justice Statistics.

Dugan, L. (1999). The effect of criminal victimization on a household's moving decision. *Criminology 37*(4), 903–930.

Duhart, D. T. (2000, October). *Urban, Suburban, and Rural Victimization, 1993–98*. NCJ 182031. Washington, DC: U.S. Department of Justice, Bureau of Justice Statistics.

Duhart, D. T. (2001, December). *Violence in the Workplace, 1993–99*. NCJ 190076. Washington, DC: U.S. Department of Justice, Bureau of Justice Statistics.

Dumond, R. W. (2006). Symposium: The impact of prisoner sexual violence: Challenges of implementing Public Law 109-79 the Prison Rape Elimination Act of 2003. *Journal of Legislation 32*(142). Accessed by LexisNexis.

Durose, M. R. and P. A. Langan (2007, July). *National Judicial Reporting Program: Felony Sentences in State Courts, 2004*. NCJ 215656. Washington, DC: U.S. Department of Justice, Bureau of Justice Statistics.

Durose, M. R., E. L. Schmitt, and P. A. Langan (2005, April). *Contacts Between Police and the Public: Findings from the 2002 National Survey*. NCJ 207845. Washington, DC: U.S. Department of Justice, Bureau of Justice Statistics.

Durose, M. R., E. L. Smith, and P. A. Langan (2007, April). *Contacts Between Police and the Public, 2005*. NCJ 215243. Washington, DC: U.S. Department of Justice, Bureau of Justice Statistics.

Eggen, D. (2005, August 25). Official in racial profiling study demoted; Justice Department denies political pressure; lawmaker demands investigation. *Washington Post*, A7.

Eisenberg, T. (2008, February 12). The need for a national civil justice survey of incidence and claiming behavior. Paper prepared for Council of Professional Associations on Federal Statistics/Bureau of Justice Statistics Data Users' Workshop.

Engel, R. S. (2005). Citizens' perceptions of distributive and procedural injustice during traffic stops with police. *Journal of Research in Crime and Delinquency 42*(4), 445–481.

Engel, R. S. and J. M. Calnon (2004). Examining the influence of drivers' characteristics during traffic stops with police: Results from a national survey. *Justice Quarterly 21*(1), 49–90.

Faggiani, D. (2007). Introduction: Special issue on research using the National Incident-Based Reporting System. *Justice Research and Policy 9*(2), 1–7.

Federal Bureau of Investigation (1989). *White Collar Crime: A Report to the Public*. Washington, DC: U.S. Government Printing Office.

Federal Bureau of Investigation (2004a). *Federal Bureau of Investigation Strategic Plan 2004–2009*. Document posted online at http://www.fbi.gov/publications/strategicplan/strategicplanfull.pdf; undated, but assumed to be 2004. Washington, DC: U.S. Department of Justice, Bureau of Justice Statistics.

Federal Bureau of Investigation (2004b). *UCR: Uniform Crime Reporting Handbook.* Washington, DC: U.S. Department of Justice, Bureau of Justice Statistics.

Felson, R. B., J. Ackerman, and C. Gallagher (2005). Police intervention and repeat domestic violence. *Criminology 43,* 563–588.

Forst, B. (2000). Prosecution's coming of age. *Justice Research and Policy* 2(1), 21–46.

Frase, R. S. (1998). Jails. In M. Tonry (Ed.), *The Handbook of Crime and Punishment,* pp. 474–508. New York: Oxford University Press.

Fyfe, J. J. (2002). Too many missing cases: Holes in our knowledge about police use of force. *Justice Research and Policy 4,* 87–102.

Glaze, L. E. and T. P. Bonczar (2007, December). *Probation and Parole in the United States, 2006.* NCJ 220218. Washington, DC: U.S. Department of Justice, Bureau of Justice Statistics.

Glaze, L. E. and L. M. Maruschak (2007, August). *Parents in Prison and Their Minor Children.* NCJ 222984. Washington, DC: U.S. Department of Justice, Bureau of Justice Statistics.

Greenfeld, L. A. (1997, February). *Sex Offenses and Offenders: An Analysis of Data on Rape and Sexual Assault.* NCJ 163392. Washington, DC: U.S. Department of Justice, Bureau of Justice Statistics.

Greenfeld, L. A. (1998, April). *Alcohol and Crime: An Analysis of National Data on the Prevalence of Alcohol Involvement in Crime.* Prepared for the Assistant Attorney General's National Symposium on Alcohol Abuse and Crime. NCJ 168632. Washington, DC: U.S. Department of Justice, Bureau of Justice Statistics.

Greenfeld, L. A. (2004). The Bureau of Justice Statistics: The statistics arm of the Department of Justice. *United States Attorneys' Bulletin 52*(5), 56–62.

Greenfeld, L. A. (2005). Larry Greenfeld steps down at BJS: Letter to JRSA members. *The Forum (Justice Research and Statistics Association) 23*(4), 7.

Greenfeld, L. A. and M. A. Henneberg (2001). Victim and offender self-reports of alcohol involvement in crime. *Alcohol Research and Health 25*(1).

Greenfeld, L. A., P. A. Langan, and S. K. Smith (1997). *Police Use of Force: Collection of National Data.* NCJ 165040. Washington, DC: U.S. Department of Justice, Bureau of Justice Statistics.

Greenfeld, L. A. and S. K. Smith (1999, February). *American Indians and Crime.* NCJ 173386. Washington, DC: U.S. Department of Justice, Bureau of Justice Statistics.

Hanson, R. A. and H. W. Daley (1995, September). *Federal Habeas Corpus Review: Challenging State Court Criminal Convictions.* NCJ 155504. Washington, DC: U.S. Department of Justice, Bureau of Justice Statistics.

Harcourt, B. E. (2006, January). From the asylum to the prison: Rethinking the incarceration revolution. Public Law Working Paper 114, University of Chicago.

Harlow, C. W. (1998, April). *Profile of Jail Inmates 1996*. NCJ 164620. Washington, DC: U.S. Department of Justice, Bureau of Justice Statistics.

Harlow, C. W. (1999, April). *Prior Abuse Reported by Inmates and Probationers*. NCJ 172879. Washington, DC: U.S. Department of Justice, Bureau of Justice Statistics.

Harlow, C. W. (2001, November). *Survey of Inmates in State and Federal Correctional Facilities: Firearm Use by Offenders*. NCJ 189369. Revised February 2002. Washington, DC: U.S. Department of Justice, Bureau of Justice Statistics.

Harlow, C. W. (2003, January). *Education and Correctional Populations*. NCJ 195670. Washington, DC: U.S. Department of Justice, Bureau of Justice Statistics.

Harlow, C. W. (2005, November). *National Crime Victimization Survey and Uniform Crime Reporting: Hate Crime Reported by Victims and Police*. NCJ 209911. Washington, DC: U.S. Department of Justice, Bureau of Justice Statistics.

Harrell, E. (2007, August). *Black Victims of Violent Crime*. NCJ 214258. Washington, DC: U.S. Department of Justice, Bureau of Justice Statistics.

Harrison, P. M. and A. J. Beck (2006, May). *Prison and Jail Inmates at Midyear 2005*. NCJ 213133. Washington, DC: U.S. Department of Justice, Bureau of Justice Statistics.

Hartford Courant (2005, September 12). Interference at Justice [editorial]. *Hartford Courant*, A8.

Heimer, K. (2008, February 12). Understanding violence against women using the NCVS: What we know and where we need to go. Paper prepared for Council of Professional Associations on Federal Statistics/Bureau of Justice Statistics Data Users' Workshop, February 12, 2008, Washington, DC.

Henriquez, M. (1999). IACP national database project on police use of force. In J. Travis, J. Chaiken, and R. Kaminski (Eds.), *Use of Force by Police: Overview of National and Local Data*, pp. 19–24. Washington, DC: U.S. Department of Justice, Bureau of Justice Statistics.

Heraux, C. G. (2007, October). The role of the National Archive of Criminal Justice Data (NACJD) in mapping and the Data Resources Program. Presentation to Ninth Crime Mapping Research Conference (National Institute of Justice), Pittsburgh, PA.

Hickman, M. J. (2003, January). *Tribal Law Enforcement, 2000*. Fact Sheet. NCJ 197963. Revised April 29, 2003. Washington, DC: U.S. Department of Justice, Bureau of Justice Statistics.

Hickman, M. J. (2005a, July). *Justice Assistance Grant (JAG) Program, 2005*. NCJ 209333. Washington, DC: U.S. Department of Justice, Bureau of Justice Statistics.

Hickman, M. J. (2005b, January). *State and Local Law Enforcement Training Academies, 2002*. NCJ 204030. Washington, DC: U.S. Department of Justice, Bureau of Justice Statistics.

Hickman, M. J. (2006, June). *Citizen Complaints About Police Use of Force*. NCJ 210296. Washington, DC: U.S. Department of Justice, Bureau of Justice Statistics.

Hickman, M. J., K. A. Hughes, K. J. Strom, and J. D. Ropero-Miller (2007, June). *Medical Examiners and Coroners' Offices, 2004*. NCJ 216756. Washington, DC: U.S. Department of Justice, Bureau of Justice Statistics.

Hickman, M. J. and B. A. Reaves (2001, February). *Community Policing in Local Police Departments, 1997 and 1999*. NCJ 184794. Washington, DC: U.S. Department of Justice, Bureau of Justice Statistics.

Hickman, M. J. and B. A. Reaves (2006a, May). *Local Police Departments, 2003*. NCJ 210118. Washington, DC: U.S. Department of Justice, Bureau of Justice Statistics.

Hickman, M. J. and B. A. Reaves (2006b, May). *Sheriffs' Offices, 2003*. NCJ 211361. Washington, DC: U.S. Department of Justice, Bureau of Justice Statistics.

Hines, J. W. and G. Engen (1992, December). BLS regional offices: 50 years of federal-state cooperation. *Monthly Labor Review*, 36–41.

Houston Chronicle (2005, September 21). Varnished truth: If the Justice Department is unhappy with the facts, it should work to change them, not to hide them [editorial]. *Houston Chronicle*, B8.

Hughes, K. A. (2006, April). *Justice Expenditure and Employment in the United States, 2003*. Revised May 10, 2006. NCJ 212280. Washington, DC: U.S. Department of Justice, Bureau of Justice Statistics.

Hughes, K. A. (2007, November). *Unidentified Human Remains in the United States, 1980–2004*. NCJ 219533. Washington, DC: U.S. Department of Justice, Bureau of Justice Statistics.

Human Rights Watch (2001). *No Escape: Male Rape in U.S. Prisons*. Accessible at http://www.hrw.org/reports/2001/prison/report.html. New York: Human Rights Watch.

Jacobs, D. and J. T. Carmichael (2001). The politics of punishment across time and space: A pooled time-series analysis of imprisonment rates. *Social Forces 80*, 61–91.

James, D. J. (2004, July). *Profile of Jail Inmates, 2002*. NCJ 201932. Revised October 12, 2004. Washington, DC: U.S. Department of Justice, Bureau of Justice Statistics.

James, D. J. and L. E. Glaze (2006, September). *Mental Health Problems of Prison and Jail Inmates* (Revised ed.). NCJ 213600. Washington, DC: U.S. Department of Justice, Bureau of Justice Statistics.

Johnson, R. and S. Raphael (2006a). How much crime reduction does the marginal prisoner buy? Working paper, Goldman School of Public Policy, University of California, Berkeley.

Johnson, R. C. and S. Raphael (2006b, June). The effects of male incarceration dynamics on AIDS infection rates among African-American women and men. Unpublished manuscript.

Joiner, R. (2005, September 5). Administration prefers spin to truth: Because he insisted on accurate reporting, a government employee is forced to change jobs [editorial]. *St. Louis Post-Dispatch*, B7.

Justice Research and Statistics Association (2003). JRSA conducts focus groups for improving crime data. *The Forum (Justice Research and Statistics Association) 21*(3), 4–5.

Justice Research and Statistics Association (2006a, December). BJS and JRSA hold 2006 national conference in Denver. *The Forum (Justice Research and Statistics Association) 24*(4), 1, 11–12.

Justice Research and Statistics Association (2006b). BJS offers crime victimization software (BJScvs). *The Forum (Justice Research and Statistics Association) 24*(1), 5.

Justice Research and Statistics Association (2008). SAC and JRSA history. Posted to http://www.jrsainfo.org/about/sac-jrsa-history.pdf.

Kane, J. and A. D. Wall (2006, January). *The 2005 National Public Survey on White Collar Crime*. Fairmont, WV: National White Collar Crime Center.

Keilitz, I. (2000). Standards and measures of court performance. See National Institute of Justice (2000), pp. 559–593.

Klaus, P. (2004, July). *Carjacking, 1993–2002*. NCJ 205123. Washington, DC: U.S. Department of Justice, Bureau of Justice Statistics.

Klaus, P. (2007, April). *National Crime Victimization Survey: Crime and the Nation's Households, 2005*. NCJ 217198. Washington, DC: U.S. Department of Justice, Bureau of Justice Statistics.

Kling, J. (2006). Incarceration length, employment, and earnings. *American Economic Review 96*(3), 863–876.

Kyckelhahn, T. and T. H. Cohen (2008, August). *Civil Rights Complaints in U.S. District Courts, 1990–2006*. NCJ 222989. Washington, DC: U.S. Department of Justice, Bureau of Justice Statistics.

LaFountain, R. C., R. Y. Schauffler, S. M. Strickland, W. M. Raftery, and C. G. Bromage (2007). *Examining the Work of State Courts, 2006: A National Perspective from the Court Statistics Project.* Williamsburg, VA: National Center for State Courts.

Langan, P. A. (1988). *Historical Statistics on Prisoners in State and Federal Institutions, Yearend 1925–86.* Washington, DC: U.S. Department of Justice, Bureau of Justice Statistics.

Langan, P. A. and J. M. Dawson (1995, September). *Spouse Murder Defendants in Large Urban Counties.* NCJ 153256. Washington, DC: U.S. Department of Justice, Bureau of Justice Statistics.

Langan, P. A. and D. Levin (1999, February). *Assessing the Accuracy of State Prisoner Statistics.* NCJ 173413. Washington, DC: U.S. Department of Justice, Bureau of Justice Statistics.

Langan, P. A. and D. J. Levin (2002, June). *Recidivism of Prisoners Released in 1994.* NCJ 193427. Washington, DC: U.S. Department of Justice, Bureau of Justice Statistics.

Langton, L. and T. H. Cohen (2007, October). *State Court Organization, 1987–2004.* NCJ 217996. Washington, DC: U.S. Department of Justice, Bureau of Justice Statistics.

Langworthy, R. H. (2002). LEMAS: A comparative organizational research platform. *Justice Research and Policy 4,* 21–38.

Larence, E. R. (2008, July 8). *Bureau of Justice Statistics Funding to States to Improve Criminal Records.* Letter to Chairmen and Ranking Members of the Judiciary Committees of the Senate and House of Representatives. GAO-08-898R. Washington, DC: U.S. Government Accountability Office.

Lattimore, P. K., S. Brumbaugh, C. Visher, C. Lindquist, L. Winterfield, M. Salas, and J. Zweig (2004, July). *National Portrait of SVORI: Serious and Violent Offender Reentry Initiative.* Research Triangle Park, NC, and Washington, DC: RTI International and Urban Institute.

Lattimore, P. K., C. A. Visher, L. Winterfield, C. Lindquist, and S. Brumbaugh (2005). Implementation of prisoner reentry programs: Findings from the Serious and Violent Offender Reentry Initiative multi-site evaluation. *Justice Research and Policy 7*(2), 87–109.

Lauritsen, J. L. and K. Heimer (2008). The gender gap in violent victimization, 1973–2004. *Journal of Quantitative Criminology 24,* 125–147.

Lauritsen, J. L. and R. J. Schaum (2005). *Crime and Victimization in the Three Largest Metropolitan Areas, 1980–98.* NCJ 208075. Washington, DC: U.S. Department of Justice, Bureau of Justice Statistics.

Lehnen, R. G. and W. G. Skogan (Eds.) (1984). *The National Crime Survey Working Papers: Volume II: Methodological Studies.* NCJ 90307. Washington, DC: U.S. Department of Justice, Bureau of Justice Statistics.

Levitt, S. D. (1996). The effect of prison population size on crime rates: Evidence from prison overcrowding litigation. *Quarterly Journal of Economics 111*, 319–351.

Lichtblau, E. (2005a, August 24). Profiling report leads to a clash and a demotion. *New York Times*, A1.

Lichtblau, E. (2005b, August 26). Democrats want official to be reinstated over report on profiling. *New York Times*, A11.

Lopoo, L. M. and B. Western (2005). Incarceration and the formation and stability of marital unions. *Journal of Marriage and the Family 67*(3), 721–734.

Love, D. A. (2005, September 2). Bush officials whitewash data on racial profiling [editorial]. *Baltimore Sun*, 21A.

Lynch, J. P. (2002). Using citizen surveys to produce information on the police: The present and potential uses of the National Crime Victimization Survey. *Justice Research and Policy 4*(1), 61–70.

Lynch, J. P. and L. A. Addington (Eds.) (2007). *Understanding Crime Statistics: Revisiting the Divergence of the NCVS and UCR.* Cambridge Studies in Criminology. New York: Cambridge University Press.

Maguire, E. R. (2002). Multiwave establishment surveys of police organizations. *Justice Research and Policy 4*, 39–59.

Maltz, M. D. (1999, September). *Bridging Gaps in Police Crime Data: A Discussion Paper from the BJS Fellows Program.* Bureau of Justice Statistics and Federal Bureau of Investigation Uniform Crime Reporting Program. NCJ 176365. Washington, DC: U.S. Department of Justice, Bureau of Justice Statistics.

Maltz, M. D. (2007). Missing UCR data and divergence of the NCVS and UCR trends. See Lynch and Addington (2007), Chapter 10.

Maruschak, L. M. (2001, July). *HIV in Prisons and Jails, 1999.* NCJ 187456. Revised October 25, 2001. Washington, DC: U.S. Department of Justice, Bureau of Justice Statistics.

Maruschak, L. M. (2006, November). *Medical Problems of Jail Inmates.* NCJ 210696. Washington, DC: U.S. Department of Justice, Bureau of Justice Statistics.

Maruschak, L. M. (2008a, April). *HIV in Prisons, 2006.* NCJ 222179. Web-only publication; PDF version accessible at http://www.ojp.usdoj.gov/bjs/pub/pdf/hivp06.pdf. Washington, DC: U.S. Department of Justice, Bureau of Justice Statistics.

Maruschak, L. M. (2008b, April). *Medical Problems of Prisoners.* NCJ 221740. Electronic-only publication and set of statistical tables. Washington, DC: U.S. Department of Justice, Bureau of Justice Statistics.

McEwen, T. (1996, April). *National Data Collection on Police Use of Force.* NCJ 160113. Washington, DC: U.S. Department of Justice, Bureau of Justice Statistics and National Institute of Justice.

Miami Herald (2005, August 28). Off message—and out of a job [editorial]. *Miami Herald.* Accessed by LexisNexis; document is dated August 29, with a note that the editorial appeared in the August 28 edition.

Minton, T. D. (2006, November). *Jails in Indian Country, 2004.* NCJ 214257. Washington, DC: U.S. Department of Justice, Bureau of Justice Statistics.

Minton, T. D. (2008, November). *Jails in Indian Country, 2007.* NCJ 223760. Washington, DC: U.S. Department of Justice, Bureau of Justice Statistics.

Motivans, M. (2006, August). *Federal Criminal Justice Trends, 2003.* NCJ 205331. Washington, DC: U.S. Department of Justice, Bureau of Justice Statistics.

Motivans, M. (2008, September). *Federal Justice Statistics, 2005.* NCJ 220383. Washington, DC: U.S. Department of Justice, Bureau of Justice Statistics.

Mumola, C. J. (2000a, August). *Incarcerated Parents and Their Children.* NCJ 182335. Washington, DC: U.S. Department of Justice, Bureau of Justice Statistics.

Mumola, C. J. (2000b, January). *Veterans in Prison or Jail.* NCJ 178888. Washington, DC: U.S. Department of Justice, Bureau of Justice Statistics.

Mumola, C. J. (2005, August). *Suicide and Homicide in State Prisons and Local Jails.* NCJ 210036. Washington, DC: U.S. Department of Justice, Bureau of Justice Statistics.

Mumola, C. J. (2007a, October). *Arrest-Related Deaths in the United States, 2003–2005.* NCJ 219534. Washington, DC: U.S. Department of Justice, Bureau of Justice Statistics.

Mumola, C. J. (2007b, January). *Medical Causes of Death in State Prisons, 2001–2004.* NCJ 216340. Washington, DC: U.S. Department of Justice, Bureau of Justice Statistics.

Murphy, L. (1976). *The Effects of the Attitude Supplement on NCS City Sample Victimization Data.* Unpublished internal document. Washington, DC: U.S. Bureau of the Census.

National Institute of Justice (2000). *Measurement and Analysis of Crime and Justice,* Volume 4 of *Criminal Justice 2000.* NCJ 182411. Washington, DC: U.S. Department of Justice, Bureau of Justice Statistics.

National Institute of Justice (2003, April). *2000 Arrestee Drug Abuse Monitoring: Annual Report*. NCJ 193013. Washington, DC: U.S. Department of Justice, Bureau of Justice Statistics.

National Research Council (1976). *Surveying Crime*. Panel for the Evaluation of Crime Surveys. Bettye K. Eidson Penick (Ed.) and Maurice E. B. Owens III (Assoc. Ed.), Committee on National Statistics, Academy of Mathematical and Physical Sciences. Washington, DC: National Academy of Sciences.

National Research Council (2001a). *Informing America's Policy on Illegal Drugs: What We Don't Know Keeps Hurting Us*. Committee on Data and Research for Policy on Illegal Drugs, Charles F. Manski, John V. Pepper, and Carol V. Petrie (Eds.), Committee on Law and Justice and Committee on National Statistics, Commission on Behavioral and Social Sciences and Education. Washington, DC: National Academy Press.

National Research Council (2001b). *What's Changing in Prosecution? Report of a Workshop*. Committee on Law and Justice. Philip Heymann and Carol Petrie (Eds.), Division of Behavioral and Social Sciences and Education. Washington, DC: National Academy Press.

National Research Council (2004a). *Fairness and Effectiveness in Policing: The Evidence*. Committee to Review Research on Police Policy and Practices. Wesley Skogan and Kathleen Frydl (Eds.), Committee on Law and Justice, Division of Behavioral and Social Sciences and Education. Washington, DC: The National Academies Press.

National Research Council (2004b). *The 2000 Census: Counting Under Adversity*. Panel to Review the 2000 Census. Constance F. Citro, Daniel L. Cork, and Janet L. Norwood (Eds.), Committee on National Statistics, Division of Behavioral and Social Sciences and Education. Washington, DC: The National Academies Press.

National Research Council (2005a). *Firearms and Violence: A Critical Review*. Committee to Improve Research Information on Data and Firearms. Charles F. Wellford, John V. Pepper, and Carol V. Petrie (Eds.), Committee on Law and Justice, Division of Behavioral and Social Sciences and Education. Washington, DC: The National Academies Press.

National Research Council (2005b). *Principles and Practices for a Federal Statistical Agency* (3rd ed.). Committee on National Statistics. Margaret E. Martin, Miron L. Straf, and Constance F. Citro (Eds.), Division of Behavioral and Social Sciences and Education. Washington, DC: The National Academies Press.

National Research Council (2006). *Once, Only Once, and in the Right Place: Residence Rules in the Decennial Census*. Panel on Residence Rules in the Decennial Census. Daniel L. Cork and Paul R. Voss (Eds.), Committee on National Statistics, Division of Behavioral and Social Sciences and Education. Washington, DC: The National Academies Press.

National Research Council (2007). *State and Local Government Statistics at a Cross-roads*. Panel on Research and Development Priorities for the U.S. Census Bureau's State and Local Government Statistics Program. Committee on National Statistics, Division of Behavioral and Social Sciences and Education. Washington, DC: The National Academies Press.

National Research Council (2008a). *Ballistic Imaging.* Committee to Assess the Feasibility, Accuracy, and Technical Capability of a National Ballistics Database, Daniel L. Cork, John E. Rolph, Eugene S. Meieran, and Carol V. Petrie (Eds.), Committee on Law and Justice and Committee on National Statistics, Division of Behavioral and Social Sciences and Education, and National Materials Advisory Board, Division of Engineering and Physical Sciences. Washington, DC: The National Academies Press.

National Research Council (2008b). *Surveying Victims: Options for Conducting the National Crime Victimization Survey*. Panel to Review the Programs of the Bureau of Justice Statistics. Robert M. Groves and Daniel L. Cork (Eds.), Committee on National Statistics and Committee on Law and Justice, Division of Behavioral and Social Sciences and Education. Washington, DC: The National Academies Press.

National Research Council (2009). *Principles and Practices for a Federal Statistical Agency* (4th ed.). Committee on National Statistics. Constance F. Citro, Miron L. Straf, and Margaret E. Martin (Eds.), Division of Behavioral and Social Sciences and Education. Washington, DC: The National Academies Press.

Noonan, M. E. and C. J. Mumola (2007, May). *Veterans in State and Federal Prison, 2004*. NCJ 217199. Washington, DC: U.S. Department of Justice, Bureau of Justice Statistics.

Pager, D. (2003). The mark of a criminal record. *American Journal of Sociology 108*(5), 937–975.

Perkins, C. (2003, September). *Weapon Use and Violent Crime*. NCJ 194820. Washington, DC: U.S. Department of Justice, Bureau of Justice Statistics.

Perry, S. W. (2004, December). *American Indians and Crime: A BJS Statistical Profile, 1992–2002*. NCJ 203097. Washington, DC: U.S. Department of Justice, Bureau of Justice Statistics.

Perry, S. W. (2005, December). *Census of Tribal Justice Agencies in Indian Country, 2002*. NCJ 205332. Washington, DC: U.S. Department of Justice, Bureau of Justice Statistics.

Perry, S. W. (2007, July). *Improving Criminal History Records in Indian Country, 2004–2006*. NCJ 218913. Washington, DC: U.S. Department of Justice, Bureau of Justice Statistics.

Peterson, J. L. and M. J. Hickman (2005, February). *Census of Publicly Funded Forensic Crime Laboratories, 2002*. NCJ 207205. Washington, DC: U.S. Department of Justice, Bureau of Justice Statistics.

Pettit, B. and B. Western (2004). Mass imprisonment and the life course: Race and class inequality in U.S. incarceration. *American Sociological Review 69*, 151–169.

Poggio, E., S. Kennedy, J. Chaiken, and K. Carlson (1985, May). *Blueprint for the Future of the Uniform Crime Reporting Program—Final Report of the UCR Study.* NCJ 098348. Washington, DC: U.S. Department of Justice, Bureau of Justice Statistics.

Police Executive Research Forum (2006, October). *Chief Concerns: A Gathering Storm—Violent Crime in America.* Washington, DC: Police Executive Research Forum.

Police Executive Research Forum (2007, November). *Critical Issues in Policing Series: Violent Crime in America: "A Tale of Two Cities".* Washington, DC: Police Executive Research Forum.

President's Commission on Law Enforcement and Administration of Justice (1967, February). *The Challenge of Crime in a Free Society.* Washington, DC: U.S. Government Printing Office. Commonly known as the Crime Commission report.

Rainville, G. A. and S. K. Smith (2003, May). *Juvenile Felony Defendants in Criminal Courts.* NCJ 197961. Washington, DC: U.S. Department of Justice, Bureau of Justice Statistics.

Ramker, G. F. and D. B. Adams (2008). Under construction: Building an information infrastructure to support BJS recidivism research. *JRSA Forum: Newsletter of the Justice Research and Statistics Association 26*(3), 1, 9–10.

Rand, M. R. (1997, August). *Violence-Related Injuries Treated in Hospital Emergency Departments.* NCJ 156921. Washington, DC: U.S. Department of Justice, Bureau of Justice Statistics.

Rand, M. R. (2008, December). *Criminal Victimization, 2007.* NCJ 224390. Washington, DC: U.S. Department of Justice, Bureau of Justice Statistics.

Rand, M. R. and S. Catalano (2007, December). *Criminal Victimization, 2006.* NCJ 219413. Washington, DC: U.S. Department of Justice, Bureau of Justice Statistics.

Rand, M. R. and C. M. Rennison (2005). Bigger is not necessarily better: An analysis of violence against women estimates from the National Crime Victimization Survey and the National Violence Against Women Survey. *Journal of Quantitative Criminology 21*(3), 267–291.

Rantala, R. R. (2004, March). *Cybercrime Against Businesses.* NCJ 200639. Washington, DC: U.S. Department of Justice, Bureau of Justice Statistics.

Rantala, R. R. (2008, September). *Cybercrime Against Businesses, 2005.* NCJ 221943. Washington, DC: U.S. Department of Justice, Bureau of Justice Statistics.

Reaves, B. A. (2006, July). *Violent Felons in Large Urban Counties.* NCJ 205289. Washington, DC: U.S. Department of Justice, Bureau of Justice Statistics.

Reaves, B. A. (2007, June). *Census of State and Local Law Enforcement Agencies, 2004.* NCJ 212749. Washington, DC: U.S. Department of Justice, Bureau of Justice Statistics.

Reaves, B. A. (2008, February). *Campus Law Enforcement, 2004–05.* NCJ 219374. Washington, DC: U.S. Department of Justice, Bureau of Justice Statistics.

Reaves, B. A. and M. J. Hickman (2002a, October). *Census of State and Local Law Enforcement Agencies, 2000.* NCJ 194066, Washington, DC: U.S. Department of Justice, Bureau of Justice Statistics.

Reaves, B. A. and M. J. Hickman (2002b, May). *Police Departments in Large Cities, 1990–2000.* NCJ 175703. Washington, DC: U.S. Department of Justice, Bureau of Justice Statistics.

Reaves, B. J. and M. J. Hickman (2004, March). *Law Enforcement Management and Administrative Statistics, 2000: Data for Individual State and Local Agencies with 100 or More Officers.* NCJ 203350. Washington, DC: U.S. Department of Justice, Bureau of Justice Statistics.

Regional Justice Information Service (2006, November). *Survey of State Procedures Related to Firearm Transfers, 2005.* NCJ 214645. Washington, DC: U.S. Department of Justice, Bureau of Justice Statistics.

Rennison, C. M. (2001, October). *Intimate Partner Violence and Age of Victim, 1993–99.* NCJ 187635. Washington, DC: U.S. Department of Justice, Bureau of Justice Statistics.

Rennison, C. M. (2002a, April). *Hispanic Victims of Violent Crime, 1993–2000.* NCJ 191208. Washington, DC: U.S. Department of Justice, Bureau of Justice Statistics.

Rennison, C. M. (2002b, August). *Rape and Sexual Assault: Reporting to Police and Medical Attention, 1992–2000.* NCJ 194530. Washington, DC: U.S. Department of Justice, Bureau of Justice Statistics.

Rosenfeld, R. (2007). Transfer the Uniform Crime Reporting program from the FBI to the Bureau of Justice Statistics. *Criminology and Public Policy* 6(4), 825–833.

Rottman, D. B. and S. M. Strickland (2006, August). *State Court Organization, 2004.* NCJ 212351. Washington, DC: U.S. Department of Justice, Bureau of Justice Statistics.

Sabol, W. J., W. P. Adams, B. Parthasarathy, and Y. Yuan (2000, September). *Offenders Returning to Federal Prison, 1986–97.* NCJ 182991. Washington, DC: U.S. Department of Justice, Bureau of Justice Statistics.

Sabol, W. J. and H. Couture (2008, June). *Prison Inmates at Midyear 2007.* NCJ 221944. Washington, DC: U.S. Department of Justice, Bureau of Justice Statistics.

Sabol, W. J., H. Couture, and P. M. Harrison (2007, December). *Prisoners in 2006.* NCJ 219416. Washington, DC: U.S. Department of Justice, Bureau of Justice Statistics.

Sabol, W. J., T. D. Minton, and P. M. Harrison (2007, June). *Prison and Jail Inmates at Midyear 2006.* Revised March 12, 2008. NCJ 217675. Washington, DC: U.S. Department of Justice, Bureau of Justice Statistics.

Scalia, J. (1996, August). *Noncitizens in the Federal Criminal Justice System, 1984–94.* NCJ 160934. Washington, DC: U.S. Department of Justice, Bureau of Justice Statistics.

Scalia, J. (1999, February). *Federal Pretrial Release and Detention, 1996.* NCJ 168635. Washington, DC: U.S. Department of Justice, Bureau of Justice Statistics.

Scalia, J. (2000, June). *Federal Firearm Offenders, 1992–98.* NCJ 180795. Revised July 17, 2000. Washington, DC: U.S. Department of Justice, Bureau of Justice Statistics.

Scalia, J. (2001, August). *Federal Drug Offenders, 1999; With Trends 1984–99.* NCJ 187285. Washington, DC: U.S. Department of Justice, Bureau of Justice Statistics.

Sedgwick, J. L. (2006, September). Bureau of Justice Statistics: environment, constraints and initiatives. *The Forum (Justice Research and Statistics Association) 24*(3), 1,6–7. Transcript of remarks to Justice Research and Statistics Executive Board, May 22, 2006.

Sedgwick, J. L. (2008). Letter to JRSA members from Jeffrey Sedgwick. *JRSA Forum: Newsletter of the Justice Research and Statistics Association 26*(3), 2.

Sedgwick, J. L. and G. F. Ramker (2007). BJS provides update on projects and priorities. *JRSA Forum: Newsletter of the Justice Research and Statistics Association 25*(2), 3–4.

Simon, T., J. Mercy, and C. Perkins (2001, June). *Injuries from Violent Crime, 1992–98.* NCJ 168633. Washington, DC: U.S. Department of Justice and U.S. Department of Health and Human Services.

Smith, S. K. and C. J. DeFrances (1996, February). *Indigent Defense.* NCJ 158909. Washington, DC: U.S. Department of Justice, Bureau of Justice Statistics.

Smith, S. K., G. W. Steadman, T. D. Minton, and M. Townsend (1999, May). *Criminal Victimization and Perceptions of Community Safety in 12 Cities, 1998.* NCJ 173940. Washington, DC: U.S. Department of Justice, Bureau of Justice Statistics and Office of Community Oriented Policing Services.

Snell, T. L. (2006, December). *Capital Punishment, 2005.* NCJ 215083. Washington, DC: U.S. Department of Justice, Bureau of Justice Statistics.

Snell, T. L. (2007, December). *Capital Punishment, 2006—Statistical Tables.* NCJ 220219. Web-only publication and set of spreadsheets; http://www.ojp.usdoj.gov/ bjs/pub/html/cp/2006/cp06st.htm. Washington, DC: U.S. Department of Justice, Bureau of Justice Statistics.

Snell, T. L. (2008, December). *Capital Punishment, 2007—Statistical Tables.* NCJ 224528. Web-only publication and set of spreadsheets; http://www.ojp.usdoj.gov/ bjs/pub/html/cp/2007/cp07st.htm. Washington, DC: U.S. Department of Justice, Bureau of Justice Statistics.

Sniffen, M. J. (2005, August 24). Study: Blacks and Hispanics fare worse than whites during traffic stops. Associated Press story, accessed by LexisNexis.

Snyder, H. N. and M. Sickmund (2006, March). *Juvenile Offenders and Victims: 2006 National Report.* Washington, DC: U.S. Department of Justice, Office of Justice Programs and Office of Juvenile Justice and Delinquency Prevention.

Solomon, A. L., V. Kachnowski, and A. Bhati (2005). *Does Parole Work? Analyzing the Impact of Postprison Supervision on Rearrest Outcomes.* Washington, DC: Urban Institute.

Spencer, B. D. (1980). *Benefit-Cost Analysis of Data Used to Allocate Funds.* New York: Springer-Verlag.

Steadman, G. W. (2002, January). *Survey of DNA Crime Laboratories, 2001.* NCJ 191191. Washington, DC: U.S. Department of Justice, Bureau of Justice Statistics.

Stephan, J. J. (2001, August). *Census of Jails, 1999.* NCJ 186633. Washington, DC: U.S. Department of Justice, Bureau of Justice Statistics.

Stephan, J. J. (2008, October). *Census of State and Federal Correctional Facilities, 2005.* NCJ 222182. Washington, DC: U.S. Department of Justice, Bureau of Justice Statistics.

Strom, K. J. (2000, February). *Profile of State Prisoners Under Age 18, 1985–97.* NCJ 176989. Washington, DC: U.S. Department of Justice, Bureau of Justice Statistics.

Strom, K. J., S. K. Smith, and H. N. Snyder (1998, September). *Juvenile Felony Defendants in Criminal Courts.* NCJ 165815. Washington, DC: U.S. Department of Justice, Bureau of Justice Statistics.

Struckman-Johnson, C., D. Struckman-Johnson, L. Rucker, K. Bumby, and S. Donaldson (1996). Sexual coercion reported by men and women in prison. *Journal of Sex Research* 33(1), 67–76.

Substance Abuse and Mental Health Services Administration (2002, August). *Drug Abuse Warning Network: Development of a New Design (Methodology Report).* DAWN Series M-4, DHHS Publication No. (SMA) 02-3754. Rockville, MD: U.S. Department of Health and Human Services.

Supreme Court of Nevada (2007, March). *Report to the 74th Regular Session of the Nevada State Legislature, 2007, Regarding the Creation of the Nevada Court of Appeals*. Carson City, NV: Supreme Court of Nevada.

Tennessee Tribune (2005, September 1). Justice Department should reward, not demote official [editorial]. *Tennessee Tribune*, A5.

Travis, J. (2008, November 11). Open letter to the American Society of Criminology. Posted to John Jay College of Criminal Justice website at http://www.jjay.cuny.edu/1918.php and http://www.jjay.cuny.edu/web_images/pres_letter.pdf.

Travis, J. and C. Visher (2005). *Prisoner Reentry and Crime in America*. New York: Oxford University Press.

Uchida, C., C. Bridgeforth, and C. Wellford (1984). Law enforcement statistics: The state of the art. Unpublished manuscript, Institute of Criminal Justice and Criminology, University of Maryland.

U.S. Census Bureau (2003). *National Crime Victimization Survey: Interviewing Manual for Field Representatives*. NCVS-550. Washington, DC: U.S. Census Bureau.

U.S. Census Bureau, Demographic Surveys Division (2007, January). *Survey Abstracts*. Washington, DC: U.S. Census Bureau.

U.S. Department of Education, Office of Postsecondary Education (2005, June). *The Handbook for Campus Crime Reporting*. Washington, DC: U.S. Department of Education.

U.S. Department of Justice (1998, March). *Comparing Case Processing Statistics*. NCJ 160274. Joint statement of Administrative Office of the U.S. Courts, Bureau of Justice Statistics, Executive Office for U.S. Attorneys, Federal Bureau of Prisons, and U.S. Sentencing Commission. Washington, DC: U.S. Department of Justice, Bureau of Justice Statistics.

U.S. Department of Justice, Bureau of Justice Statistics (2004, October). *Department of Justice: The Nation's Two Crime Measures*. NCJ 122705. Washington, DC: U.S. Department of Justice, Bureau of Justice Statistics.

U.S. Department of Justice, Criminal Division, Fraud Section (2006, May). Questionnaire on fraud and the criminal misuse and falsification of identity (identity fraud): Response of the United States delegation to the Intergovernmental Expert Group. Response to questionnaire issued by the United Nations Crime Commission Intergovernmental Expert Group on Fraud and the Criminal Misuse and Falsification of Identity. Posted at http://www.usdoj.gov/criminal/fraud/docs/05-06questionnaire.pdf.

U.S. Department of Justice, Office of Justice Programs (2006, August 28). *Office of Justice Programs Strategic Plan, Fiscal Years 2007–2012*. U.S. Department of Justice, Bureau of Justice Statistics.

U.S. General Accounting Office (2004, February). *National Criminal History Improvement Program: Federal Grants Have Contributed to Progress*. GAO-04-364. Washington: U.S. General Accounting Office.

U.S. Government Accountability Office (2007, March). *Bureau of Justice Statistics: Quality Guidelines Generally Followed for Police-Public Contact Surveys, but Opportunities Exist to Help Assure Agency Independence*. GAO-07-340. Washington, DC: U.S. Government Accountability Office.

U.S. House of Representatives, Committee on the Judiciary (1991). *Oversight Hearings on Emerging Criminal Justice Issues*. Serial No. 126. Hearings Before the Subcommittee on Criminal Justice, March 27, April 19, July 23, and August 15, 1990. Washington, DC: U.S. Government Printing Office.

U.S. House of Representatives, Committee on the Judiciary (2003). *Prison Rape Reduction Act of 2003*. Number Serial No. 36, 86-706 PDF. Hearing Before the Subcommittee on Crime, Terrorism, and Homeland Security, 108th Congress, First Session, on H.R. 1707, April 29, 2003. Washington, DC: U.S. Government Printing Office.

U.S. Office of Management and Budget (2007). *Statistical Programs of the United States Government: Fiscal Year 2008*. Washington, DC: U.S. Government Printing Office.

Useem, B., A. M. Piehl, and R. V. Liedka (2001). The crime control effect of incarceration: Reconsidering the evidence. Final Report to the National Institute of Justice.

Wainer, H. (1996). Depicting error. *American Statistician 50*(2), 101–111.

Weisburd, D., S. D. Mastrofski, R. Greenspan, and J. J. Willis (2004, April). *The Growth of COMPSTAT in American Policing*. Washington, DC: Police Foundation.

Weisburd, D., S. D. Mastrofski, A. M. McNally, R. Greenspan, and J. J. Willis (2003). Reforming to preserve: COMPSTAT and strategic problem solving in American policing. *Criminology and Public Policy 2*(3), 421–455.

Western, B. (2006). *Punishment and Inequality in America*. New York: Russell Sage Foundation.

Western, B., M. Kleykamp, and J. Rosenfeld (2006). Did falling wages and employment increase U.S. imprisonment? *Social Forces 84*, 2291–2312.

Wildeman, C. (2009). Parental imprisonment, the prison boom and the concentration of childhood disadvantage. *Demography 46*(1), 265—280.

Willis, J. J., S. D. Mastrofski, and D. Weisburd (2003). *COMPSTAT in Practice: An In-Depth Analysis of Three Cities*. Police Foundation Report. Washington, DC: Police Foundation.

Wilson, D. J. (2000, May). *Drug Use, Testing, and Treatment in Jails*. NCJ 179999. Washington, DC: U.S. Department of Justice, Bureau of Justice Statistics.

Wilson, J. M. (2005, August). *Determinants of Community Policing: An Open Systems Model of Implementation*. Working Paper, National Institute of Justice. NCJ 211975. Washington, DC: U.S. Department of Justice, Bureau of Justice Statistics.

Xie, M. and D. McDowall (2008). The effects of residential turnover on household victimization. *Criminology 46*, 539–575.

Xie, M., G. Pogarsky, J. P. Lynch, and D. McDowall (2006). Prior police contact and subsequent victim reporting: Results from the NCVS. *Justice Quarterly 23*(4), 481–501.

Zawitz, M. W. and K. J. Strom (2000, October). *Firearm Injury and Death from Crime, 1993–97*. NCJ 182993. Washington, DC: U.S. Department of Justice, Bureau of Justice Statistics.

Appendixes

– A –

Findings and Recommendations

This appendix lists the panel's findings and recommendations in this final report for ease of reference.

Finding 2.1: The data on crime currently collected by BJS are primarily focused on street crime. This focus on certain forms of violent and property crime does not account for important or emerging types of crime—notably, many forms of white-collar crime such as corporate fraud, health care fraud, financial institution fraud, money laundering, government fraud, consumer fraud, public corruption, and Internet crimes. The broad area of civil justice proceedings—distinct from criminal justice—is represented by one principal data series in BJS's extensive portfolio, and is limited by its construction to cover only completed court cases (and not out-of-court settlements). BJS's slate of cross-sectional series also does not readily provide for comprehensive analyses of contextual factors such as drugs and their impact on crime and violence.

Finding 2.2: Responsibility within the U.S. Department of Justice for coordinating and organizing data collections on juveniles is generally assumed by the Office of Juvenile Justice and Delinquency Prevention (OJJDP), instead of BJS. Though BJS's series do cover some segments of the juvenile population (e.g., juveniles housed in adult correctional facilities), the results of BJS and OJJDP studies are not well integrated. Within both BJS's and OJJDP's statistical coverage, there remain substantial gaps in data for juvenile offenders and victims with respect to their processing through the justice system "funnel."

Recommendation 2.1: Consistent with its legal mandate to collect, analyze, and disseminate statistical information on all aspects of the justice system, BJS should (a) document and organize the available statistics on forms of crime not covered by the NCVS, the FBI's UCR and NIBRS data systems, and other major data series maintained by other statistical agencies, (b) pursue research on what new statistics could be feasibly and usefully developed, and (c) propose such new data collections as the research suggests to be both feasible and useful. BJS should strive to function as a clearinghouse of justice-related statistical information, including reference to data not directly collected by BJS.

Recommendation 2.2: In line with its original charge and to better document and understand the contribution of juveniles to street crime and violence, the victimization of youth, and the consequences for youth and society of their victimization and offending, BJS should develop juvenile victimization, crime, and justice statistical series suitable for describing the patterns of offending and victimization of youth, longitudinal progression of youth through the juvenile and criminal justice systems, and reentry into the community and criminal system. Taking on this responsibility would require additional resources.

Finding 3.1: BJS currently gathers data about the criminal justice system but it does so on an institution-by-institution basis (police, courts, corrections) using varying units of analysis (crimes, individuals, cases) and sometimes varying time periods and samples. This approach provides good cross-sectional assessments of parts of the system, but makes it difficult or impossible to answer questions about the flow of individuals from arrest through eventual exit from the system. Yet people exit the system at many different stages in ways that are ill-understood but consequential for the effectiveness and fairness of criminal justice system processes. The cross-sectional approach misses the interfaces between the institutions, such as the large but unknown number of individuals who are arrested but not prosecuted.

Recommendation 3.1: BJS's goal in providing statistics from basic administrative data on corrections should be the development of a yearly count of correctional populations capable of disaggregation and cross-tabulation by state, offense categories, and demographic groups (age, race, gender, education).

Recommendation 3.2: BJS should produce yearly transition rates between steps in the corrections process capable of disaggregation and cross-tabulation by state, offense categories, and demographic groups.

Recommendation 3.3: BJS should explore the possibilities of increasing the utility of their correctional data collections by facilitating the linkage of records across the data series. For example, the ability to link records from the Recidivism Studies or from NCRP to the Census of Adult Correctional Facilities (CACF) would increase the ability to understand how correctional facilities contribute to recidivism.

Recommendation 3.4: BJS should develop an approach to measure the experiences of individuals through the criminal justice system on a prospective, longitudinal basis, beginning as early as practicable in the process (arrest) and ending with their eventual exit (ranging from early dismissal of charge through completion of sentence).

Recommendation 3.5: BJS should develop an approach to measure the victimization experiences of individuals on a prospective, longitudinal basis, beginning from a focal victimization and following the victim forward in time measuring subsequent victimizations and possible consequences of victimization. The NCVS may be used to recruit respondents to a panel survey of crime victims.

Recommendation 3.6: BJS should develop a panel survey of people under correctional supervision to understand how individuals move between institutional and community settings, and to understand the social contexts of correctional supervision.

Recommendation 3.7: To be useful, a BJS strategic plan must articulate a blueprint of interrelated data collection and product activities, including both current and potentially new data products. This blueprint would be used to evaluate new opportunities.

Recommendation 3.8: BJS should make supplements a regular feature of the NCVS. Procedures should be developed for soliciting ideas for supplements from outside BJS and for evaluating these supplements for inclusion in the survey.

Finding 3.2: The multitude of scattershot "census" studies of specific law enforcement agency types (e.g., campus law enforcement, medical examiners, training academies) detracts from the appearance of a coherent measurement program in the area of law enforcement. Instead, the impression left is that these "censuses" are sporadic inventories or catalogs of particular agency types with no obvious internal consistency.

Recommendation 3.9: To maximize both utility and timeliness of information, the LEMAS survey should be conducted as core-supplement design in the context of a continuous data collection.

Recommendation 3.10: To improve the utility of censuses of law enforcement agencies, BJS should develop an integrated conceptual plan for their periodicity, publish a 5-year schedule of their publication, and integrate their measurement into the LEMAS as supplements.

Recommendation 3.11: The NCVS (and its supplements) should be more effectively used as a tool for studying law enforcement, both in terms of the types of crime that are reported (and not reported) to police and the action that results from the reporting of a crime (e.g., the Police-Public Contact Survey).

Finding 3.3: BJS's current approach to data collection in adjudication lacks an effective basis in sampling.

Recommendation 3.12: As court records become more accessible through computerized case management systems, BJS should implement more rigorous methods of probability sampling in its adjudication series.

Recommendation 3.13: To inform future revisions to its adjudication portfolio and to more efficiently acquire and work with court data in the future (including longitudinal analysis), BJS should develop a research program to build representative samples of courts and to assess strategies for collection of case records.

Recommendation 3.14: BJS should mount a feasibility study of the flow of individuals between correctional supervision and community settings. Repeated interviews of samples of about-to-be-released prisoners that track their successes and failures in reintegrating with the community would enhance understanding of this critical policy issue.

Finding 4.1: BJS's state Statistical Analysis Center (SAC) program has cultivated a strong federal-state relationship, relative to other federal statistical agencies. Development of the SAC network—which provides points of contact across the justice system to facilitate research on individual data series, dissemination of BJS information, and coordination of activities—has involved forging unique relationships adapted to state environments (for instance, whether the SAC is part of a state law enforcement department or is housed at a university).

Recommendation 4.1: Through its Statistical Analysis Center and State Justice Statistics programs, BJS should continue to develop its ties with the states, and more fully exploit the potential for using states as partners in data collections.

Recommendation 4.2: Developments toward longitudinal and small-area measurement systems should involve state partners who are active in data collection and knowledgeable about state justice systems.

Finding 4.2: The National Criminal History Improvement Program (NCHIP) is a grantmaking program but not directly a statistical collection, even though it is administered by BJS. However, improved criminal history records are important for the prospects of longitudinal analysis of the criminal justice system. Analysis of the National Instant Background Check System serves as one approach to provide the data necessary to evaluate national policy on regulation of firearms purchases.

Recommendation 4.3: **BJS should actively utilize the NCHIP program to improve criminal history records necessary for longitudinal studies of crime.**

Finding 4.3: A full-fledged NIBRS would be a source of basic information on police responses to public complaints (911 calls), including whether or not a case is "cleared" by police through an arrest.

Recommendation 4.4: **To improve the timeliness of crime statistics, BJS should explore the development of a crime reporting system based on a probability sample of police administrative records. The goals of such a system would be national representativeness, high response, high data quality, timeliness and flexibility in terms of crime classification and analysis, and national statistics for the monitoring of crime trends.**

Finding 5.1: Under the terms of the Prison Rape Elimination Act of 2003, BJS was required to release the identity of selected responding institutions (i.e., facilities with the highest and lowest rates of sexual violence against inmates) for later regulatory action as part of a statistical program.

Recommendation 5.1: **Congress and the Department of Justice should not require, and BJS should not provide, individually identified data in support of regulatory functions that compromise the independence of BJS or require BJS to violate any of the principles of a federal statistical agency.**

Finding 5.2: The appearance of political interference in release of statistical information undermines public trust in that information and in the entire agency.

Recommendation 5.2: **The Department of Justice review of any BJS statistical product and related communications should not require changes to the content, the release schedule, or the mode of dissemination planned by BJS.**

Finding 5.3: The placement of BJS within the Office of Justice Programs has harmed the agency's ability to innovate in data collections and expand the efficiency of achieving its statistical mission. It suffers from a zero-sum game in competition with programs of direct financial benefit to states and localities.

Recommendation 5.3: **BJS should be administratively moved out of the Office of Justice Programs, reporting to the attorney general or deputy attorney general.**

Finding 5.4: Under current law, the director of the Bureau of Justice Statistics serves at the pleasure of the president; the director is nominated to an unspecified term by the president, with the advice and consent of the Senate (42 USC § 3732(b)).

Recommendation 5.4: **Congress and the administration should make the BJS director a fixed-term presidential appointee with the advice and consent of the Senate. To insulate the BJS director from political interference, the term of service should be no less than 4 years.**

Recommendation 5.5: **The BJS director needs to reach out to other agencies within DOJ, forming partnerships to propose initiatives for information collection that are relevant to policy needs.**

Recommendation 5.6: **The Department of Justice should build provisions for BJS collection of data and statistical information into its program initiatives aimed at crime reduction. These are not intended as program evaluation funds, but rather as funds for the basic monitoring and assessment of the phenomena targeted by the initiative.**

Finding 5.5: BJS enjoys high credibility but often is critiqued for missing fine-grained data by geography or time.

Recommendation 5.7: **To effectively get input on contemporaneous topics of interest, BJS should regularly convene ad hoc stakeholder workshops to suggest areas of immediate data needs.**

Recommendation 5.8: **BJS should establish an Advisory Group under the Federal Advisory Committee Act to provide guidance to BJS on the addition of new data collection efforts and the modification of current ones in light of needs identified by the group. Membership in the group should include, at a minimum, leaders and practitioners from each of the major subject matters covered by BJS data, as well as those with statistical and other types of academic expertise in these subject matters. The members of the group should be selected by the BJS director and the group should provide the director with at least two reports each year that contain its recommendations.**

Recommendation 5.9: DOJ should take steps to ensure that congressional staff are aware of BJS data that could be used in developing legislation; DOJ and BJS should learn from congressional staff how their data are needed to inform/support legislation so that they can improve the utility of their current data and so that they can develop new data sets that could enhance policy development.

Recommendation 5.10: To improve the utility and accuracy of the National Corrections Reporting Program (NCRP), BJS should work with correctional agencies to develop their own internal records to promote consistent data collections and expand coverage beyond the 41 states covered in the most recent NCRP.

Finding 5.6: A recurring criticism of BJS data products is that their quality is highly valued but that they are not sufficiently timely to meet user needs. All statistical agencies are attempting to grapple with new data collection designs that offer more timely estimates.

Recommendation 5.11: BJS should evaluate each of its data programs to inquire whether more timely estimates might be obtained by (a) making discrete data collections into more continuous operations and (b) issuing preliminary estimates, to be followed by final estimates.

Finding 5.7: The credibility of BJS's products is a function of its quality review procedures.

Recommendation 5.12: BJS should articulate why some data collections are housed on external websites and describe the process by which links to external websites are allowed. BJS should articulate and justify the use of its insignia on external websites.

Finding 5.9: The active investigation of new ways of measuring and understanding crime and criminal justice issues is a critical responsibility of BJS. The agency has lacked the resources needed to fully meet this responsibility and, for some issues, has fallen behind in developing such innovations.

Finding 5.10: BJS has lacked the resources to sufficiently produce new topical reports with data it currently gathers. It also lacks the resources and staff to routinely conduct methodological analyses of changes in the quality of its existing data series and to fully document those issues. Instead, the BJS production portfolio primarily is limited to a routine set of annual, biannual, and periodic reports and for some topics, the posting of updated data points in online spreadsheets.

Recommendation 5.13: BJS should carefully study changes in the NCVS survey design before implementing them.

Recommendation 5.14: BJS should study the measurement of emerging or hard-to-reach groups and should develop more appropriate approaches to sampling and measurement of these populations.

Recommendation 5.15: BJS must improve the technical skills of its staff, including mathematical statisticians, computer scientists, survey methodologists, and criminologists.

Finding 6.1: Relatively little is understood about how BJS data are used in developing policy and could be used in improving policy development.

Recommendation 6.1: BJS must ensure that the nation has quality annual estimates of levels and changes in criminal victimization.

Recommendation 6.2: Congress and the administration should ensure that BJS has a budget that is adequate to field a survey that satisfies the goal in Recommendation 6.1.

Recommendation 6.3: More information about the needs of victims is essential to the compensation and assistance goals of the Office of Victims of Crime. Congress should allow additional funding for the collection and improvement of victimization data to be obtained from funds obtained through the Victims of Crime Act.

Recommendation 6.4: Additional resources made available for the NCVS should be used not only to increase the reliability of annual estimates but also to supplement the survey in ways that increase our understanding of criminal victimization.

– B –

Summary of *Surveying Victims: Options for Conducting the National Crime Victimization Survey*

> This appendix reprints the executive summary of the panel's interim report *Surveying Victims: Options for Conducting the National Crime Victimization Survey* (National Research Council, 2008b), which included all of that report's findings and recommendations. The only change to the text as it appeared in the interim report is to change the numbering scheme for the findings and recommendations to include the prefix "Int-" to avoid confusion with finding and recommendation numbers in this report.

The Bureau of Justice Statistics (BJS) of the U.S. Department of Justice, Office of Justice Programs (OJP), requested that the Committee on National Statistics (in cooperation with the Committee on Law and Justice) convene this Panel to Review the Programs of the Bureau of Justice Statistics. The panel has a broad charge to:

> examine the full range of programs of the Bureau of Justice Statistics (BJS) in order to assess and make recommendations for BJS' priorities for data collection. The review will examine the ways in which BJS

statistics are used by Congress, executive agencies, the courts, state and local agencies, and researchers in order to determine the impact of BJS programs and the means to enhance that impact. The review will assess the organization of BJS and its relationships with other data gathering entities in the Department of Justice, as well as with state and local governments, to determine ways to improve the relevance, quality, and cost-effectiveness of justice statistics. The review will consider priority uses for additional funding that may be obtained through budget initiatives or reallocation of resources within the agency. *A focus of the panel's work will be to consider alternative options for conducting the National Crime Victimization Survey, which is the largest BJS program.* The goal of the panel's work will be to assist BJS to refine its priorities and goals, as embodied in its strategic plan, both in the short and longer terms. The panel's recommendations will address ways to improve the impact and cost-effectiveness of the agency's statistics on crime and the criminal justice system. [emphasis added]

BJS specifically requested that the panel begin its work by providing guidance on options for conducting the National Crime Victimization Survey (NCVS), one of many data series sponsored by BJS and one that consumes a large share (as much as 60 percent) of the agency's annual appropriations. This interim report responds to this request.

Since the survey began full-scale data collection in the early 1970s, the NCVS has become a major social indicator for the United States. Serving as a complement to the official measure of crimes reported to the police (the Uniform Crime Reporting [UCR] program administered by the Federal Bureau of Investigation), the NCVS has been the basis for better understanding the cost and context of criminal victimization. However, and particularly over the course of the last decade, the effectiveness of the NCVS has been undermined by the demands of conducting an increasingly expensive survey in an effectively flat-line budgetary environment. In order to keep the survey going in light of tight resources, BJS has reduced the survey's sample size over time, and other design features have been altered. When the survey began in 1972, the sample of addresses for interviewing numbered 72,000; in 2005, the NCVS was administered in about 38,600 households, yielding interviews with 67,000 people. Although this sample size still qualifies the NCVS as a large data collection program, occurrences of victimization are essentially a rare event relative to the whole population: many respondents to the survey do not have incidents to report when they are contacted by the survey. At present, the sample size is such that only a year-to-year change of 8 percent or more in the NCVS measure of violent crime can be deemed statistically to be significantly different from no change at all. In its reports on the survey, BJS has to combine multiple years of data in order to comment on change over time, which is less desirable than an annual measure of year-to-year change.

In approaching this work, the panel recognizes the fiscal constraints on the NCVS, but we do not intend to be either strictly limited by them or completely indifferent to them. Rather, our approach is to revisit the basic goals and objectives of the survey, to see how the current NCVS program meets those goals, and to suggest a range of alternatives and possibilities to match design features to desired sets of goals.

PRESERVING THE VICTIMIZATION MEASURE

There are no nationally available data on crime and victimization—collected at the incident level, with extensive detail on victims and the social context of the event—except those collected by the NCVS. It is this basic fact that is the strongest argument for the continuation and maintenance of the survey. Certainly, one option for the future of the NCVS—and the ultimate cost-reducing option—is to suspend or terminate the survey. It is an option that would have to be considered, if budget constraints require further reductions in sample size. To be clear, though, abandonment of the NCVS is not an option that we favor in any way.

Annual national-level estimates from the NCVS are routinely used in conjunction with the UCR to describe the volume and nature of crime in the United States. There is great value in having two complementary but nonidentical systems—the NCVS and the UCR—addressing the same phenomenon, for the basic reason that crime and victimization are topics that are too broad to be captured neatly by one measure. The police are not a disinterested party when it comes to characterizing the crime problem, and it is unwise to have data generated by the police as a sole measure of crime nationally. The UCR tells us little about the victims of crime; although its National Incident-Based Reporting System (NIBRS) has the potential to capture some of the detail currently measured by the NCVS, NIBRS has substantial limitations and remains incapable of providing national-level estimates after 20 years of implementation. Moreover, it is clear that a substantial proportion of crime is not reported fully and completely to law enforcement authorities. Thus, there remains a vital role for a survey-based measure that sheds light on unreported crime.

> *Recommendation Int-3.1:* **BJS must ensure that the nation has quality annual estimates of levels and changes in criminal victimization.**

The current design of the NCVS has benefited from years of experience, methodological research, and evaluation; it is a good and useful model that has been adopted by international victimization surveys as well as subnational surveys within the United States. The principal fault of the current NCVS is not a design flaw or methodological deficiency, or even that the design inherently costs too much to sustain, but rather—simply—that it costs more than is tenable under current budgetary priorities. In its present size and configuration, the NCVS can permit insights into the dynamics of victimization. However, in our assessment, the current NCVS falls short of the vibrant measure of annual change in crime that was envisioned at the survey's outset.

> **Finding Int-3.1:** As currently configured and funded, the NCVS is not achieving and cannot achieve BJS's legislatively mandated goal to "collect and analyze data that will serve as a continuous and comparable national social indication of the prevalence, incidence, rates, extent, distribution, and attributes of crime . . ." (42 USC § 3732(c)(3)).

By several measures—comparison with the expenditures of foreign countries for similar measurement efforts or with the cost of crime in the United States—the NCVS is underfunded. Accordingly, the panel recommends that BJS be afforded the budgetary resources necessary to generate accurate measures of victimization, which are as important to understanding crime in the United States as the UCR measure of crimes reported to the police.

> *Recommendation Int-3.2:* **Congress and the administration should ensure that BJS has a budget that is adequate to field a survey that satisfies the goal in Recommendation Int-3.1.**

OVERALL GOAL AND DESIGN CONSIDERATIONS

In considering historical goal statements of the NCVS, as well as new ones, we find three basic goals to be particularly prevalent and important, in addition to the previously expressed goal of maintaining annual national-level estimates of victimization that are independent of official reports to the police:

- Flexibility, in terms of both content (capability to provide detail on the *context and etiology of victimization* and to assess *emerging crime problems*, such as identity theft, stalking, or violence against and involving immigrants) and analysis (providing *informative metrics beyond basic crime rates*);
- Utility for gathering information on crimes that are not well reported to police or on *hard-to-measure constructs* (e.g., crimes against adolescents, family violence, and rape); and
- Small-domain estimation, including providing *information on states or localities*, which we think will be crucial to maximizing the utility of the NCVS and to building and maintaining constituencies for the survey.

In this report, we describe various design possibilities and their implications relative to these goals; however, we do not suggest one single path as the ideal for a redesigned NCVS. In part, this is because it is difficult to justify the case that our preferred set of NCVS goals is correct to the exclusion of all others; in part, it is because of the short time frame and the sequencing of this report (since it is inherently difficult to try to consider NCVS in isolation from the balance of BJS programs). But in large part we refrain from expressing a single, unequivocal path because the potential effectiveness and cost implications of some major design choices are simply unknown at this time.

We do think that it is critical to emphasize that even small changes to the design of a survey can have significant impacts on resulting estimates and the errors associated with them. Design changes made in the name of fiscal expedience, without grounding in testing and evaluation, are highly inadvisable. They risk unexplained changes in the time series and confusion among users.

> *Recommendation Int-4.1:* **BJS should carefully study changes in the NCVS survey design before implementing them.**

One potential cost-saving design choice is to change from asking respondents to recall and describe crime incidents in a 6-month window to using a 12-month window. This would entail contacting households once a year rather than twice (and, presumably, only 3 or 4 times if one chose to keep with the current regime of keeping households in the sample for 3.5 years). This would reduce the per-unit interviewing cost and free up resources to add additional sample addresses within each single year; 12 months is also the common reference period in victimization surveys in other countries. However, it could also increase problems of recall error by making respondents search their memories over a longer period. On its conceptual strengths and its use in comparable crime surveys in other western nations, we prefer a switch to a 12-month reference period as a cost-saving mechanism over options that would simply reduce the total sample size. That said, the empirical case for implementing this change is not completely clear and warrants up-to-date research. We note that such a move requires an overlap of designs over time to safely incorporate the change to 12 months.

> *Recommendation Int-4.2:* **Changing from a 6-month reference period to a 12-month reference period has the potential for improving the precision per-unit cost in the NCVS framework, but the extent of loss of measurement quality is not clear from existing research based on the post-1992-redesign NCVS instrument. BJS should sponsor additional research—involving both experimentation as well as analysis of the timing of events in extant data—to inform this trade-off.**

It is also the case that cost savings might be achieved by refining the NCVS sample stratification schemes. The current multistage cluster design of the NCVS automatically includes households sampled from counties and other geographic regions with large population sizes, clustering the remaining geographic areas by social and demographic information to produce similar strata from which the remaining sample is drawn. The composition of the sample is relatively slow to change with each decennial census, although effort is made to include some new housing stock by sampling from housing permit data. If the NCVS continues to be conducted by the Census Bureau (see "Collecting the Data," below), particular insight for altering the basic sample design and modifying sample strata based on an up-to-date sampling frame could come from interaction with the new American Community Survey (ACS). But, again, quantitative methodological research that could suggest exactly what benefits might or might not accrue is lacking.

> *Recommendation Int-4.7:* **BJS should investigate changing the sample design to increase efficiency, thus allowing more precision for a given cost. Changes to investigate include:**
>
> (i) **changing the number or nature of the first-stage sampling units;**
>
> (ii) **changing the stratification of the primary sampling units;**
>
> (iii) **changing the stratification of housing units;**
>
> (iv) **selecting housing units with unequal probabilities, so that probabilities are higher where victimization rates are higher; and**
>
> (v) **alternative person-level sampling schemes (sampling or subsampling persons within housing units).**

As early as 1980, the NCVS began the use of multiple response modes. Face-to-face personal interviews after the first contact with a sample household were replaced with interviews conducted by telephone, and—after the 1992 implementation of the full NCVS redesign—some interviewing began to be done by Census Bureau computer-assisted telephone interviewing (CATI) centers using a fully automated survey instrument. The NCVS path to automation has been somewhat complicated: full conversion to nonpaper survey questionnaires was achieved only in 2006, and—as part of the most recent round of cost reductions—BJS and the Census Bureau abandoned the use of the centralized CATI centers for NCVS interviews because anticipated cost savings never occurred. However, as redesign possibilities are considered, it is important that BJS continue to seek automation possibilities and not be limited to the NCVS traditional interview formats. A particular area of focus should be self-response options, such as computer-assisted self-interviewing (effectively, turning the interviewer's laptop around so that the respondent answers questions directly) or Internet response for interviews after several visits. As with the central CATI centers, cost savings from new modes of data collection are not guaranteed, but they may put the survey in good stead for implementing new topical modules and promoting high respondent cooperation. They can also serve to reduce overall respondent burden.

> *Recommendation Int-4.8:* **BJS should investigate the introduction of mixed mode data collection designs (including self-administered modes) into the NCVS.**

The NCVS is subject to the same pressures facing all household surveys in modern times, whether federal or private. It is increasingly difficult (and expensive) to obtain survey responses from persons or households in an age of cell phones, call waiting, and Internet chat. A significant fraction of survey costs are incurred to contact the most hard-to-find respondents. In considering design possibilities, it is important that BJS try to develop schemes that are relatively robust to declines in response rate, as such declines are virtually certain.

> *Recommendation Int-4.9:* **The falling response rates of NCVS are likely to continue, with attendant increasing field costs to avoid their decline. BJS should sponsor nonresponse bias studies, following current OMB guidelines, to guide trade-off decisions among costs, response rates, and nonresponse error.**

BUILDING AND REINFORCING CONSTITUENCIES

A continuing challenge for the NCVS is the development of constituencies with a strong interest in the data and their quality. The public is aware of the NCVS mainly due to one regular constituency—the media—and the spate of crime uptick or downtick stories that accompanies each year's release of NCVS and UCR estimates. Likewise, findings from topical supplements (such as racial dimensions of traffic stops, measured by the Police-Public Contact Survey supplement) typically get prominent press coverage. Official statistics, like other societal infrastructures, are often highly valued but rarely passionately promoted by day-to-day users. However, the long-term viability of the survey depends crucially on building and shoring up constituencies for NCVS products and on cultivating the survey's user base among researchers.

Small-Domain Estimates

The world has changed since the mid-1970s—computers are more powerful, data users are more sophisticated, and the demand for small-area geographic data is more insatiable. It is too strong to say that the NCVS can remain relevant *only* if it provides estimates for areas or populations smaller than the nation as a whole: state and local governments, which are among the most prodigious of NCVS users, continue to find national benchmarks very valuable. However, the survey will increasingly grow out of step with potential constituencies if it cannot be used to provide estimates for smaller areas.

> *Recommendation Int-4.5:* BJS should investigate the use of modeling NCVS data to construct and disseminate subnational estimates of major crime and victimization rates.

This recommendation runs counter to the principal effect of one of our predecessor National Research Council (1976) panel's recommendations— that the separate "impact city" victimization surveys that were originally part of the National Crime Surveys suite should be terminated. However, it is very much consistent with that previous recommendation's focus on an integrated set of estimates, including subnational geographies. These subnational estimates need not be exhaustive: expanding the sample to support estimates for the largest metropolitan statistical areas is a more sensible and cost-effective approach than a system for generating estimates for all 50 states. But they should permit insight on victimization for some smaller units than the nation as a whole. Small-domain estimates also refer to estimates by other social or demographic constructs, such as urbanicity (urban, suburban, or rural), in addition to the basic disaggregation by major race-ethnicity groups that is currently done.

With particular regard to the generation of small-domain estimates, it should be noted that enhancing the NCVS to better serve constituencies is not strictly a process of addition, in terms of sample size or implementation of a full supplemental questionnaire. In some important respects, user constituencies may best be served by more creative use of the current NCVS design. In the years since National Research Council (1976) advocated eliminating the city surveys, statistical developments in small-domain estimation techniques have been considerable; hence, some small-domain estimates may be possible through modest investment by BJS in technical infrastructure for statistical modeling tasks.

In addition to small-domain modeling using NCVS data, it may also be useful to explore ways to strengthen victimization surveys conducted by states and localities. Currently, BJS operates a program under which it develops victimization survey software and provides it to interested local agencies; however, those agencies must supply all the resources (funds and manpower) to conduct a survey. An approach to strengthen this program would be to make use of BJS's organizational position within the U.S. Department of Justice. The bureau is housed in the Office of Justice Programs, the core mission of which is to provide assistance to state and local law enforcement agencies; it does so through the technical research of the National Institute of Justice and the grant programs of the Bureau of Justice Assistance (BJA), among others. We suggest that OJP consider ways of dedicating funds—like BJA grants, but separate from BJS appropriations—for helping states and localities bolster their crime information infrastructures through the establishment and regular conduct of state or regional victimization surveys. Such surveys would most likely involve cooperative arrangements with research organizations or local universities and make use of the existing BJS statistical analysis center infrastructure. This approach is analogous to the Behavioral Risk Factor Surveillance System (BRFSS) of the Centers for Disease Control and Prevention, and it is similar in its partnership arrangements to the Federal-State Cooperative Program for Population Estimates (FSCPE) of the Census Bureau.

> ***Recommendation Int-4.6:*** **BJS should develop, promote, and coordinate subnational victimization surveys through formula grants funded from state-local assistance resources.**

We discuss an extreme interpretation of this approach—wherein the "national" victimization survey would be effectively be the combination of the subnational surveys—in Chapter 4 [of the interim report]. However, we emphasize that we suggest that this BRFSS/FSCPE approach should be considered independent of (and as a complement to) the chosen design of the NCVS.

Topic Constituencies

The NCVS first added a topic supplement to the survey questionnaire in 1977, querying respondents on their perceptions of the severity of crime. Particularly since 1989, supplements have been an irregular part of the NCVS structure; the School Crime Supplement on school safety has been repeated six times and the Police-Public Contact Survey three times, with other supplements being (to date) one-time efforts.

A strong program of topic supplements is an important part of the NCVS, both because of the breadth of topics that may be handled and because the ability to quickly field questions on new topics of interest is a key advantage of survey-based collection compared with official records.

> ***Recommendation Int-4.3:*** **BJS should make supplements a regular feature of the NCVS. Procedures should be developed for soliciting ideas for supplements from outside BJS and for evaluating these supplements for inclusion in the survey.**

What is necessary regarding NCVS supplements is a more structured plan for their implementation, better exploration (and marketing) of sponsorship opportunities by other state and federal agencies, and greater transparency in real costs of conducting a supplement. Regardless of the overall design of the NCVS, the British Crime Survey offers an attractive model: a streamlined core set of questions combined with a planned, regular slot for topical content.

> **Recommendation Int-4.4:** BJS should maintain the core set of screening questions in the NCVS but should consider streamlining the incident form (either by eliminating items or by changing their periodicity).

This would reduce respondent burden and allow additional flexibility for adding items to broaden and deepen information about prevalent crimes.

ATTENTION TO DATA QUALITY AND ACCESS

We make a series of recommendations that are agency-level in focus, aimed at better equipping BJS to understand its own products and to interact with its users. They are presented here in initial form because they are pertinent to the NCVS. We expect to expand on them in our final report on the full suite of BJS programs and products.

First, BJS currently receives periodic advice from the Committee on Law and Justice Statistics of the American Statistical Association (ASA). Although this input is certainly valuable, we think that BJS—and the NCVS in particular—would benefit from the commissioning of an ongoing scientific technical advisory board, such as is in place for other statistical agencies. This board should include subject-matter, survey methodological, and statistical expertise; spots on the board are also a vehicle for strengthening stakeholder constituencies for the NCVS.

> **Recommendation Int-5.1:** BJS should establish a scientific advisory board for the agency's programs; a particular focus should be on maintaining and enhancing the utility of the NCVS.

Several of our recommendations listed earlier identify gaps in existing research that must be filled to accurately inform trade-offs in design choices. More generally, the NCVS developmental work in the 1970s and the research conducted as part of the 1980s redesign effort are extensive, but we think that there is a paucity of recent methodological research making use of the post-1992-redesign NCVS instrument and techniques. BJS has already made some strides in fostering methodological research with its fellowship program, operated in conjunction with the ASA. We urge BJS to continue this work and to explore other creative ways to foster internal and extramural research using the NCVS and other BJS data sets, including graduate fellowships, as part of continuous efforts to assess the quality of NCVS estimates.

Recommendation Int-5.3: BJS should undertake research to continuously evaluate and improve the quality of NCVS estimates.

Conceptually, the survey-based NCVS is ideally suited (as the official record-based UCR is not) to study the dynamics of crimes that are emotionally or psychologically sensitive, such as violence against women, violence against adolescents, and stalking or harassment. We urge BJS to develop lines of research to ensure that such crimes are accurately measured on the NCVS instrument; these might include the testing of self-response options, such as audio computer-assisted interviewing.

Recommendation Int-3.3: BJS should continue to use the NCVS to assess crimes that are difficult to measure and poorly reported to police. Special studies should be conducted periodically in the context of the NCVS program to provide more accurate measurement of such events.

The quality of NCVS data and its scientific rigor in measuring crime should always be the survey's primary goal and acknowledged as its principal benefit. However, for the purpose of cultivating constituencies and users for the survey, attention to the accessibility and the ease of use of NCVS data is also vitally important. Part of this work involves reevaluation of basic products and reports from the NCVS and expansion of the range of analyses based on the data, and it involves both in-house research by BJS and effective ties with other users and researchers.

Recommendation Int-5.2: BJS should perform additional and advanced analysis of NCVS data. To do so, BJS should expand its capacity in the number and training of personnel and the ability to let contracts.

A necessary consequence of this recommendation is that the agency must expand its capacity, both in the number and training of personnel and the agency's ability to let contracts for external research.

Recommendation Int-5.4: BJS should continue to improve the availability of NCVS data and estimates in ways that facilitate user access.

Recommendation Int-5.5: The Census Bureau and BJS should ensure that geographically identified NCVS data are available to qualified researchers through the Census Bureau's research data centers, in a manner that ensures proper privacy protection.

In the case of this last recommendation, we understand that arrangements to place detailed NCVS data at the research data centers are under development; we state it here as encouragement to finalize the work.

COLLECTING THE DATA

It is important to note that some of the resource constraints on the NCVS are common to those on other important federal surveys, which have faced difficulties carrying out basic maintenance tasks like updating samples to reflect new census and address list information. The country needs a mechanism to alert itself to budget cuts that undermine the basic purposes of key federal statistical products.

> *Recommendation Int-5.6:* **The Statistical Policy Office of the U.S. Office of Management and Budget is uniquely positioned to identify instances in which statistical agencies have been unable to perform basic sample or survey maintenance functions. For example, BJS was unable to update the NCVS household sample to reflect population and household shifts identified in the 2000 census until 2007. The Statistical Policy Office should note such breakdowns in basic survey maintenance functions in its annual report *Statistical Programs of the United States Government.***

Any review of a major survey program—particularly one carried out with an eye toward cost reduction—must inevitably raise the question of the agent that collects the data: could survey operations be made better, faster, or cheaper by getting some other organization to carry out the survey? In this case, the U.S. Census Bureau's involvement with the NCVS predates the formal establishment of the survey, as the Census Bureau convened planning discussions and conducted NCVS pilot work.

The optimal decision on who should do the data collection for the NCVS will depend on the weight that one puts on desired objectives for the survey. For instance, an extremely strong weight on flexibility and quick response to emerging trends might argue against the Census Bureau, where implementation of a supplement can be made time-consuming through detailed cognitive testing (which ultimately improves the quality of the questions but can be slow) and passage through bureaucratic channels (e.g., clearance by the Office of Management and Budget, as required of all federal surveys). However, dominant weight on maintaining high response rates and drawing from the experience of other large, ongoing surveys would suggest that staying with the Census Bureau is the best course. Just as we do not offer a single design path for the NCVS, we do not find justification for offering a conclusion on "Census Bureau" or "not Census Bureau." Based on the advantages and disadvantages, we suggest that "privatizing" the NCVS is not the panacea for high survey costs that some may believe it is. We have been provided no way of estimating the various costs associated with switching NCVS data collection agents; however, it is altogether appropriate to consider means of getting detailed and specific answers to these questions.

In the interim, we suggest that the Census Bureau would benefit both BJS and itself itself by providing greater transparency in true survey costs.

> *Recommendation Int-5.7:* **Because BJS is currently receiving inadequate information about the costs of the NCVS, the Census Bureau should establish a data-based, data-driven survey cost and information system.**

We further suggest that BJS consider a design competition—providing some funds for bidders to specify in detail how they would conduct a victimization survey. This design competition would effectively compensate bidders for their time in developing proposal specifications, but it should be run with a statement that a formal request for proposals *may* result from the competition (and not that it will definitely occur).

> *Recommendation Int-5.8:* **BJS should consider a survey design competition in order to get a more accurate reading of the feasibility of alternative NCVS redesigns. The design competition should be administered with the assistance of external experts, and the competition should include private organizations under contract and the Census Bureau under an interagency agreement.**

– C –

Biographical Sketches of Panel Members and Staff

Robert M. Groves *(Chair)* is director of the U.S. Census Bureau, having been confirmed to that position on July 13, 2009. During this panel's work, he was professor of sociology and director of the Survey Research Center in the Institute for Social Research at the University of Michigan. He is the author of *Survey Errors and Survey Costs* and the coauthor of *Nonresponse in Household Surveys*. A National Associate of the National Academies, he has served on seven National Research Council committees and is a former member of the Committee on National Statistics. From 1990 to 1992, he served as associate director for statistical design, standards, and methodology at the U.S. Census Bureau. He is a fellow of the American Statistical Association and an elected member of the International Statistical Institute, and he has received the Innovator Award and an award for exceptionally distinguished achievement from the American Association for Public Opinion Research. He has an M.A. in statistics, an M.A. in sociology, and a Ph.D. in sociology, all from the University of Michigan.

William G. Barron, Jr., is a consultant to Princeton University and has served as consultant to the U.S. Department of Commerce and the U.S. Census Bureau. After a 30-year career at the Bureau of Labor Statistics—serving as deputy commissioner (1983–1988) and acting commissioner (once for a 23-month period)—he moved to the U.S. Census Bureau in 1998. There he served as deputy director and chief operating officer. Heavily involved in the conduct and completion of the 2000 census and the development

347

of plans for the 2010 census, he served as acting director of the Census Bureau in 2001 and early 2002. Prior to his consultancy at Princeton, he was visiting lecturer and Frederick H. Shultz Class of 1951 professor of international economic policy, and later the John L. Weinberg/Goldman Sachs and Company visiting professor and lecturer at the university's Woodrow Wilson School of Public and International Affairs. He has served as senior vice president for economic studies at the National Opinion Research Center at the University of Chicago and as senior client executive at Northrop Grumman Corporation. He has a B.A. from the University of Maryland.

William Clements is dean of the School of Graduate Studies and professor of criminal justice at Norwich University. Prior to assuming the role of dean, he was director and creator of the Master of Justice Administration program (2002–2005) and executive director of the Vermont Center for Justice Research (1994–2005), Vermont's Bureau of Justice Statistics–affiliated Statistical Analysis Center. He has been involved in bringing Norwich's curriculum to the online environment and developing the online graduate program model. His professional research interests and experience include a variety of criminal justice system studies in program evaluation, data systems development, and adjudication patterns. He has worked on and published in the areas of incident-based crime data, juvenile justice, the operation of the courts, and sentencing trends. He has served in various capacities and as president of the Northeast Academy of Criminal Justice Sciences, and he is a past president and executive committee member of the Justice Research and Statistics Association. He is coeditor of *Justice Research and Policy*. He has a Ph.D. in sociology from the University of Delaware.

Daniel L. Cork *(Study Director)* is a senior program officer for the Committee on National Statistics (CNSTAT), currently serving as study director of the Panel to Review the Programs of the Bureau of Justice Statistics and co-study director of the Panel on the Design of the 2010 Census Program of Evaluations and Experiments. He previously served as study director of the Panel on Residence Rules in the Decennial Census and program officer for the Committee to Review the Feasibility, Accuracy, and Technical Capability of a National Ballistics Database, as well as work with other CNSTAT census panels. His research interests include quantitative criminology, particularly space-time dynamics in homicide; Bayesian statistics; and statistics in sports. He has a B.S. in statistics from George Washington University and an M.S. in statistics and a joint Ph.D. in statistics and public policy from Carnegie Mellon University.

Janet L. Lauritsen is professor of criminology and criminal justice at the University of Missouri–St Louis. Much of her research is focused on understanding individual, family, and neighborhood sources of violent victimization as well as racial and ethnic differences in violence. She served as chairperson of the American Statistical Association Committee on Law and Justice Statistics from 2004 to 2006 and as visiting research fellow at the Bureau of Justice Statistics from 2002 to 2006. During her fellowship, she assembled two expert meetings on major options for the National Crime Victimization Survey, several of the participants of which are also members of this panel. She currently serves on the editorial boards of *Criminology* and the *Journal of Quantitative Criminology* and on the National Research Council's Committee on Law and Justice. She has a Ph.D. in sociology from the University of Illinois at Urbana-Champaign.

Colin Loftin is co-director of the Violence Research Group, a research collaboration with colleagues at the University at Albany and the University of Maryland that conducts research on the causes and consequences of interpersonal violence. The major themes of the research are (1) understanding violence as a social process extending beyond individual action, (2) improving the quality of data on the incidence and nature of crime, (3) design and evaluation of violence prevention policies, and (4) investigation of population risk factors for violence. The Violence Research Group published the *Statistical Handbook on Violence in America*. A past member of the National Research Council's Committee on Law and Justice, he previously served on the Panel on Understanding and Preventing Violence. He has a Ph.D. in sociology from the University of North Carolina.

James P. Lynch is distinguished professor at John Jay College of Criminal Justice in New York. At the Bureau of Social Science Research in the 1980s, he served as manager of the National Crime Survey redesign effort for the bureau. He became a faculty member in the Department of Justice, Law, and Society at American University in 1986, where he remained as associate professor, full professor, and chair of the department until leaving for John Jay in 2005. He has published 3 books, 25 refereed articles, and over 40 book chapters and other publications. He was elected to the executive board of the American Society of Criminology in 2002 and has served on the editorial boards of *Criminology* and the *Journal of Quantitative Criminology* and as deputy editor of *Justice Quarterly*. He has also chaired the American Statistical Association's Committee on Law and Justice Statistics. He has a Ph.D. in sociology from the University of Chicago.

Ruth D. Peterson is professor of sociology and director of the Criminal Justice Research Center at Ohio State University, where she has been on the faculty since 1985. She is also a fellow of the National Consortium of Violence Research, where she coordinates the Race and Ethnicity Research Working Group. She has conducted research on legal decision making and sentencing, crime and deterrence, and most recently, patterns of urban crime. She is widely published in the areas of capital punishment, race, gender, and socioeconomic disadvantage. Her current research focuses on the linkages among racial residential segregation, concentrated social disadvantage and race-specific crime, and the social context of prosecutorial and court decisions. She has a Ph.D. in sociology from the University of Wisconsin.

Carol V. Petrie *(Senior Program Officer)* is director of the Committee on Law and Justice at the National Academies. She also served as the director of planning and management at the National Institute of Justice, U.S. Department of Justice, responsible for policy and administration. In 1994, she served as the acting director of the National Institute of Justice. She has conducted research on violence and public policy, and managed numerous research projects on the development of criminal behavior, domestic violence, child abuse and neglect, and improving the operations of the criminal justice system. She has a B.S. in education from Kent State University.

Trivellore E. Raghunathan is professor of biostatistics and research professor at the Institute for Social Research at the University of Michigan. He also teaches in the Joint Program in Survey Methodology at the University of Maryland. He is the director of the Biostatistics Collaborative and Methodology Research Core, a research unit designed to foster collaborative and methodological research with the researchers in other departments in the School of Public Health and other allied schools. He is an associate director of the Center for Research on Ethnicity, Culture and Health and a faculty member of the Center of Social Epidemiology and Population Health; he is also affiliated with the University of Michigan Transportation Research Institute. Before joining the University of Michigan in 1994, he was on the faculty in the Department of Biostatistics at the University of Washington. His research interests are in the analysis of incomplete data, multiple imputation, Bayesian methods, design and analysis of sample surveys, small-area estimation, confidentiality and disclosure limitation, longitudinal data analysis, and statistical methods for epidemiology. He has a Ph.D. in statistics from Harvard University.

Steven R. Schlesinger is chief of the Statistics Division at the Administrative Office of the U.S. Courts (AO). He was director of the Bureau of Justice Statistics from 1983 to 1988 and was deputy director of the U.S. Department of Justice Office of Policy and Communications from 1991 to 1993.

He has also taught on the political science faculties of Rutgers University and the Catholic University of America. He is the author of 2 books and over 25 articles on legal topics. Among his professional awards are the O.J. Hawkins Award for Innovative Leadership and Outstanding Contributions to Criminal Justice Systems, Policy and Statistics in the United States, the U.S. Attorney General's Award for Excellence in Management, and AO's Meritorious Service Award. He has a Ph.D from the Claremont Graduate School.

Wesley G. Skogan has been a faculty member at Northwestern University since 1971 and holds joint appointments with the political science department and the University's Institute for Policy Research. His research focuses on the interface between the public and the legal system, crime prevention, victim services, and community-oriented policing. He has written four books on policing; all are empirical studies of community policing initiatives in Chicago and elsewhere. His 1990 book *Disorder and Decline* examined public involvement in these programs, their efficacy, and the issues involved in police-citizen cooperation in order maintenance. Another line of his research concerns neighborhood and community responses to crime. He has edited a series of technical monographs on victimization research and authored a technical review of the National Crime Victimization Survey that was published in *Public Opinion Quarterly*. He served as a consultant to the United Kingdom Home Office, developing and analyzing the British Crime Survey. He has been a visiting scholar at the Max-Planck-Institut (Freiburg), the Dutch Ministry of Justice, the University of Alberta, and Johns Hopkins University. He spent 2 years as a visiting fellow at the National Institute of Justice. At the National Research Council, he has served on the Committee on Law and Justice and chaired the Committee on Research on Police Policies and Practices. He has a B.A. in government from Indiana University, an M.A. in political science from the University of Wisconsin, and a Ph.D. in political science from Northwestern University.

Bruce D. Spencer is professor of statistics and faculty fellow in the Institute for Policy Research at Northwestern University. His interests include the interactions between statistics and policy, demographic statistics, and sampling. He chaired the statistics department at Northwestern from 1988 to 1999 and 2000 to 2001. He directed the Methodology Research Center of the National Opinion Research Center (NORC) at the University of Chicago from 1985 to 1992. From 1992 to 1994 he was a senior research statistician at NORC. At the National Research Council he served as a member of the CNSTAT Panel on Formula Allocations, as well as the Mathematical Sciences Assessment Panel and the Panel on Statistical Issues in AIDS Research; as a staff member he served as study director for the Panel on Small Area Estimates of Population and Income. He has a Ph.D. from Yale University.

Bruce Western is professor of sociology at Harvard University and director of the Multidisciplinary Program in Inequality and Social Policy at the Kennedy School of Government. Previously, he was professor of sociology at Princeton University and faculty associate in the Office of Population Research. His research interests broadly include political and comparative sociology, stratification and inequality, and methodology. More specifically, he has studied how institutions shape labor market outcomes. Work in this area has developed along two tracks: the growth and decline of labor unions and their economic effects in the United States and Europe; and the impact of the American penal system on labor market inequality. His methodological work has focused on the application of Bayesian statistics to research problems in sociology. He is the author of *Punishment and Inequality in America* and, with Mary Patillo and David Weiman, of *Imprisoning America: The Social Effects of Mass Incarceration*, both publications of the Russell Sage Foundation. He has a Ph.D. in sociology from the University of California, Los Angeles.

COMMITTEE ON NATIONAL STATISTICS

The Committee on National Statistics was established in 1972 at the National Academies to improve the statistical methods and information on which public policy decisions are based. The committee carries out studies, workshops, and other activities to foster better measures and fuller understanding of the economy, the environment, public health, crime, education, immigration, poverty, welfare, and other public policy issues. It also evaluates ongoing statistical programs and tracks the statistical policy and coordinating activities of the federal government, serving a unique role at the intersection of statistics and public policy. The committee's work is supported by a consortium of federal agencies through a National Science Foundation grant.